Biology of Populations

THE BIOLOGICAL BASIS OF PUBLIC HEALTH

Biology of Populations

THE BIOLOGICAL BASIS OF PUBLIC HEALTH

Edited by

BRENDA K. SLADEN AND FREDERIK B. BANG

The Johns Hopkins University
School of Hygiene and Public Health
Baltimore, Maryland

1969

AMERICAN ELSEVIER PUBLISHING COMPANY, INC.

NEW YORK

AMERICAN ELSEVIER PUBLISHING COMPANY, INC.

52 Vanderbilt Avenue, New York, N.Y. 10017

ELSEVIER PUBLISHING COMPANY LTD.

Barking, Essex, England

ELSEVIER PUBLISHING COMPANY

335 Jan Van Galenstraat, P.O. Box 211, Amsterdam, The Netherlands

Standard Book Number 444–00063–1
Library of Congress Card Number 68–28347

PRINTED IN THE UNITED STATES OF AMERICA

Contributors*

FREDERIK B. BANG
M.D.
Department of Pathobiology and Department of Medicine, The Johns Hopkins University School of Medicine
(*pages* 1–12, 168–179, 234–246, 334–371, 380–389, 414–430)

BERNICE H. COHEN
Ph.D., M.P.H.
Department of Chronic Diseases
(*pages* 180–205)

JEROME D. FRANK
M.D.
Psychiatry and Behavioral Sciences, Johns Hopkins Hospital
(*pages* 322–331)

ROGER M. HERRIOTT
Ph.D.
Department of Biochemistry
(*pages* 128–153)

WILLIAM D. HILLIS
M.D.
Department of Pathobiology
(*pages* 26–34, 221–231)

HARVEY RABIN
Ph.D.
Department of Pathobiology. Currently with the Zoological Society of San Diego
(*pages* 372–379)

ANTHONY J. READING
M.D., M.P.H., Sc.D.
Department of Pathobiology; and Psychiatry, and Behavioral Sciences, Johns Hopkins Hospital
(*pages* 311–321)

LLOYD E. ROZEBOOM
Sc.D.
Department of Pathobiology
(*pages* 206–220, 390–402)

R. BRADLEY SACK
M.D., Sc.D.
Department of Pathobiology
(*pages* 35–46)

GEORGE B. SCHALLER
Ph.D.
Department of Pathobiology. Currently with the New York Zoological Society
(*pages* 116–126)

KEERTI V. SHAH
M.B., B.S., Dr.P.H.
Department of Pathobiology
(*pages* 403–413)

BRENDA K. SLADEN
B.Sc.
Department of Pathobiology
(*pages* 47–100, 180–205)

WILLIAM J. L. SLADEN
M.D. (London), D.Phil.
Department of Pathobiology
(*pages* 247–283)

CHARLES H. SOUTHWICK
Ph.D.
Department of Pathobiology
(*pages* 101–115, 284–310)

BYRON S. TEPPER
Ph.D.
Department of Pathobiology
(*pages* 14–25, 154–167)

FRANCIS S. L. WILLIAMSON
Sc.D.
Department of Pathobiology and the Smithsonian Institution
(*pages* 206–220)

*Unless otherwise indicated, all contributors are with The Johns Hopkins University School of Hygiene and Public Health.

Contents

Chapter

Editors' Preface

During the last few decades the explosion of information in biology has included population dynamics and other aspects of ecology, population genetics, ethology, and the nature of interactions of populations in disease. At the same time the applied field of public health has been forced by public need to move into population dynamics and control, medical care (particularly concerning chronic disease), the effects of the environment on man, and other broad areas. Most workers in the newly expanding fields of public health and welfare have had no training in population biology and are frequently unaware of its principles. Yet many of these principles apply directly to immediate problems. This divergence has developed despite pioneer efforts of people such as Raymond Pearl and Lowell Reed, who worked for many years in the Johns Hopkins School of Hygiene and Public Health. Although there is a scattering of texts containing some of the general ideas of population biology, we have found none that brings together the multifaceted aspects and makes them understandable to people in related fields.

For twelve years the Department of Pathobiology has taught a course in Population Biology to Master of Public Health students in the Johns Hopkins School of Hygiene and Public Health. The lecturers and students have included specialists in ecology, animal behavior, psychiatry and other fields of medicine, microbiology, genetics, and sanitary engineering. In this compilation of the lectures there has been no attempt to describe all the applications of population biology to public health. The fundamental ideas from the various component disciplines have been given, illustrated by some of the best studied examples. We hope that this bringing together of several disciplines will be useful for many biologists, and that the richness of the data will be apparent for all interested students of man.

When a text is compiled from a repeatedly taught course, many debts are incurred. Dr. Bentley Glass gave a number of lectures during the first years and he encouraged the development of the course. Drs. David Davis and Clarke Read, formerly members of the Department, contributed to the teaching of ecology and disease. Dr. Paul Lemkau, of the Department of Mental Health, gave earlier lectures relating animal and human behavior. Drs. Everett Schiller and Phillip Grant lectured in parasitology and embryology. Dr. Helen Abbey, Professor of Biostatistics, repeatedly helped staff and students with statistical analyses. All these people helped in the development of laboratory exercises to illustrate the principles of population biology. Their ideas became part of this attempt to describe population biology for all who are interested in man's welfare in a densely populated and very rapidly changing world.

A Quotation from Raymond Pearl (1937)* Introducing Population Biology

"A population is a group of living individuals set in a frame that is limited

*Pearl, R. (1937) On biological principles affecting populations: human and other. *Am. Naturalist* **71**: 50–68.

and defined in respect of both time and space. The biology of populations is consequently a division or department of group biology in general. The essential and differentiating feature of group biology is that it considers groups as wholes. It aims to describe the attributes and behavior of a group as such, that is as an entity in itself, and not as the simple sum of the separate attributes and behavior of the single individual organisms that together make up the group. The concept of group attributes, separate and different from the attributes of the component individuals, is a familiar one in other fields. For example the familiar measures on variation such as the standard deviation and coefficient of variation, are quantitative expressions of a group attribute, namely, the shape of the distribution of frequency of the component individuals in respect of the character measured. Similarly, birth rates and death rates are quantitative expressions of group attributes, meaningless relative to any individual. . . .

"The purely biological attributes or characteristics of populations as whole groups, whether of men or mice, that are directly relevant to the discussion and elucidation of significant population problems are not numerous. In the first line of importance there are but three of them. These are size, growth, and quality. Each may be briefly considered separately.

"*Size* of populations may be considered in either absolute or relative terms. It is the total absolute numbers of individuals composing a given population at a particular time, or the number of individuals at that time in proportion to the size of the spatial frame-work in which the population exists. The most significant method of measuring relative population size or population density will of course differ according to the habits and ecology of the species concerned.

"*Growth* of population is measured by the change of size with time in either the positive or negative direction. The increase or decline of populations in size with the passage of time, to the special student of population problems, is easily the most interesting biological phenomenon they exhibit. It rests upon the fundamental variables of fecundity, fertility and life duration, and takes the investigator at once into the fascinating realm of the physiology of reproduction, that has been until recently the most neglected part of all physiology. In human populations the rate of growth in a particular period is closely correlated with social and economic behavior and in extremely complex ways. History plainly shows that the rate of population growth—positive or negative—has been a factor of no mean importance in shaping national policies and influencing political conduct. In infra-human populations in Nature the rate of population growth at a particular time may be associated with abrupt changes in habits and behavior of great significance evolutionarily, particularly in relation to what is called the "Balance of Nature." This expression "Balance of Nature" is, of course, only a figurative way of speaking in general terms about the relative sizes of the populations of different species existing in the same spatial framework at the same time.

"*Quality* is a convenient term to designate the genetic constitution of the population, or, put in another way, the frequency distribution in absolute or relative terms of the qualitatively different genes represented in the group. In human populations this attribute has come during the last half century to be regarded as of prime importance. It is what the eugenists fight and bleed about, as well as preach and pray over. It is like the weather in at least the respect that people discuss it a lot but seldom really do much about it. The reason why is at bottom about the same in the two cases. The problem presented is somewhat too big a bite for our feeble masticatory powers. Here the laboratory experimenter with animal populations has one of his greatest advantages. He can make the genetic constitution of his populations precisely what he wants it to be and hold it there. In the long run much useful knowledge to the whole enterprise of population study will surely accrue from this advantage.

"It is evident that these three biological attributes of populations are not separate and disparate things. On the contrary, they are in fact integrally interconnected with each other. The size of a population at any given time depends upon its past growth—is in fact the static aspect of the dynamic phenomenon. Both are reflections of the biological quality of the population —its genetic constitution. But this interconnection and integration is nothing peculiar to population biology. It is one of the most characteristic features of all biological events and living organisms.

"There are other group attributes of populations that might be listed besides these three. But none seems to be of the same order of significance, so far as may be judged in the present state of knowledge. *Group behavior* or psychology is the one omitted that most insistently demands to be discussed."

Brenda K. Sladen
Frederik B. Bang

Baltimore, Maryland
Spring, 1969

Acknowledgments

Some of the assistance given in developing the course upon which this text is based has been mentioned in the Preface. For the text, Mrs. R. Kimmerer transcribed tape recordings of lectures. Dr. P. Marler kindly read Chapter 18, and Dr. N. Hairston read Chapter 27. Dr. G. S. Watson and G. Chase made calculations for Figure 14.4. Dr. A. H. Schultz provided Plate 30.2 and Dr. D. S. Borgdonkar Plate 13.1. Dr. R. S. Peterson allowed Figure 19.3 to be used from his unpublished doctoral thesis. Reproduction of the self-portraits by Rembrandt for Plate 30.1 was permitted by The Glasgow Art Gallery and Museum; the National Gallery, London; The National Gallery of Art, Washington, D.C.; and the Wallraf-Richartz Museum, Cologne. The Rijksmuseum, Amsterdam, gave permission for reproduction of the print, "The Fighting Stags," at the beginning of Section III.

The following copyright holders kindly allowed their material to be used: Abelard-Schuman, Publishers (Table 19.2); Academic Press (Figures 3.1, 3.2, 12.3, 24.2, 30.4, Tables 1.1, 5.2, 24.2); American Association for the Advancement of Science (Figure 11.3, copyright 1964); American Genetics Association (Figure 15.2); *American Journal of Epidemiology* (Table 24.1); American Ornithologists Union (Table 19.1); American Physiological Society (Figure 30.7, copyright 1954); *American Scientist* (Table 5.8); American Society for Microbiology (Figure 26.1, Tables 26.3, 26.4); Annual Reviews, Inc. (Figures 15.1, 15.2); Blackwell Scientific Publications (Figures 5.5, 5.7, 7.1, 7.2, Tables 5.4, 5.6., 7.1); British Ornithologists' Union (Figures 18.3, 18.6); Cambridge University Press (Figures 6.1, 25.5, 25.6, Table 25.1); *Cancer Research* (Tables 26.1, 26.2); Clarendon Press (Figures 8.3, 8.5, 8.8, 18.7); Cold Spring Harbor Laboratory of Quantitative Biology (Figures 6.5, 14.2, Table 14.2); Commonwealth Scientific and Industrial Research Organization (Figures 19.4, 19.5); Controller of Her Brittanic Majesty's Stationery Office (Figure 25.3); Duke University Press (Figures 6.3, 6.6, 8.2, 25.1, 25.2); E. P. Dutton and Co., Inc. and Hutchinson and Co., Ltd., Publishers (plate of Tassili Frescoes at the beginning of Section II); Forestry Branch, Canadian Department of Fisheries and Forestry (Figures 5.1, 5.6); Holt, Rinehart, and Winston, Inc., Publisher (Figures 16.1, 30.2); *Human Biology* (Figure 1.2); Harcourt, Brace, and World, Publishers (Figures 18.4, 18.5, copyright 1953); Harper and Row, Publishers (quotation on page 356); Indian Council of Medical Research (Tables 29.2, 29.3); J. and A. Churchill, Ltd., Publishers (Tables 27.1, 27.2); *Journal of Wildlife Management* (Figure 8.1); A. A. Knopf, Inc., Publishers (Figure 6.2); *Lancet* (Figure 25.4); Lewis State Bank, executor of the estate of M. F. Boyd (Figure 1.2); J. B. Lippincott Co., Publishers (Figure 3.3); Little, Brown, and Co., Publishers (Figure 30.3, copyright 1959); Liverpool School of Tropical Medicine (Table 28.2); Macmillan (Journals), Ltd., and *Nature* (Tables 14.3, 15.4); National Academy of Sciences (Figures 9.1, 23.1, Tables 9.2, 9.3); Oliver and Boyd, Publishers (Figure 14.6); *Physiological Zoology* (Figure 6.3); Prentice-Hall, Inc., Publishers (Figures 2.2, copyright 1963, 11.2 and 11.6, copyright 1964, 13.1, copyright 1969); *Quarterly Review of Biology* (Figures 1.1, 20.1); Rockefeller Press (Figures 4.2, 4.3, Table 23.1); Reinhold Publishing Co. (Figure 7.4); The Royal Society (London) (Figure 6.7); W. B. Saunders Co., Publishers (Figures 2.4, 12.1, 12.2); Scientific American, Inc. (Figures 18.1, 18.2, copyright 1958); *Transactions of the Royal Society of Tropical Medicine and Hygiene* (Table 28.3); Charles H. Thomas, Publisher (Table 16.2); University of California Press (Figure 17.2, Table 6.1); University of Chicago Press (Table 14.2 and quotation on pages xvii–xix); Wayne State University Press (Figure 1.2); John Wiley and Sons, Inc., Publishers (Figure 14.7); Williams and Wilkins Co., Publishers (Figures 30.5, 30.6, 30.8, copyright 1952); World Health Organization (Figures 27.1, 27.2, Tables 15.3, 28.4); The Zoological Society of London (Figure 8.4).

CHAPTER 1

Introduction

FREDERIK B. BANG

POPULATION BIOLOGY AND PUBLIC HEALTH

The practice of medicine is largely applied to individuals whereas public health and hygiene are applied to populations. For effective practice and research in public health one needs to understand why a population should be considered as more than a sum of the characteristics of its members. Populations should be studied as entities in themselves and their study involves several integrated aspects (Pearl, 1937, quoted on page xv). The main purpose of this text is to present four major areas in which the characteristics of a population assume importance not apparent in the individuals—ecology, genetics, behavior, and disease. It will also introduce some of the methods used in population research, as well as many of the problems and some recent advances. The value of this general biological approach to public health may be seen by taking two important diseases and discussing their places as population problems. The first disease is arthropod-borne encephalitis in the United States, the second is sickle-cell anemia.

THE ARTHROPOD-BORNE ENCEPHALITIDES

A variety of arthropod-borne viral encephalitides is known in North America but for simplicity the discussion will be limited to eastern and western equine encephalomyelitis.

Vertebrate Hosts

It was early recognized that the disease in horses and the disease in man were caused by the same agents, but that any relationship between numbers of cases was not direct. In one year the incidence of infection with equine encephalitis might be high in horses and low in man and in another year the reverse would occur. The disease might even occur in man in the absence of apparent disease of the horse. Moreover, times of occurrence within the season might not be the same.

Recognition of the eastern virus in pigeons and pheasants led to the question of whether there was a reservoir of infection from which the disease was transmitted to man or the horse. Serum neutralization tests showed that various species of birds had experienced the infection. Whether they might in nature have served as sources of virus was another matter. Then outbreaks of the disease occurred in various pheasant colonies in New Jersey, affecting birds in one pen and not in the next even though a vector insect could fly from

1

one to the other through coarse chicken wire. Thus there were unidentified events occurring in this epidemic.

Mosquitoes as Vectors

Japanese workers had shown that an encephalitis of man in Japan was transmitted by mosquitoes. In the United States the distribution of the horse encephalitis disease on the eastern shore of Virginia showed that the infection occurred primarily along the salt marshes, and salt-marsh mosquitoes came under suspicion.

If mosquitoes transfer the virus from one vertebrate to another, then the biology of mosquitoes is important. How many mosquitoes are there? How many are infected in nature? How long does the mosquito live? Where does it feed? Does it feed on man most of the time or can it survive on cattle, horses, and chickens? Does it feed often on a particular species that carries the virus? Does it serve as a frequent link between the reservoir of infection and man, or between the reservoir and the horse? If it is a vector, what is its efficiency?

The extensive studies of Reeves and Hammon (1962) examined these questions for *Culex tarsalis*, the vector of western equine and St. Louis encephalitis viruses in California. This mosquito develops from egg to larva to pupa in standing water in fields. Reeves showed that perhaps one in a thousand, sometimes more, of the wild-caught *C. tarsalis* were infected and that mosquitoes acquired their virus by feeding on infected birds. The ingested virus passed through the wall of the gut and multiplied and persisted in large amounts in the tissues, including the salivary glands. Then the virus was inoculated with saliva into the vertebrate host.

How many mosquitoes are there in nature? Various trapping methods are available, for instance a light-trap uses light at night as an attractant and a fan sweeps the assembled insects into a killing jar. After some hours the dead mosquitoes are removed, identified, and counted. This method does not indicate absolute numbers because attractiveness of the light varies with species and sex. Other methods include collecting all mosquitoes that alight on exposed skin or are attracted to an animal in a trap in given times. These methods allow comparisons of prevalence at different places and times, but there are few, if any, reliable calculations of total numbers.

The Lincoln index can be used for estimating population size (Chapter 5). Captured mosquitoes can be marked with fluorescent dusts, for instance, and then released and later their dilution in a second sample of the population can be measured. If the total population of mosquitoes be taken as x and the number first caught marked and released be a, the total caught the second time, b, and the number of those of b that are marked be c, then with many qualifications:

$$\frac{a}{x} = \frac{c}{b}.$$

This equation is solved for x, the total number of mosquitoes. However many factors can disturb these ratios. The marked individuals can be damaged in handling, there can be recruitment and loss between samplings because of emergence, death, and migration. Temperature and wind influence flying and distribution, and so on.

From a marking-recapture study of the invasion of adult *Culex tarsalis* into an area of 29 square miles that had been heavily treated with larvicides (Dow *et al.*, 1965) the following very rough estimates of population size can be made, according to the formula given above:

$$\text{Test } 1 : \frac{110{,}301}{x} = \frac{132}{26{,}776} \; ; \; x = 22 \times 10^6$$

$$\text{Test } 2 : \frac{73{,}763}{x} = \frac{75}{71{,}086} \; ; \; x = 70 \times 10^6.$$

The authors themselves did not calculate x, the estimated total population size, but it is done here in order to show the type of data that are available.

In order to evaluate the efficiency of mosquito populations as vectors, knowledge of the number infected in nature should be combined with laboratory investigations of vector efficiency. Mosquitoes are reared in the laboratory and fed on experimentally infected vertebrates or on virus suspensions, and the amount of virus needed for mosquito infection is measured. Some are more readily infected than others and some carry more virus in the salivary glands and are better transmitters.

Reservoir and Amplifier Populations

The idea of a reservoir from which virus is transmitted by mosquitoes to man has been already introduced. Such a reservoir is not a large body of infection which merely overflows to man, but must be considered in terms of the amount of active infection present at a given time and its availability to mosquitoes. This depends on the number of birds with viremias high enough for infection of mosquitoes, and on the duration of the viremias.

The numbers of infected birds and their proportions in whole populations of particular areas must be known. These also are ecological considerations. Birds can be caught in mist nets, banded, and recaptured for estimates of population size and for tracing local movements and long-distance migrations. Birds occupying territory are the least difficult to count and various methods are available.

In Maryland the redwing blackbird, *Agelaius phoeniceus*, may be one part of the ecological chain involved in the maintenance and irruption of the disease. These birds occupy breeding territories in marshy places. Meanley and Webb (1963) found as many as 40 or 50 pairs per 100 acres in certain types of marshes of the Chesapeake Bay. In such areas infection may be present all the year, or, because these populations of the species migrate away for the winter, the infection is perhaps brought in anew each spring. Banded redwing blackbirds of the Chesapeake Bay populations have been recaught during the winter all along the Atlantic coast from Virginia to Florida. The birds gather in flocks in winter, and roosts may consist of many thousands of individuals. At the present time this ecological information cannot be translated into terms of quantity of natural reservoir of infection.

If most birds are infected one year, develop antibodies, and survive to the next year, they cannot then serve as a source of further infection. Therefore host mortality rates must be known. Survivorship curves typical of wild birds are shown in Figure 1.1.

What percentage of birds is infected? Data will be used from the Yamika Valley of Washington State (Reeves and Hammon, 1962) concerning western equine encephalitis virus. About 50 percent of chickens, ducks, geese, pigeons,

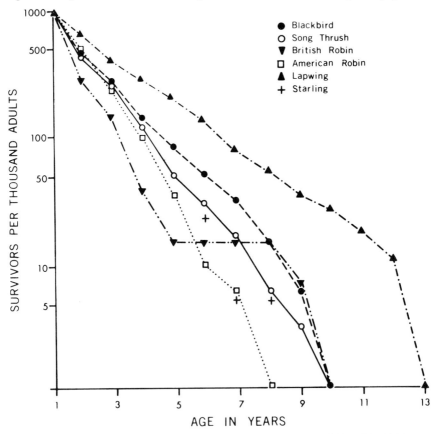

Figure 1.1. Survivorship (l_x) curves for natural populations of the British robin, *Erithacus rubecula*; song thrush, *Turdus ericetorum*; blackbird, *Turdus merula*; starling, *Sturnus vulgaris*; lapwing, *Vanellus vanellus*; and American robin, *Turdus migratorius*, age being expressed in years. (After Deevey, 1947.)

and turkeys were demonstrated to have neutralizing antibodies in their blood; about 17 percent of flickers, killdeers, owls, pheasants, quails, and robins and other wild birds had antibodies.

Perhaps the young bird in spring is more susceptible than others and has a viremia high enough to infect mosquitoes, persisting for a few days or weeks. Buescher *et al.* (1959) made a further hypothesis concerning Japanese B encephalitis virus. They suggested that since a young bird that has recovered from the infection maintains its antibodies and transmits them through the egg yolk to the offspring, the fledglings are protected during the next season. These birds lose their passive immunity and produce susceptible young. Thus there might be a series of fluctuations of the numbers of susceptibles and thereby a variation in the frequency of infections.

What is a reservoir of infection? Is a reservoir necessary for seasonal persistence of the virus? Or is it the total mass of infection, the virus populations that exist in nature, that can infect the mosquito which may also bite man? This leads to the idea of amplifier populations. These populations are not able to maintain infection from year to year, but serve as masses of tissue susceptible to virus from other populations. In them the population of infectious agent is multiplied to the level where it itself can serve as a source of infection of a large number of mosquitoes. Under these conditions the infection might become so great that it is carried over to man.

Finally, it may well be that this infection passes from one group of animals to another at different seasons of the year, not just going from bird to bird by way of mosquito and occasionally infecting man and horse, but perhaps going somewhere else. How many, then, of all these animals of various species are there in a particular area? It is only when the total ecological problem presented by such a source of infection is understood, that there can begin to be some understanding of transmission of the disease.

Survival of the Virus through the Winter

Culex tarsalis is able to keep the virus of western equine encephalitis in its body throughout the winter. It would seem, then, that in one of the encephalitides the continuing source of virus from year to year might be the overwintering mosquito. In another case, or another place, the overwintering mechanism could be something else.

In the continual search for overwintering mechanisms investigators have recently turned to animals that have usually been ignored. The garter snake was shown to be susceptible to western equine encephalitis virus. Virus persisted in artificially inoculated garter snakes for 70 days or more. It was recovered from some 23 out of 26 snakes infected in the laboratory. Thomas and Eklund (1960) showed that these snakes were infectious to mosquitoes, and that the snakes maintained titers in their blood as high as 10^6 per milliliter. Then garter snakes were artificially infected and put into hibernation in artificial conditions and on emerging they became viremic again. Here was a potentially excellent mechanism for the overwintering of virus. Does it occur in nature? Gebhardt *et al.* (1964) caught 84 wild snakes of the genera *Thamnophis*, *Coluber*, and *Pituophis*, of which 37 had virus of western equine encephalitis in the blood. There was a cyclical viremia; it disappeared and reappeared with changes in the temperature of the environment. Some of the baby snakes born of infected mothers were viremic. A viremia titer of 2.7×10^2 per milliliter in adult snakes infected 31 percent of *Culex tarsalis* that were fed on them, and a viremia of 2.7×10^7 infected 79 to 88 percent of the mosquitoes. These findings need confirmation.

Summary

All these considerations lead to the inference that there are four conditions of the presence of this virus in populations.

In some infected species there is no evidence of further transmission of virus. These are dead-end infections and by overt disease or by development of antibodies may serve as *indicators* of the presence of infection. Examples of these for the encephalitis viruses are human and horse populations.

Reservoir populations maintain virus from season to season and sometimes a vector is required. From these populations virus is available for transmission to other populations. Perhaps snake populations are one of the true reservoirs of western equine encephalitis in the United States.

Amplifier populations are large susceptible populations in which the total amount of infectious agent increases to large quantities. Deaths may occur, as in plague in rats, or the agent may be present asymptomatically in the blood of many individuals, as in encephalitis in nestling birds. The size of the pool of the agent may then allow transmission or overflow into other populations.

Vector populations transmit the agent. Their sizes play crucial roles in transmission among reservoir, amplifier, and indicator populations.

SOME RELATIONSHIPS OF DISEASE INCIDENCE TO GENETICS

"Survival of the fittest."—This phrase of Darwin's is a circular statement and yet it is inherent in all thought about populations. It means "Survival of the individuals, groups, or species that do best survive." Factors contributing to survival from generation to generation include not only the survival of sexually mature competitors but also differential survival rates of the youngest stages up to the end of the reproductive period, differential reproductive rates, and all those attributes contributing to the survival of whole groups and species, operating over long periods of time. The survival value for a population of a behavior pattern may frequently involve cooperative procedures.

All populations alter their environments to certain extents. If a population "improves" the environment, it probably has a greater chance of survival. As the dangers of the environment are eliminated, those dangers no longer have selective influence. For instance, when man emerged from life in a cave he was developing environmental control greater than the presence of fire in the cave, but while he has changed he has undoubtedly lost some characteristics that enhanced his survival in a cave. Because of such loss man cannot be said to be weaker; he is not less adapted to his environment; he has changed his adaptations in response to changing situations.

Sometimes the accusation is made against public health workers that by eliminating the selective action of certain diseases they are weakening the human race. Medicine changes the environment, and the effects of this change upon the population genetics of man are complex and should be thoroughly examined. Two relevant situations will be discussed here, and the topic will be specifically raised again later.

GENETICALLY CONTROLLED RESISTANCE TO SPECIFIC DISEASE AGENTS IN MICE

The genetic basis of resistance to some infections in laboratory animals is examined in Chapter 23, and one example will be repeated here. Following the work of Sabin (1952), Goodman and Koprowski (1962) found that C3H laboratory mice were highly susceptible to arbovirus B infection, whereas the Princeton strain was highly resistant. These two strains were crossbred and the offspring were mated with each other or backcrossed with the parents. Resistance and susceptibility to the virus segregated into the simple Mendelian ratios that are determined by two autosomal alleles, and resistance was shown to be dominant over susceptibility. But the same strains of mice showed a reversed pattern when tested with mouse-hepatitis virus (Bang and Warwick, 1960). C3H mice had recessive resistance and Princeton mice had dominant susceptibility but this was controlled by another pair of alleles at another locus. Thus if there had been selection for mice resistant to one disease, arbovirus B infection, little or nothing would have been done thereby about resistance to the other disease, hepatitis. As far as is known, there is no general factor of resistance to specific viral disease. Thus as the selective effects of one disease agent are removed from populations there may be no change in susceptibility to other disease agents, and there is no reason to believe that this sort of relationship is different for most diseases of man.

SICKLE-CELL DISEASE AND GENETIC POLYMORPHISM

Sickle-cell anemia is a disease primarily of the inhabitants of Africa, and it is present also in some parts of India. Probably its highest prevalence is in West Africa, and it is a serious problem of Negroes of the United States. The disease has a genetic basis, being determined by alleles at one locus. People homozygous for alleles Hb^S/Hb^S suffer from sickle-cell anemia and have low rates of survival to adulthood. Thus they are heavily selected against. The heterozygous individuals of genetic constitution Hb^S/Hb^A, have some sickle-cell hemoglobin and have the sickle-cell trait but not the actual disease. Normal people, Hb^A/Hb^A, have no hemoglobin of the sickle-cell type. In West Africa between 10 and 20 percent of the human population carry an Hb^S

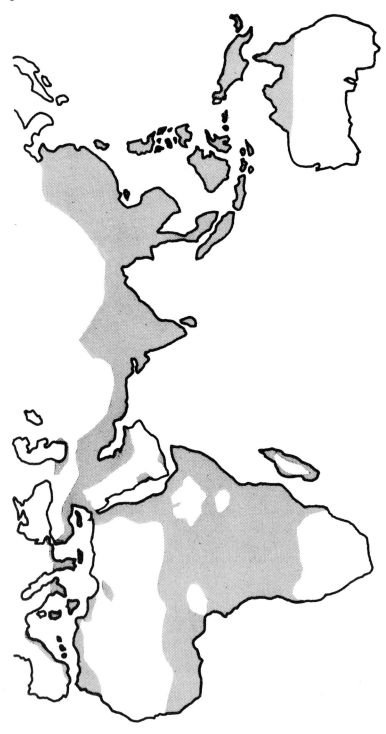

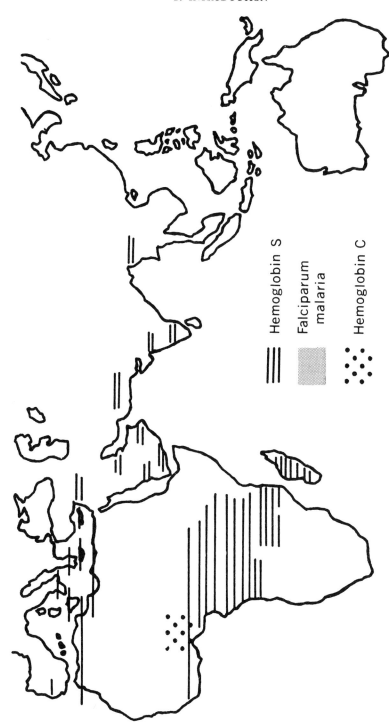

Figure 1.2. The distribution of falciparum malaria (upper figure) and the distribution of hemoglobin S and hemoglobin C (lower figure). Negro populations in some parts of the world, particularly the United States, are not included. (After Boyd, 1949, and Motulsky, 1960.)

Hemoglobin S

Falciparum malaria

Hemoglobin C

allele. Individuals with other hemoglobins due to other alleles of the same locus are present in the same population; another allele, Hb^C, will be mentioned below. This discontinuous distribution, at high relative frequencies, of characters determined at the same genetic locus is genetic polymorphism.

The Correlation of Sickle-Cell Disease with Falciparum Malaria

The continuing presence in the populations of the allele Hb^S that has deleterious effects was a puzzle for a long time, and Allison (1957) and Raper (1959) were largely responsible for the solution. They recognized that the sickle-cell trait must have some selective advantage, perhaps in resistance to a disease. The most common serious disease of the areas is falciparum malaria (Figure 1.2). Therefore the question arose, "Is the Hb^S/Hb^A individual protected against malaria?" This introduces the idea of a heterozygote having selective advantage over both homozygotes. Fisher (1929) stated that the most general basis of all genetic polymorphisms is a balance of opposed advantage and disadvantage so that the heterozygote is favored compared with either homozygote.

Allison estimated survival values of the various hemoglobin types determined by combinations of the alleles Hb^S, Hb^A, Hb^C. Survival values were calculated from comparison of proportions of the various types in the very young members of the population, with proportions of the various types in the adult members of the population (Table 1.1).

TABLE 1.1.

Survival Values of Various Hemoglobin Types in Africa[a]

Genetic factors for hemoglobin type	Estimated survival value
Hb^A/Hb^S	1.13
Hb^A/Hb^C	1.03
Hb^A/Hb^A	0.976
Hb^C/Hb^C	0.55
Hb^S/Hb^S	0.192
Hb^S/Hb^C	0.40

[a]Allison (1957).

The survival values of the individuals Hb^C/Hb^C, Hb^S/Hb^S, Hb^S/Hb^C were low, and if these were the only values operating the alleles Hb^S and Hb^C should reach very low relative frequencies. However, the selective advantages of the heterozygotes Hb^A/Hb^S and Hb^A/Hb^C operate for large numbers of individuals over long periods of time. The greater survival of these heterozygotes in the presence of falciparum malaria must have largely contributed to the maintenance of the Hb^S and Hb^C alleles.

Mechanism of Protection of the Heterozygotes against Falciparum Malaria

A large proportion of persons heterozygous for the Hb^S allele, that is, of those people who have sickle-cell trait, do contract malaria. But the numbers of parasites in their blood are at about one-third the level found in normal persons with malaria. So the sickling trait is associated with resistance to malaria. This may again be demonstrated from autopsy data from Nigerian children. Of 40 children that died of falciparum malaria, not one carried an allele for sickle-cell hemoglobin. Yet, as stated above, the prevalence of the gene in the population at large is from 10 to 20 percent.

The protective mechanism is not known. It is not necessarily based on the presence of the sickle-cell hemoglobin. On the basis of some genetically controlled mechanism associated with possession of the allele for sickling, individuals can survive the onslaughts of malaria.

Removal of Populations from the Environment of Malaria

When the environment no longer contains falciparum malaria, the selective advantage of the heterozygote is removed. The homozygote Hb^S/Hb^S continues to exert its deleterious effect and the allele Hb^S should decrease in frequency, and perhaps eventually almost disappear. There is evidence that in the Negro population of the United States, a country where there is now no significant amount of falciparum malaria, the allele for sickling is indeed becoming less frequent. Thus this is an example of how elimination of a particular disease could cause decline of another disease. It is hardly an adequate example to support the idea that removing the selective effect of a disease may weaken the human species.

CONCLUSIONS

The equine encephalitides are unsolved problems in the sizes and distribution of animal and virus populations—they are problems in ecology. Sickle-cell disease illustrates that disease can be a major factor in the natural selection of man.

It is recognized that extensive areas have been covered in this chapter. Many parts of the accounts have been left unfinished, but this was done purposely. It is hoped that the population approach will stimulate closer examination of the ideas of ecology, population genetics, and comparative behavior. It will then be clearer how the basic data of public health problems must be examined in order to frame specific questions, and, hopefully, to answer them.

REFERENCES

Allison, A. C. 1957. Malaria in carriers of the sickle-cell trait and in newborn children. *Exptl. Parasitol.* **6**: 418.

Allison, A. C. 1964. Polymorphism and natural selection in human populations. *Cold Spring Harbor Symposia* **27**: 137–149.

Audy, J. R. 1958. The localization of disease with special reference to the zoonoses. *Trans. Roy. Soc. Trop. Med. Hyg.* **52**: 308–328.

Bang, F. B., and Warwick, A. 1960. Mouse macrophages as host cells for the mouse hepatitis virus and the genetic basis of their susceptibility. *Proc. Natl. Acad. Sci. U.S.* **46**: 1065–1075.

Boyd, M. F. (ed.) 1949. *Malariology.* Saunders, Philadelphia, Penn.

Buescher, E. L., Scherer, W. F., McClure, H. E., Moyer, J. T., Rosenberg, M. Z., Yoshii, M., and Okada, Y. 1959. Ecologic studies of Japanese encephalitis virus in Japan. IV. Avian infection. *Am. J. Trop. Med. Hyg.* **8**: 678–688.

Deevey, E. S. 1947. Life tables for natural populations of animals. *Quart. Rev. Biol.* **22**: 283–314.

Dobzhansky, T. 1962. Mankind Evolving—the Evolution of the Human Species. Yale Univ. Press, New Haven, Conn. and London.

Dow, R. P., Reeves, W. C., and Bellamy, R. E. 1965. Dispersal of female *Culex tarsalis* into a larvicided area. *Am. J. Trop. Med. Hyg.* **14**: 656–669.

Fisher, R. A. 1929. On some objections to mimicry theory: statistical and genetic. *Trans. Roy. Ent. Soc. London*, **75**: 269–278.

Gebhardt, L. P., Stanton, G. J., Hill, D. W., and Collett, G. C. 1964. Natural overwintering hosts of the virus of western equine encephalitis. *New Engl. J. Med.* **271**: 172–177.

Goodman, G. T., and Koprowski, H. 1962. Study of the mechanism of innate resistance to virus infection. *J. Cellular Comp. Physiol.* **59**: 333–373.

Jennings, H. S. 1927. Public health progress and race progress—are they compatible? *Trans. Meet. National Tuberculosis Assoc.* **23**: 125–142.

Meanley, B., and Webb, J. S. 1963. Nesting ecology and reproductive rate of the redwinged blackbird in tidal marshes of the upper Chesapeake Bay region. *Chesapeake Science* **4** (2): 90–100.

Motulsky, G. A. 1960. Metabolic polymorphisms and the role of infectious diseases in human evolution. *Human Biol.* **32**: 28–62.

Pearl, R. 1937. On biological principles affecting populations: human and other. *Am. Naturalist* **71**: 50–68.

Raper, A. B. 1959. Further observations on sickling and malaria. *Trans. Roy. Soc. Trop. Med.* **53**: 110–117.

Reeves, W. C., and Hammon, W. M. 1962. Epidemiology of the arthropod-borne viral encephalitides in Kern County, California, 1943–1952. Univ. of California Publications in Public Health 4.

Sabin, A. B. 1952. Nature of inherited resistance to viruses affecting the nervous system. *Proc. Natl. Acad. Sci. U.S.* **38**: 540–546.

Thomas, L. A., and Eklund, C. M. 1960. Overwintering of western equine encephalitis in experimentally infected garter snakes and transmission to mosquitoes. *Proc. Soc. Exptl. Biol. Med.* **105**: 52–55.

I

POPULATION GROWTH AND ECOLOGY

SECTION I

POPULATION GROWTH AND ECOLOGY

A population is defined here as a reproducing group of individuals of one species, living in a certain area or volume. No population exists by itself in nature; populations are very interdependent. Single species cannot take from the physical environment all the energy and raw material that they need, and on their own return inorganic material. Rather, species are adapted for small roles in the whole series of biochemical processes of the total living organization, the biosphere. Species specialize also in ability to withstand only a small range of the conditions of the total environment. The role of a species in the total biological process, and the small range of conditions in which it operates, is its ecological niche. The gross kind of place in which it lives is its habitat—for instance, soil of deciduous woodland, sheep carcasses in open grassland, or the bottom of shallow temperate seas.

Food energy enters the biosphere almost entirely by photosynthesis, with sunlight as the energy source. Then this energy is distributed through consumers of living and dead organic material. Thus most living populations are harvested in many ways by other populations. Eventually a very small portion of the energy is left to accumulate as unused organic materials in the soil and ocean bottoms, for example, as peat, coal, and petroleum.

Long-term survival of a population implies that it does not overdamage its food resources, that some of its habitat is always available to it, and that it is able to withstand various onslaughts. These onslaughts include harvesting and competition by other populations, and all the physical forces of the environment.

A population has an input and loss of individuals and a distribution of individual ages. Mathematical and statistical description of these conditions is the subject of demography. Demographers and ecologists study the numerical and biological properties of single-species populations, of pairs of dependent or competing populations, of larger groups of interrelated populations (ecological communities), and the relation of all of these to the total environment. Human populations are also cogs in the wheels of the total biological organization, and can be examined by the same criteria as other populations.

CHAPTER 2

Population Growth of Bacteria

BYRON S. TEPPER

The growth of a population is continually influenced by the environment and it, in turn, modifies the environment. The nature of the forces acting on a population may be simple or highly complex, and different populations may vary in response to the same environmental factors. In attempts to understand the principles involved, some of these forces can be controlled in experiments.

Bacteria, as described in this chapter and Chapter 4, offer unusual conditions for controlled population study. These unicellular organisms can be cultivated with relative simplicity, they multiply rapidly, and attain enormous population sizes. Furthermore, sexual fusion is not obligatory, and one unit is sufficient to start a population. The individuals of the resulting clone are then of similar genetic makeup, and thus one of the most important variables in populations is almost eliminated.

In bacteria there is growth of the individual as well as growth of the population. At a characteristic size the growth of the individual ceases, and the individual divides into two so that the population grows in number. Bacterial population growth is expressed in numbers or in mass. Cell concentration is the number of cells per unit volume; bacterial density is total material per unit volume. The number of viable organisms is determined by culture and the total number is calculated from sample microscopic counts. The bacterial density may be measured turbidimetrically, the optical density representing a product of cell size and the number; by volume or weight of the bacterial mass after centrifugation; and by total cellular nitrogen. In certain cases, cellular activities such as rates of acid or carbon dioxide production, oxygen, or nutrient uptake are employed as indices of total bacterial protoplasm. Each method has limitations and must be proved to be applicable.

MATHEMATICAL DESCRIPTION OF BACTERIAL POPULATION GROWTH

One bacterial cell divides, producing two new cells, and a bacterial population can develop by geometric progression. The daughter cells grow and divide at approximately the same rate as the parent cell and the number of cells doubles at essentially regular intervals. If the number of bacteria is b,

at the 1st generation, $b = 1 \times 2 = 2$

at the 2nd generation, $b = 1 \times 2^2 = 4$

and at the nth generation, $b = 1 \times 2^n$

In practice the initial number of bacteria is probably several thousand and is represented by a, so that

$b = a \times 2^n$

The logarithms of the numbers can be substituted for the numbers themselves and the equation for population size becomes

$$\log_{10}b=\log_{10}a+n\log_{10}2$$
$$=\log_{10}a+0.301\ n.$$

To determine the number of generations the equation is solved for n.

$$n=\frac{\log_{10}b-\log_{10}a}{0.301}$$

The average generation time, G, is equal to t, the time for a to develop to b, divided by the number of generations, n.

$$G=\frac{t}{n}$$

Substituting for n

$$G=\frac{0.301\ t}{\log_{10}b-\log_{10}a}$$

THE GROWTH CURVE

It has been assumed that G is fairly constant, but this is so during only a part of normal population growth. In reality at least five phases of population growth can be recognized.

The Lag Phase, or Phase of Adjustment (A in Figure 2.1)

In this phase increase in the number of cells is not demonstrable but there is intense growth activity in the individual cell. After a short time in a new medium there may be two- to threefold increase in cell size, indicating intense synthetic activity. These cells contain more of most constituents and have higher metabolic activities than cells from any other phase of the growth cycle. There is increased susceptibility to disinfectants and to physical agents such as heat, cold, and osmotic pressure. Changes in the electrochemical state of the surface can be demonstrated by a diminution in the electrophoretic charge of the cell and by decreased susceptibility to nonspecific agglutination. The phase of adjustment has been referred to as a period of rejuvenation. Cells can overcome toxic effects of their previous environment; they accumulate nutrients and attain the synthetic and metabolic capabilities necessary for multiplication. Activities build up to some steady state which must be attained before division can occur. Once these levels are reached the individual cell divides. As more and more cells divide the population growth rate increases.

The Exponential Growth, or Logarithmic, Phase (B)

Most bacteria are growing at a fairly constant rate in this phase. The logarithms of the numbers plotted against time make a straight line. The rate of increase in protoplasm is constant and maximal. Under favorable conditions mortality is very low with more than 90 percent of the cells continuing to

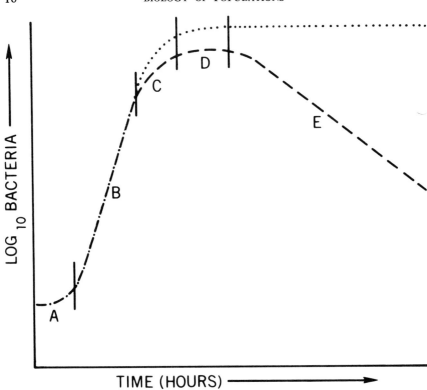

Figure 2.1. Typical growth curve, showing total numbers (dotted line) and numbers of viable bacteria only (dashed), plotted as logarithms. The phases of the curve are described in the text.

divide. The rate of multiplication is independent of the concentrations of nutrients and toxic metabolites. However at critical concentrations the toxic metabolites increase sharply and contribute in large part to deceleration at the end of the phase. The exponential phase is a steady state. Enzyme and substrate concentrations are maintained at such levels that the rate of synthesis of cell protoplasm proceeds at the maximum for the strain of bacteria in that environment.

Decreasing Growth Phase (C)

The decreasing growth rate of the population may depend on various factors, such as exhaustion of food, accumulation of toxic metabolites, and rates of diffusion of oxygen into the medium. If growth is accompanied by a decreasing concentration of food, the rate falls when food is reduced to limiting levels. In the presence of excess food the accumulation of toxic metabolites eventually reaches concentrations high enough to inhibit growth. Since decreases in available nutrients or accumulations of toxic products are gradual there is no sharp change in the growth curve at the end of the exponential phase, but rather a gradual deceleration.

The Maximum Stationary Phase (D)

Exhaustion of nutrients and accumulation of a toxic environment, together or singly, eventually halt population growth and the death rate becomes equal to the multiplication rate. The viable count becomes constant at its maximal value. For those organisms that are very susceptible to an unfavorable environment the stationary phase is brief and is followed rapidly by death of cells. However, in most cases the culture may stay in the stationary phase for hours, even days, before a decline in viable cells becomes noticeable.

The population of viable bacteria per unit volume of medium at the maximum stationary phase tends to be constant for a given organism and given conditions of culture. This population of bacteria is the *M-concentration*.

The factors limiting the population growth at the M-concentration are not clearly defined. Frequently, one can remove the organisms by centrifugation or filtration and find that a fresh inoculum grows in the medium, so that cessation of growth by the original culture could not have been due to exhaustion of nutrients or accumulation of toxic metabolic products. When a fresh medium, rich in nutrients, is inoculated with a concentration of organisms greater than the M-concentration, there is a decrease in the number of cells to the M-concentration and no further growth occurs.

Such observations have suggested that a certain amount of "biological space," is required to support the growth of a bacterium. On the other hand it is likely that a critical concentration of nutrients per organism must be exceeded for growth and multiplication. This is suggested by findings that in very dilute media, in spite of the presence of some nutrients, no growth and multiplication may occur. Probably the concentration gradients and quantity of metabolites diffusing into the cell are quite small and these minimal quantities permit low metabolic rates in the cell but are insufficient for the increases in protoplasm and metabolic activities necessary for multiplication.

Under ordinary methods of cultivation the available nutrient concentration declines continuously and multiplication ceases when a critical level is reached. Even in this case, when the organisms are removed, and a few inoculum organisms are added, growth ensues. This second growth does not achieve as large a population as the first growth. Yet growth of the first population was halted because of nutrient limitation and the nutrients should have been consumed and there should have been no second growth. Apparently some unidentified factor is responsible for the phenomenon.

The Phase of Decline, or Death Phase (E)

In this phase the viable count decreases. Decline of the population begins and the rate of decline increases and eventually reaches a maximum that is usually constant, and the culture dies exponentially.

Deviations from the exponential death rate are observed. Frequently, after the majority of the population has died, the decline decreases rapidly. A small number of survivors may persist using nutrients released by the decomposition of the dead individuals.

FACTORS INFLUENCING THE GROWTH CURVE

The Organism

Different genera and species have different generation times, varying from less than 20 minutes to hours and days. The slope of the exponential phase of the growth curve reflects these differences. The maximum population density in a given medium, the length of the stationary phase, and the rate of decline of the population are determined by the inherent metabolic capabilities of the bacteria and their susceptibility to toxic environments.

The Medium (Figure 2.2)

The composition and concentration of the medium control the initiation of growth, the rate, and the final population size. Those metabolites which the cell cannot synthesize must be supplied. These include growth factors, nitrogen, and carbon sources which the cells use in the synthesis of protoplasm, and the substrates which furnish energy for the synthetic processes. The absence from the medium of a metabolite which the cell synthesizes more slowly than necessary for maximal rates of growth will affect the rate at which the final level of growth is reached.

Within limits the rate of growth of a bacterial cell during the exponential phase may be inversely proportional to the concentration of an essential nutrient. Above a concentration specific for each nutrient further additions may have no effect or may become inhibitory.

The concentration of nutrients has already been shown to influence the maximum population density. The exhaustion of a limiting nutrient decreases the growth rate of the exponential phase and establishes the maximum population. Under appropriate conditions it is possible to establish a direct linear relationship between the initial concentration of the limiting nutrient and the maximal population of viable organisms attained.

Oxygen and carbon dioxide tensions, hydrogen ion concentration, and physical conditions such as surface tension and osmotic pressure influence the initiation and rate of growth.

Temperature

Temperature influences bacterial growth by two means. It influences the rate of chemical reactions and at the higher levels it influences destruction of proteins, especially enzymes. The rates of chemical reaction within a certain range are approximately doubled for each 10°C rise in temperature. As the temperature approaches the maximum compatible with growth, the rate at which enzymes are destroyed increases. At these levels even small increments

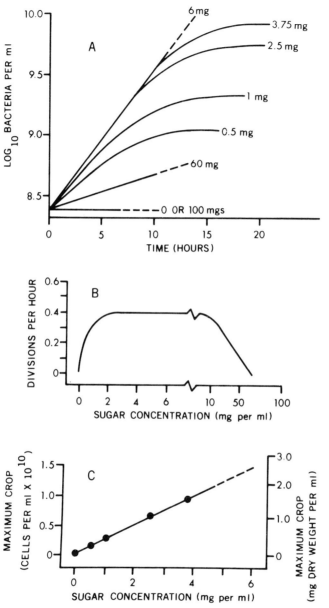

Figure 2.2. Growth of *Pseudomonas* populations when fructose (given in milligrams per milliliter) is sole carbon source, and other environmental factors are constant. All curves are based on the same data.   A. Growth as a function of time at various concentrations of fructose.   B. Growth rate as a function of fructose concentration. C. Maximum crop as a function of fructose concentration. Note the direct relationship. (After Stanier *et al.*, 1963.)

of temperature result in very marked acceleration in the rate of destruction of
cellular components resulting in complete cessation of growth. A curve
indicating the influence of temperature on bacterial growth is shown in
Figure 2.3. Three major influences are shown: lower temperatures at which
the organism will not grow or multiply, higher temperatures above which the

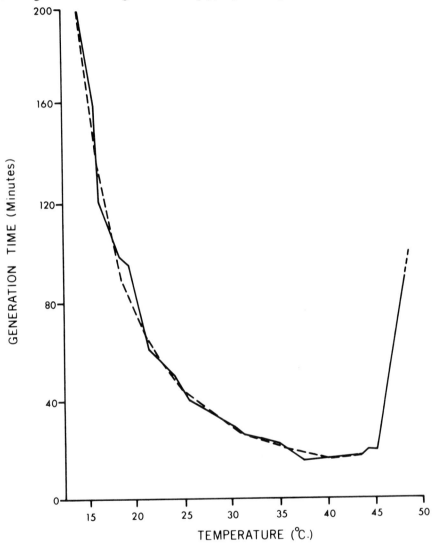

Figure 2.3. Influence of temperature on generation time of *Escherichia coli*. The solid line
connects points that represent data; the broken line is a smoothed curve. (After Barber,
1908.)

organism will not grow, and the optimum temperature at which the organism has minimal generation time and maximal growth rate.

Three temperature optima can be determined for a given culture, optima for growth rate, population size, and metabolic activities. The optimum for growth rate may be as much as 10°C above the temperature at which maximum population size is produced.

LABORATORY CONTROL

Bacterial populations can be maintained for long periods in the exponential phase if the supply of nutrients is maintained and the toxic products of metabolism are eliminated. This can be achieved by repeated transfer of large numbers of cells into fresh media every few hours, but then the population size fluctuates markedly at each transfer. Activated carbon may be added to the culture for adsorption of toxic metabolites. Cultures may be periodically brought to neutral pH. Under appropriate conditions aeration of the culture may increase the population attained. Another successful method is growing bacteria in dialysis bags surrounded by large quantities of sterile medium. Consumed nutrients are replenished from the external fluid while at the same time toxic metabolites are lost by outward diffusion. However, in all of these manipulations in closed systems, the growth cannot be extended over long periods.

The chemostat provides a method for keeping a culture growing exponentially for an indefinite time at a constant population size and at a controlled rate. It is a culture vessel with a continuous automatic supply of fresh medium, that simultaneously removes an equal flow of culture fluid containing organisms. The removal of bacteria compensates exactly for population growth. Two variables are manipulated, the concentration of the introduced nutrient and its flow through the system. The medium contains excess of nutrients required for growth, except for one whose concentration is held at a limiting level. The multiplication rate is controlled by the rate of flow. When the system reaches the steady state for a particular fixed flow rate, the concentration of the limiting nutrient determines the population density.

MICROBIAL ECOLOGY

Pure cultures of microorganisms rarely occur thus in nature. In most natural environments bacteria associate with yeasts, molds, actinomycetes, viruses, protozoa, algae, and other multicellular organisms.

Mutualism

When organisms mutually benefit from an obligatory association the relationship is called mutualism. For instance the growth of anaerobic bacteria in the presence of aerobic forms is frequently observed. The aerobic species use oxygen and thus create conditions favorable for the growth of anaerobes. The aerobes in turn may oxidize the waste products of the anaerobes. One of the most important mutualistic relationships in the soil is that between the nitrogen-fixing rhizobia and leguminous plants. The bacteria grow in root nodules of the plants and provide nitrogenous nutrients to the plants and the bacteria derive nutrients from the plants.

Synergism

Two or more species in association can produce changes which are not produced by either alone. For example, *Proteus vulgaris* and *Staphylococcus aureus* each ferments lactose, producing acid but no gas. However when the species are together acid and gas are produced. Certain infections of plants and animals are caused by synergistic organisms. Swine influenza is caused by the combined action of swine influenza virus and the bacterium *Hemophilus influenza suis* (Chapter 24).

Metabiosis

One organism can produce substances which favor the growth of another organism. In its broadest sense, metabiosis is the simultaneous growth of one organism on waste products of another, which is not injured thereby. When throat swabs are streaked into blood agar plates it is frequently observed that colonies of *Hemophilus influenza* grow more rapidly and bigger when they are near staphylococcus colonies. Staphylococci produce a diffusible material which is essential to the growth of *Hemophilus* and is deficient in the blood medium. Other examples of metabiosis are fermentation of grape and apple juices by yeasts, which prepare for the development of acetic acid bacteria.

Commensalism

One organism can live in, with, or on another, partaking usually of the same food without harming or helping the host. Many of the bacteria of the intestines of warm-blooded animals are thought to be commensals.

Antibiosis

One microorganism can produce a substance or condition inimical to the growth or survival of other organisms. Organisms that produce antibiotics are an example. The formation of acids by fermentative bacteria, which inhibit the growth of putrefactive species, is an example of an inhibitory change in the environment, a nonspecific antibiosis.

Parasitism

Parasitism is an intimate association between two organisms in which only one member profits from the association.

Parasites live in or on other organisms, hosts, sometimes to the detriment of the hosts. Staphylococci are parasitic on the skin and mucous membranes of man. When they penetrate into, and multiply in, deeper tissues they

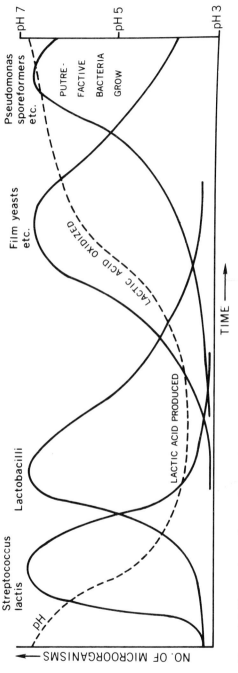

Figure 2.4. Changes in numbers and types of microorganisms in raw milk at room temperature. For explanation see text. (After Carpenter, 1961.)

damage the host and become destructuve parasites or pathogens. Parasitism, with or without pathogenicity, is widespread in nature.

Two examples of cooperation and conflict among naturally occurring microbial populations will be briefly considered.

Fermentation of Raw Milk at Room Temperature (Figure 2.4)

Milk is a good culture medium with a nearly neutral pH. It contains lactose, a fermentable sugar. *Streptococcus lactis* or other lactose fermenters grow and form acids, mostly lactic acid, curdle the milk, and cease growth when a limiting amount of acid has been developed. Next to grow are the more acid-tolerant lactobacilli, which develop still more acidity. This high acidity eventually inhibits the growth of other bacteria, but permits the growth of molds and film yeasts, which are not sensitive to high acidity and can, in fact, use lactic and similar organic acids as foods. These aerobic organisms grow over the surface of the curd or on the whey and destroy enough acid to permit further lactic acid fermentation of the remaining lactose. Finally the molds and film yeasts may use all the acid with the result that the reaction becomes neutral or even slightly alkaline. This gives the protein-splitting bacteria, hitherto inhibited by the presence of acid, an opportunity to decompose the milk proteins. This series of events is a good example of ecological succession (Chapter 7).

The Decomposition of Organic Matter in Soils

One gram of soil may contain 10^9 bacteria, 5×10^6 actinomycetes, 5×10^5 protozoa, 2×10^5 algae, and 2×10^5 molds, the numbers being dependent on moisture, temperature, type of soil, nutrients, and other factors. Soil is composed of inorganic particles coated with colloids, mainly of organic matter—waste and dead plant and animal materials, chiefly proteins, fats, and carbohydrates. Soils with high moisture content are subject to smaller temperature variations than are dry soils. Saturated soils may become anaerobic and drier soils permit greater oxygen diffusion. Thus moisture influences temperature and oxygen content, and consequently the microbes and their activities.

The organic matter is broken down into increasingly simpler compounds. The main end products in well cultivated soil are water, carbon dioxide or carbonates, nitrates, and traces of other salts. This process of reduction of complex organic matter to salts is called mineralization. It results from numerous complex sequences of microbial activities. The sequences that are particularly vital to the living world are involved in the cycling of carbon, nitrogen, and sulfur.

Though a wide range of species may enter an environment, many appear to be unable to survive in competition with the established forms, while at times one form may enter and establish itself at the expense of other forms. Much remains to be learned about activities of organisms in their natural environment in mixed cultures.

CONCLUSIONS

The growth of bacterial populations in pure culture as described in this chapter is an example of the logistic pattern that occurs in some other populations. Bacteria have more homogeneous populations than higher organisms since reproduction occurs by binary fission and sexual recombinations are rare. Many environmental conditions influence the rate and extent of population growth and limits are imposed by, among other things, exhaustion of nutrients and production of toxic substances. That a given environment will support a certain maximal population (at the M-concentration) compares with the concept of carrying capacity that is applied to habitats supporting higher organisms. In mixed populations a bacterial population may be influenced by favorable or unfavorable associations with other organisms and a few of these associations have been described. Chapter 4 describes what is known of interactions of populations in the ecology of the bacterial flora of the gastrointestinal tract.

REFERENCES

Alexander, M. 1964. Biochemical ecology of soil microorganisms. *Ann. Rev. Microbiol.* **18:** 217–252.

Barber, M. A. 1908. The rate of multiplication of *Bacillus coli* at different temperatures. *J. Infect. Diseases* **5:** 379.

Carpenter, P. L. 1961. *Microbiology.* Saunders, Philadelphia, Penn.

Hungate, R. E. 1962. Ecology of bacteria. In *The Bacteria*, Vol. 4. I. C. Gunsalus and R. Y. Stanier (eds.). Academic Press, New York.

Lamanna, C., and Mallette, M. F. 1965. *Basic Bacteriology.* Williams and Wilkins, Baltimore, Md.

Oginski, E. L., and Umbreit, W. W. 1959. *An Introduction to Bacterial Physiology.* Freeman, San Francisco, Calif.

Pelzar, M. J., and Reid, R. D. 1958. *Microbiology.* McGraw-Hill, New York.

Sarles, W. B., Frazier, W. S., Wilson, J. B., and Knight, S. C. 1951. *Microbiology, General and Applied.* Harper and Bros., New York.

Society for General Microbiology. 1957. *Microbial Ecology.* Seventh Symposium. Cambridge Univ. Press, Cambridge, England.

Stanier, R. Y., Doudoroff, M., and Adelberg, E. A. 1963. *The Microbial World.* Prentice-Hall, Englewood Cliffs, New Jersey.

CHAPTER 3

Population Growth of Viruses

WILLIAM D. HILLIS

Viruses, the smallest and biochemically least complex of all replicating organisms, offer a unique opportunity for the study of population growth. Not only can population clones composed of great numbers of individuals result from infectivity by a single viral particle, but of even greater fundamental biological interest is the fact that viral progeny of considerable numbers can often result from the infection of living cells by highly purified, virus-derived nucleic acid of a single molecular species. Thus, entire populations are available for study, the ancestry of which can be reduced to the molecular level. The gene pool of the resulting population is uncomplicated by differences in parent genes and further simplified by the absence of the variations that accompany sexual processes. Genetic variation thus arises only as the result of mutations, which in viruses can be considered simply as rare miscopying events in the replication of molecules in nucleic acid templates. Furthermore, maximal population sizes are achieved in relatively short periods of time.

Virus populations are still further simplified by the absence of increase in size or mass of individual particles. Every viral particle of the same species is the same size and mass as every other particle, provided it is not deformed, incomplete, or immature. It can be inferred, then, that no time is expended for increase in mass of individual units. Reproduction, or better, replication, continues as long as the original parent molecule is supplied with adequate building materials and synthesizing and assembly processes, at least up to certain practical limitations.

However, population studies with viruses are complicated by their obligate parasitic nature. The environment of the replicating viral population must be the interior of a living host cell. In fact, in the absence of suitable living host cells viruses can hardly be considered biological entities, only a little more than relatively simple mixtures of organic chemical compounds. Their inclusion in the biological sphere is largely dependent upon their capability of demonstrating the most fundamental of population characters, the ability to increase in numbers. This character might best be described for viruses as the ability to direct an increase in numbers. In the final analysis this ability is highly sensitive to the viability of the host cell, but if an adequate host-supplied metabolic machinery continues in operation, growth proceeds in a characteristic pattern for any given viral type.

Specific environmental requirements of any given virus are beyond the scope of this chapter. Assuming that a suitable, viable environment can be supplied for a sufficient time, attention can then be given to the pattern of growth of a population arising from host cells which are infected with identical viral particles. In general, all of the mature progeny of such infections should then be identical, if mutations do not occur.

Human poliomyelitis virus type 1, infecting monkey kidney cells in suspension cultures, has been selected as a prototype virus-host system. It should not be construed from this example that all virus populations grow in an identical manner, but the principles of growth demonstrated by this animal virus type hold generally true for many or most animal viruses, if allowances are made for relatively minor differences in the time required for population increases. Such temporal differences are to be expected when considering the wide differences in the basic chemical composition of different virus types.

MEASUREMENT OF POPULATION GROWTH

Sizes of mature virus populations at any point in time following the initial infection event may be measured in several ways. For the purposes of this presentation, determinations of the number of infective units of virus present will serve to represent the number of whole or mature poliomyelitis virus particles formed. Infective units are here measured as particles which give rise to definitive plaques when exposed to monolayers of monkey kidney cell cultures and overlaid with an agar gel, allowing the cells to continue active metabolism but preventing the free movement of viral particles from cell to cell. Measurement of total virus formed within the suspended cells can be achieved by first removing aliquots of the infected mixture at given intervals, then freezing and thawing the cells quickly to disrupt the cells and release intracellular virus, and finally mechanically dispersing the harvest to obtain individual viral infective units in the suspension. Other technical aspects of the study are unimportant to the present discussion, but details can be found in the original report upon which this example is based (Howes, 1959).

THE PATTERN OF POPULATION GROWTH

Growth of such a viral population is illustrated in Figure 3.1, in which time, in hours, is plotted on the horizontal axis and the size of the virus population, in terms of logarithms of plaque-forming units, PFU, per infected cell is plotted on the vertical axis. The curve illustrates the population which results when each monkey kidney cell in the suspension is infected with an average of 4.9 type 1 poliovirus particles. Although not shown here, a similar curve would result from individual cells each infected with single viral particles.

The Lag Phase Including the Period of Eclipse

As can be seen in Figure 3.1, a period of inactivity of some 4 hours' duration follows the infection of the cells at time zero. The period corresponds to the lag phase of other biological populations, and numerous significant events occur during this initial interval. First, the infecting viral particles must gain

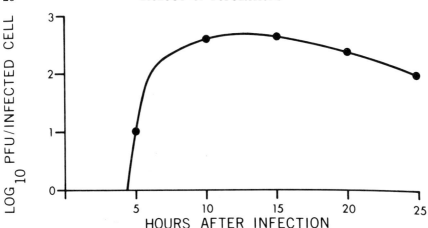

Figure 3.1. Growth curve of human poliomyelitis virus Type 1 in suspension cultures of monkey kidney cells, as measured by changes in numbers of infective units (plaque-forming units, PFU), at a viral multiplicity of 4.9 units per cell. (After Howes, 1959.)

entrance into the interior of the cell. This is accomplished by two recognizable events: the viral particles are adsorbed onto the surface of the cell, and the cell membranes are penetrated by the adsorbed particles. Having now gained entrance into the interior of the cell, the particles can be shown to be in-accessible to the inhibitory effect of specific type 1 antibody. The remainder of the period of inactivity, from penetration until the first new viral particles begin to be assembled, is known as the eclipse period. The infectivity of the penetrated particles is reduced or altogether lost, resulting from enzymatic breakdown by host processes. Eventually, the stripped particles of viral nucleic acid, in this case RNA (ribonucleic acid) associate themselves inti-mately with the endothelial reticulum of the host cell cytoplasm and there begin to direct the synthesis of required enzymes and virus-specific proteins. Other viral types may localize on or within different cellular components, but wherever the localization, the direction or redirection of syntheses begins during this period in which no actual growth can be detected. Simultaneously, or somewhat later, viral nucleic acid molecules replicate into multiple units of new RNA or DNA (deoxyribonucleic acid) in the case of some other viruses. By the end of the eclipse period the infected cells have produced in limited quantities all building blocks necessary for the assembly of new viral particles.

Logarithmic Phase to Cessation of Growth

At the beginning of logarithmic growth, some 4 hours after infection, self-assembly of the newly available building blocks begins and newly formed infectious units are detectable in virus assays. During the following 2 hours or so the increase in virus population proceeds at a logarithmic rate and begins to level off during the next 2 hours. Peak yields occur at 8 to 10 hours after infection and vary between 100 and 200 infectious units per infected cell. Thereafter no further replication occurs, despite the fact that excess viral building blocks are still available. A population saturation point has been achieved, and from this time no new increases in population can be detected. Instead, over the next several hours the infectious titer decreases somewhat, owing probably to thermal inactivation of some of the mature viral particles.

Release of Virus

Infectious virus is released, in this case as the result of lysis of some of the infected cells beginning some 6.5 hours after infection. The resulting free virus, detectable in the supernatant culture medium and distinct from intracellular virus, increases as shown in Figure 3.2. With increasing time more and more

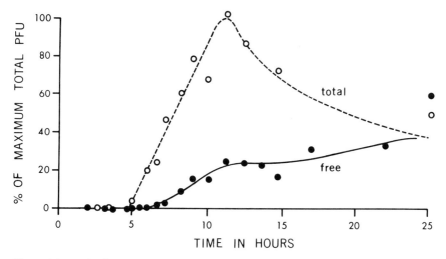

Figure 3.2. Production of total and free (released) Type 1 poliovirus in suspension cultures of monkey kidney cells at a multiplicity of infection of 4.9. Population size is expressed as percentage of the maximum production of plaque-forming units, PFU, as a function of time, in hours. (After Howes, 1959.)

cells are lysed with the resulting release of increasing numbers of free virus particles. In Figure 3.2 time is plotted against the percent of the maximum number of total infectious units produced for both total and free virus. Virus newly released from lysed cells is now fully infective and can attach to and penetrate other susceptible cells and thus initiate a new cycle of growth for another generation.

FACTORS INFLUENCING THE GROWTH CURVE

As with bacterial populations, virus growth can be strongly influenced by a large number of factors, many of which are necessarily mediated through changes in the host cell physiology. Thus, temperature, by regulating the rate of cellular metabolism, as well as thermal inactivation of newly formed virus, exerts a marked effect on the shape of the growth curve. Other intracellular host factors are known to regulate the rate of growth of infecting viral populations, including pH and the inhibiting action of certain cellular products. Of the latter, one of the most thoroughly studied is interferon, a highly cell-specific basic protein which is capable of interfering with the multiplication of certain virus types. Synthesis of this substance may be stimulated by both infectious and noninfectious particles of various viruses, and the interferon thus produced is inhibitory to both homologous and heterologous viruses. The exact mechanism of inhibition is poorly understood, but it is known that interferon inhibits the production of viral nucleic acid and also viral RNA polymerase activity.

The Virus Species

Different species of viruses, as mentioned above, vary widely in the growth pattern each characteristically exhibits under optimal conditions. Even the same virus type may show highly variable growth behaviour in different species of susceptible cells. Thus, variations in growth curves can be seen in comparative studies of a wide range of animal viruses, as shown in Figure 3.3. Six prototype viruses, representing the major virus classes, exhibit total growth curves as shown by the upper curve in each set of coordinates, during a single cycle of replication. The lower curves in each graph represent the virus which is spontaneously released from infected cells, the free virus production. In each case the fundamental characteristics of growth as described above for poliomyelitis virus are seen to obtain, despite differences in growth rates and in the maximal population sizes that can be detected. Even in the declining phase of growth, which is not illustrated for most of the viruses included in Figure 3.3, the different viruses exhibit widely divergent patterns, owing in large part to differences in thermal lability and susceptibility to toxic products.

Resistance of the Host

The highly selective host requirements shown by many viruses are in part due to the innate susceptibility or resistance of the host cells determined by genetic mechanisms. As is elsewhere discussed (Chapters 1 and 23), an excellent illustration of genetically determined susceptibility and resistance to the same agent is afforded by mouse hepatitis virus. This virus grows readily in *in vitro* cultures of peritoneal exudate cells from inbred Princeton mice, which are likewise highly susceptible to natural infection *in vivo*. Similar cultures from C3H mice, which *in vivo* are highly resistant to infection with the identical virus, fail to replicate the agent to any significant degree *in vitro*. Many other examples of such host-determined genetic susceptibility or resistance to viral multiplication could be cited.

As has been shown by Holland and Hoyer (1962), failure of poliovirus to

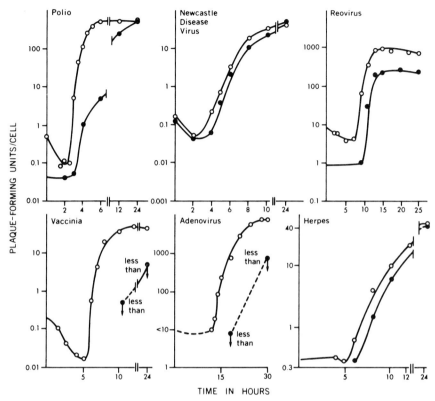

Figure 3.3. One-step growth curves of various viruses in cell cultures measured by plaque assay. Upper curves represent total virus production, whereas lower curves indicate free virus released from infected cells. (After Darnell, 1965.)

replicate in a particular host cell type may be based, not upon the cell's inadequate provision of a satisfactory replication mechanism, but upon the failure of the virus to adsorb onto the cell's membrane, which in turn is the consequence of the absence of appropriate cell receptors. This mechanism may be the basis for many host systems displaying so-called innate resistance to given viruses, but other mechanisms are also undoubtedly at play. The latter include known host barriers and filtering systems at the organ level, the production of host cell antiviral substances such as interferon, the absence of certain required intracellular uncoating enzymes, and the presence of destructive intracellular enzymes.

The age of the host supplying the cells in which virus is to multiply is a highly significant determinant of virus growth. As an illustration, newborn mice provide cells which are highly productive of mouse hepatitis virus or certain enteroviruses, even though cells from adult mice of the same genetic type are highly unproductive of virus (Bang and Warwick, 1960). A similar ontogeny of resistance has been noted with numerous arboviruses.

The Culture Medium

A given virus infecting a specific host cell type may show marked variations of its growth curve related to differences in the chemical composition of the culture medium utilized for the growth or maintenance of the cells. Variations in virus production thus result from differences in donor animals, even of the same species or litter, which provide the serum utilized in different culture media. More specific substances may substantially contribute to virus yield. Poliomyelitis virus, for example, grows to much higher titers when excess L-glutamine is included in the growth medium of its host cells, even though the metabolism of the latter is not appreciably enhanced by such additions. It can thus be seen that variations in nutritional states of host cells may strongly affect the growth curves of their infecting viruses.

A virus in many ways changes the host cell in which it replicates. It is remarkable that the changes in the host appear to be as great as, even greater than, the changes brought about by various hosts in the growth of the parasitic virus. The sequence of multiplication of viral populations, however, is not appreciably different in broad terms from that exhibited by other biological populations, even in the absence of sexual processes and of individual increases in mass.

VIRAL PERSISTENCE

The above description is concerned with a growth cycle for a virus which produces destruction of its host cell. Destruction may occur at the time of virus release, or in other prototype viruses destruction may occur at some later stage after progeny virus particles have been fully or partially released. Of perhaps even greater importance are the infectious cycles in which virus persists in its host cells without producing any evidence of destruction concurrent with or immediately following replication, or in some cases even in the absence of any cellular destruction. Such latent virus infections are of considerable importance in the maintenance of viral infectivity potential. In these circumstances the viral genome may be preserved intracellularly until such time as conditions allow its multiplication and the subsequent release of progeny, sometimes continually throughout the remaining life of the host cell. In this manner viral infectivity, or potential for infectivity, may persist intracellularly even in the presence of high titers of neutralizing antibody or other antiviral substances in the extracellular fluid. Of especial interest are the cases in which viruses are capable of stimulating the production of neoplasms (Chapter 26). The viruses may be concomitantly produced and released, or else infectivity, and/or evidence of multiplication of recognizable viral progeny, are altogether obscured. Frequently new antigens that the host neoplastic cells produce under the influence of the infecting virus are the only evidence of continued viral activity. Under certain circumstances, not always understood, the neoplastic host cells may renew their support of viral multiplication, and progeny of the original agent may be produced and released. In certain other viral infections host cells may be stimulated to produce syncytia by combining with one another, and huge multinucleated cells result. A wide variety of other variations is known among animal viruses. In summary, the pattern of the infectious cycle of a given virus may vary considerably from the typical cycle for poliovirus that is described above. Also the responses of the host cell environment to viral infection are similarly almost as varied as the multiple virus types that are currently recognized.

COMPLEMENTATION
Mutation may give rise to two different genotypes of a given virus, each of which would be defective in that neither of the genotypes could alone direct satisfactory replication. Under certain circumstances when the same host cell is infected by both of the genotypes, they may both multiply. Such replication apparently occurs because the different genotypes complement each other, perhaps by each providing a defective genetically determined character of the other. Such complementation has been observed not only between viruses of identical species and types, but also between different classes of viruses, and also in bacteria. Hybrid viruses may possess a genetic machinery which does not correspond to the phenotype of the capsule or surface structure. This occurs, for example, in a hybrid virus resulting from simultaneous infections of appropriate host cells by papovavirus SV-40 and any of several types of adenoviruses. Here, the nucleoid of the hybrid virus can be demonstrated to be identical with that of the papovavirus, whereas the viral caspsule is composed of surface units which can be shown to be of adenovirus origin (Rapp *et al.*, 1964). Such a combination is referred to as a phenotypic mixing. Mixing of this nature is not inherited by progeny from single infections of cells by such hybrids.

Several double viral infections are now recognized in which viruses of one of the types may be defective, that is to say, they lead to host cell infections which fail to produce new infectious virus. However, virus of the second type simultaneously infecting the same host cell serves as a helper virus, and the resulting progeny are phenotypically mixed. The best known illustration of this joint association of two different viral types, discussed further in Chapter 26, is that of the defective Rous sarcoma virus, RSV, which in the absence of its helper virus, Rous-associated virus, RAV, fails to produce infectious virus under certain conditions in infected cells. But simultaneous infections with both agents leads to production of mixtures of RAV and of RSV–RAV hybrid populations (Hanafusa *et al.*, 1963).

CONCLUSIONS
A simple description of population growth of viruses has been presented here,

with some indications of the sources of variety in the patterns and extent of growth, and it has been indicated also that reactions of the host cells are also extremely varied. A variation in viral types may arise during population growth by mutation, recombination, phenotypic mixing, and complementation. These phenomena and their biochemical bases are further discussed in Chapters 10 and 11. The extent to which all these possible sources of variation occur during infection, replication, and population growth outside the laboratory is unknown.

REFERENCES

Bang, F. B., and Warwick, A. 1960. Mouse macrophages as host cells for the mouse hepatitis virus and the genetic basis of their susceptibility. *Proc. Nat. Acad. Sci.* **46**: 1065–1075.

Darnell, J. E. 1965. Biochemistry of animal virus reproduction. In *Viral and Rickettsial Infections of Man*, 4th ed., Chapter 9. F. L. Horsfall, Jr. and I. Tamm. (eds.). Lippincott, Philadelphia, Penn.

Hanafusa, H., Hanafusa, T., and Rubin, J. 1963. The defectiveness of Rous sarcoma virus. *Proc. Nat. Acad. Sci.* **49**: 572–580.

Hirst, G. K. 1965. Genetics of viruses. In *Viral and Rickettsial Infections of Man*, 4th ed. Chapter 11. F. L. Horsfall, and I. Tamm (eds.). Lippincott, Philadelphia, Penn.

Holland, J. J., and Hoyer, B. H. 1962. Early stages of enterovirus infection. *Cold Spring Harbor Symp. Quant. Biol.* **27**: 101–111.

Howes, D. W. 1959. The growth cycle of poliovirus in cultured cells. II. Maturation and release of virus in suspended cell populations. *Virology* **9**: 96–109.

Rapp, F., Melnick, J. L., Butel, J. S., and Kitahara, T. 1964. The incorporation of SV40 genetic material into adenovirus 7 as measured by intranuclear synthesis of SV40 tumor antigen. *Proc. Nat. Acad. Sci.* **52**: 1348–1352.

CHAPTER 4

The Ecology of the Gastrointestinal Tract

R. BRADLEY SACK

The body surfaces of animals harbor mixtures of populations of lower forms of life, outside on the integument and inside in the gastrointestinal tract. These populations frequently provide the host with distinctive characteristics that seem to be inseparable from the host itself. The populations vary considerably with the type, age, and geographical location of the host. For instance, enteroviruses are relatively common in the gastrointestinal tracts of children in many parts of the world, but are rare in adults from these same areas. Hookworm and ascaris are an integral part of the intestinal flora of man in many parts of the world, but in others they are virtually absent because of sanitary measures.

Since the latter part of the nineteenth century, physicians have queried the significance of the numerous and varied fecal bacteria in the intestine of man. In the absence of scientific data, many speculations were made. For instance, peasants who ate milk products containing large numbers of lactobacilli frequently lived to be quite old. Therefore, Metchnikoff surmised that longevity might be increased by ingestion of large numbers of lactobacilli (Dubos, 1965). Another example is that today many laymen assume that constipation is the cause of many ill-defined symptoms.

As bacteriological and sampling methods have improved, knowledge of the bacterial components of the mammalian gastrointestinal tract has markedly increased, particularly in the last 10 years. However, the information is still very incomplete and there are undoubtedly organisms in the stool that have not yet been cultured. This discussion will be confined to bacteria because these are best known and are the major component of the gastrointestinal flora.

THE HABITAT

The stomach, small intestine, and large intestine have characteristic physical and chemical properties which help determine the type of microbial life that can be supported. These environments, although anatomically separate, are functionally closely related. Intestinal contents normally move distally in an orderly fashion, making the flora of each segment dependent on that of the adjacent segments. These environments are never static, because of digestion and absorption by the host.

The best understood of these properties are diagrammatically shown in Figure 4.1. The *stomach* contents normally have a low pH, which is a formidable barrier to many microorganisms. The extent to which the pH is main-

35

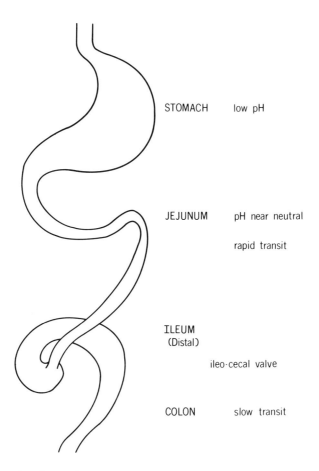

STOMACH   low pH

JEJUNUM   pH near neutral

               rapid transit

ILEUM
(Distal)

               ileo-cecal valve

COLON   slow transit

Figure 4.1 continued opposite.

tained largely determines the microflora. Only bacteria that are acid-resistant or are protected within food particles are able to survive passage through this environment. When this selective barrier is removed, for example by gastric resection, microbial populations in distal areas are considerably altered.

Bacteria entering the *small bowel* are normally propelled fairly quickly from duodenum through the jejunum to terminal ileum. This rapid transit seems to be a primary antibacterial mechanism of this segment of bowel (Dixon, 1960). Bacteria proliferate when this normal function is inhibited experimentally

Region		Mice	Dogs	Man
Stomach	Coliforms	2.0	4.4	0
	Clostridia	1.7	5.2	0
	Streptococci	6.0	5.5	0.3
	Lactobacilli	7.7	4.2	0.9
	Yeasts	6.7	0	0
	Bacteroides	0	0	0
Jejunum	Coliforms	1.7	2.7	0
	Clostridia	0	4.5	0
	Streptococci	5.0	5.9	0.6
	Lactobacilli	6.3	3.3	0.5
	Yeasts	6.2	0	0.7
	Bacteroides	0	0	0
Ileum	Coliforms	4.0	5.2	4.0
	Clostridia	0	6.2	0
	Streptoccoci	6.7	7.0	1.2
	Lactobacilli	8.0	5.8	2.4
	Yeasts	6.7	0	2.8
	Bacteroides	0	0	4.2
Colon	Coliforms	6.0	7.2	6.8
	Clostridia	1.7	7.8	3.2
	Streptococci	6.7	9.3	5.5
	Lactobacilli	6.6	9.0	7.1
	Yeasts	6.6	0	1.3
	Bacteroides	9.4	9.3	9.3

Figure 4.1. Bacterial populations of regions of the gastrointestinal tract of laboratory mice and dogs (Smith, 1965) and man (Gorbach *et al.*, 1967). Some properties of some of these regions are given. The numbers are estimated total population size of the region in mean $\log_{10}$.

with morphine or by disease such as scleroderma. In contrast, the very slow transit time of the *large bowel* is thought to be the primary reason why bacterial growth is abundant there. Dividing the small and large intestine is the ileocecal valve, an anatomical barrier that normally prevents reflux of colonic contents. That it acts as a true barrier is indicated by the marked differences in bacterial flora from one side to the other (Gorbach *et al.*, 1967).

Many other physical and chemical properties of these environments undoubtedly play important roles in the regulation of bacterial flora, but most of these are poorly understood.

THE BACTERIA INVOLVED

The microbial populations inhabiting the various segments of the gastrointestinal tract of the laboratory mouse, dog, and man are listed in Figure 4.1. The differences are striking. Except for the large intestine, the flora in man is considerably less than in the dog or mouse. Only in recent years was it appreciated that man had any normal bacterial flora in the small bowel and the complete explanation for the paucity of this flora is not known. Lactobacilli and streptococci are the most numerous in the upper gastrointestinal tract of mice. In dogs large numbers of clostridia are also present in these areas.

The flora of the large intestine is more comparable among these species. *Bacteroides*, anaerobic gram-negative rods, are the predominant organisms here and frequently account for over 90 percent of the flora of feces. *Bacteroides*, moreover, are completely absent from other parts of the gastrointestinal tract except the terminal ileum. Fungi on the other hand are found in similar concentrations throughout the gastrointestinal tract.

So far in this discussion the bacterial flora of any one part of the tract has been treated as a unit. However, Dubos *et al.* (1965) demonstrated two distinct types of flora in the intestine of NCS mice. These mice are raised with a known flora in sanitary but not germ-free conditions. They described an "autochthonous" flora that was primarily anaerobic, lived in the mucous layer, and was thought to have merely a mutualistic relationship with the host. Another distinctive flora was primarily aerobic, found only in the lumen, and was felt to probably represent infection (parasitism) from which the host had recovered. Such distinctions have not been made clearly for other animal hosts, including man (Plaut *et al.*, 1967).

STABILITY OF THE SPECIES COMPOSITION

The microbial flora in mammals has been shown to be fairly stable throughout the life of the individual and unrelated to the sex of the host. Repeated sampling of the same healthy animals on constant diets yields remarkably similar results. The flora is so stable that when other enteric organisms are experimentally introduced they usually pass through the gastrointestinal tract without becoming permanent residents.

DEVELOPMENT OF THE MICROBIAL FLORA

Animals in the uterus are normally sterile and become colonized with bacteria in early infancy. The colonization begins almost immediately after birth and is nearly complete several weeks later. Studies of mice have indicated that the intestinal flora of the young closely resembles that of the mother. In some young animals there are temporary antibacterial substances that limit intestinal colonization. Suckling infant rabbits have an antibacterial substance in the stomach which is produced by stomach enzymes acting on a factor in the mother's milk (Smith, 1966). This defensive measure is present for only the first few weeks of life, and it retards only the early colonization.

Schaedler's group (1965a), employing the NCS strain of mice, enumerated the bacterial populations in various gut segments as they develop from the

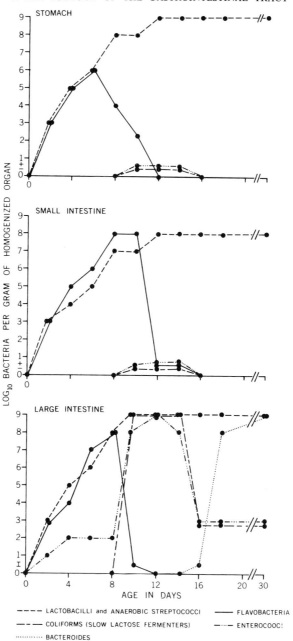

Figure 4.2. The development of enteric flora in NCS mice—mice raised with a known bacterial flora in sanitary but not germ-free conditions. (After Schaedler *et al.*, 1965a.)

time of birth (Figure 4.2). These experiments demonstrated an orderly systematic colonization of all segments of the gastrointestinal tract. Lactobacilli and anaerobic streptococci establish themselves quickly throughout the gastrointestinal tract, whereas coliforms, enterococci and, considerably later, *Bacteroides* are confined to the large intestine. Flavobacteria reach high concentrations in all parts until day 10, after which they suddenly disappear. Once the flora is established it remains stable for life.

The same investigators (Schaedler *et al.*, 1965b) fed cultures of lactobacilli, streptococci, and *Bacteroides* to germ-free mice and noted the same colonization patterns that occurred in NCS mice. However, when coliforms and enterococci were fed in a similar manner they grew throughout the intestinal tract. Only after feeding the previously germ-free mice with feces from NCS mice did these abnormal patterns of colonization revert to those of the NCS mice.

These studies indicated that during the development of the bacterial flora there is regulation by factors depending not only on gastrointestinal location, but also on characteristics of the mixed flora itself.

INTERACTIONS OF THE FLORA

The large mixtures of bacterial cells in the intestine interact in many ways. The roles of various anatomical features—size and shape, flagellar appendages, and capsular components—are poorly understood but presumably have ecological significance. Certain interactions among the species have been demonstrated in the laboratory with pure cultures, and it is assumed that the interactions occur also in nature and are important. For instance, the oxidation-reduction potential in the environment is lowered by the growth of aerobic organisms, resulting in a favorable environment for the growth of anaerobic bacteria. Also, some organisms utilize metabolic products given off by other organisms during growth, for instance some fungi grow in lactic acid produced by *Streptococcus lactis* (Brock, 1966). Two or more organisms may compete for limited nutrients. Some species produce lethal substances—antibiotics killing a wide variety of other strains and bacteriocines killing only closely related ones. Exchange of either chromosomal or extrachromosomal DNA (episomes) occurs among bacteria of the same or different species. Episomal transfer has been shown to be important in the rapid development of resistance to a wide spectrum of antibiotics (Smith, 1967; and Chapter 12).

INTERACTIONS OF THE BACTERIAL FLORA AND THE HOST
Morphological Effects on the Host

Many of the normal morphological features of the mammalian bowel may actually represent a reaction of the host to its intestinal flora. Germ-free rodents usually develop very large cecums, which may account for 20 percent of their body weight. When these animals are subsequently fed enteric bacteria the cecum decreases to normal size. The intestinal mucosa of germ-free mice is quite different histologically from the normal in that the crypts are shallow and the lamina propria poorly developed. When these animals are colonized with enteric organisms a normal picture appears with deepening of the crypts and an increase in cellularity of the lamina propria (Abrams *et al.*, 1963). Thus the normal architecture of the mucosa in these animals is in reality an inflammatory response to the bacterial flora.

Antibody Responses of the Host

One of the regulatory factors imposed by the host on its bacterial flora may be its production of antibodies. Bacteria and their products within the lumen are known to stimulate the host to produce systemic antibodies but whether these antibodies are active within the lumen of the bowel is not certain. It has been postulated that one reason for resistance of the host to enteric organisms, such as *Vibrio cholerae*, is the presence of intraluminal antibodies called coproantibodies.

Within the past few years a secretory gamma globulin found in many secretions has been identified and it seems to be directed against bacteria and viruses (Thomasi, 1967). This antibody, Gamma A, contains an extra protein piece that gives it a distinctive character. Plasma cells containing Gamma A have been identified in the lamina propria of the entire gastrointestinal tract (Crabbe and Heremans, 1966) and Gamma A has been demonstrated in the luminal edge of epithelial cells lining the intestinal tract (Gelzayd *et al.*, 1967). It has been postulated that Gamma A antibody is formed by the plasma cells, after which the extra protein piece is added in the epithelial cell before discharge of the antibody into the intestinal lumen. The function of these antibodies within the lumen is not understood.

Antibodies have been found in intestinal fluid following oral or parenteral vaccination with *Vibrio cholerae* but their molecular structure has not been determined (Freter and Gangarosa, 1963). In germ-free mice orally inoculated with *V. cholerae* (Miller and Sack, 1968) the *Vibrio* populations within the intestine regularly undergo serotype changes coincident with the appearance of serum antibodies. These serotypic changes can be delayed by immunosuppression of the animals and hastened by previous vaccination. These phenomena are best explained by assuming that antibody is active within the intestinal lumen, selectively inhibiting growth of homologous serotypes, thus allowing mutant serotypes to become predominant in the population. This indirect evidence suggests that intraluminal antibodies within the intestinal tract may be one host factor regulating microbial populations.

Favorable Effects of the Flora on the Host

A few benefits of the bacterial flora to the host have been well defined. They primarily concern the flora of the large intestine. Mammals which subsist primarily on grasses and other plants are assisted in the utilization of this food source by specialized microbes which break down cellulose. Ruminants have an especially highly developed enzymatic factory in the first portion of the stomach, the rumen, in which large populations of microbes anaerobically digest cellulose.

Synthesis of vitamin K has been shown to occur in pure cultures of enteric organisms. When fed to germ-free rats kept on a diet deficient in vitamin K these bacteria are able to prevent the occurrence of deficiency symptoms (Lev, 1963).

The stability of the normal flora has been demonstrated to be of considerable protective value to the host against invasion by pathogenic enteric bacteria. This phenomenon is directly comparable with the protection of other undisturbed ecological communities against invasion by new populations. Guinea pigs normally are resistant to *Shigella* infections whereas starved guinea pigs or germ-free guinea pigs are quite susceptible. Normal guinea pigs are resistant to oral inoculation of *Vibrio cholerae*. However, they are susceptible after starvation, alkalinization of the stomach, intraperitoneal morphine, or streptomycin treatment (Freter, 1955). *Vibrio cholerae* do not become established in the intestine of normally raised mice whereas they luxuriantly colonize the intestinal tracts of germ-free mice (Miller and Sack, 1968).

Extensive investigations of mice infected with *Salmonella* have elegantly demonstrated and clarified this protective phenomenon. Although mice are normally resistant to *Salmonella* they become very susceptible after antibiotic administration. Meynell (1963) demonstrated that whereas the lethal dose of *Salmonella typhimurium* was normally greater than 10^6 organisms, the lethal dose was reduced to one organism when the mice were pretreated with streptomycin. Bohnhoff and Miller (1962) found that whereas 10^6 *Salmonella enteritidis* were needed to infect 50 percent of normal mice, this infecting dose was reduced to 10^3 organisms after fasting or morphine treatment and to less than 10 organisms after streptomycin treatment. This marked increase in susceptibility is shown in Figure 4.3. The explanation of this remarkable phenomenon was further clarified by demonstrations that the elimination of *Bacteroides* from the flora of the cecum was the critical determinant in increasing the susceptibility of the mice (Meynell, 1963; Bohnhoff *et al.*, 1964). Moreover it was not only the presence of *Bacteroides*, but more importantly of their metabolic products, identified as butyric and acetic acids. These acids were bacteriocidal for *Salmonella* and thus protected normal mice against infection.

Exogenous antibiotics not only render the host more susceptible to known pathogenic organisms, but they also disturb the flora so that species resistant to the antibiotic may overgrow their usually more numerous competitors. This phenomenon has clearly been shown to occur in man as well as in laboratory animals. Staphylococci and yeasts are most often the antibiotic-resistant opportunist organisms.

Unfavorable Effects of the Flora on the Host

The functions of the flora of the small intestine are even less well understood.

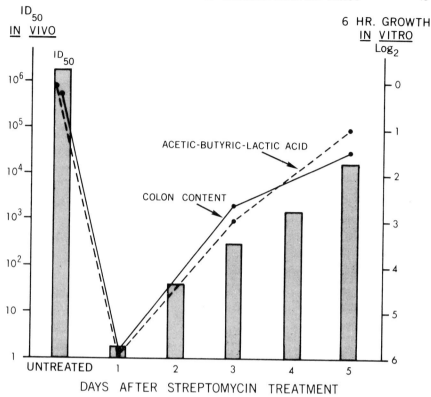

Figure 4.3. Streptomycin was given orally to laboratory mice. The acid content of the lumen of the colon was measured on following days, and broth media were made with added corresponding amounts of lactic, acetic, and butyric acids. Also, at various days after streptomycin administration, colon contents were extracted and used as a further set of culture media. *Salmonella enteritidis* was added to the media, and the *M* concentrations supported are shown by the dashed line for acid-broth and the solid line for colon content. The bars on the graph show doses of the bacterium required for infection of 50 percent of batches of mice before streptomycin administration, and on various days after. (After Bohnhoff, Miller, and Martin, 1964.)

However in certain instances when the small bowel flora is markedly altered a pronounced effect can be produced on the normal physiology of the host. An increase in the bacterial flora of the upper small bowel has been demonstrated in patients with partial gastrectomies. The growth of enteric bacteria in the small bowel is even more marked in humans or animals with surgically produced blind intestinal loops. Intestinal malabsorption of vitamin B_{12} and fat occurs in both of these disturbances. This malabsorption can be improved by lowering the small intestinal bacterial populations with antibiotic treatment. Several explanations have been given for these absorption defects (Donaldson, 1967). The vitamin B_{12} malabsorption may be explained by the fact that enteric bacteria which take up this vitamin compete with the absorptive epithelial cell. In normal fat absorption, conjugated bile salts are important because they aid in lipolysis of fats and they associate in the small bowel with products of fat hydrolysis, that is, they form micelles. Fat malabsorption occurs if these bile salts are not present. It has been shown that *Bacteroides* in the small bowel can split conjugated bile salts into free bile acids (Drasar *et al.*, 1966). Free bile acids normally present only in the cecum have also been identified in the small bowel in such cases of malabsorption. Free bile salts are ineffective in fat digestion and absorption and are even possibly damaging to the intestinal mucosa.

The presence of a very numerous gastrointestinal bacterial flora may be responsible in some way for an overall inhibition of growth of the host. The administration of antibiotics in food increases the growth rates in many commercially raised animals such as calves, chickens, and pigs. These effects are most striking when the animals are raised in insanitary conditions and on less than optimal diets. Such growth-stimulating effects of antibiotics have also been shown in ordinary laboratory mice but not in germ-free mice (Dubos *et al.*, 1963).

FUTURE STUDIES

It is clear from this discussion that the intestinal bacterial flora is poorly understood. Current studies that are still incomplete indicate some fields of investigation that may increase understanding of this flora and its relationship to the host. Bacterial products such as ammonia and vasoactive amines have been implicated in the causation of certain clinical symptoms in man but these phenomena are yet to be explored more fully.

Morphological changes in the intestinal mucosa have been found in persons from the United States living in tropical countries (Lindenbaum *et al.*, 1966). Biopsy specimens have shown inflammation of the jejunal mucosa and blunted villi. Similar inflammatory mucosal changes have been noted in healthy residents of tropical areas (Sprinz *et al.*, 1962). More severe changes of mucosal inflammation are found in persons with tropical sprue, a malabsorptive disease which is responsive to antibiotic treatment. All of these observations suggest that the small bowel flora, not yet studied in these conditions, may be responsible for some of these histological and absorptive abnormalities.

CONCLUSIONS

Some of the known components of the gastrointestinal bacterial flora have been discussed. Although present knowledge concerning the many interrelationships in this system is still incomplete, observations of significance have been made. The bacterial intestinal flora is characteristic for the species of the host. Although usually quite stable within an individual, the flora can be markedly altered by diet, starvation, and antibiotics. A stable flora seems to have certain beneficial effects on the host, whereas a disturbed one may interfere with normal host functions. Finally, bacterial flora may determine what has previously been regarded as normal intestinal mucosal architecture.

REFERENCES

Abrams, G. D., Bauer, H., and Sprinz, H. 1963. Influence of the normal flora on mucosal morphology and cellular renewal in the ileum. *Lab. Invest.*, **12**: 355–364.

Bohnhoff, M., and Miller, C. P. 1962. Enhanced susceptibility to *Salmonella* infection in streptomycin-treated mice. *J. Infect. Dis.* **111**: 117–127.

Bohnhoff, M., Miller, C. P., and Martin, W. R. 1964. Resistance of the mouse's intestinal tract to experimental *Salmonella* infection. II. Factors responsible for its loss following streptomycin treatment. *J. Exptl. Med.* **120**: 817–828.

Brock, T. D. 1966. *Principles of Microbial Ecology*. Prentice-Hall, Englewood Cliffs, New Jersey.

Crabbe, P. A., and Heremans, J. F. 1966. The distribution of the immuno-globulin-containing cells along the human gastrointestinal tract. *Gastroenterology* **51**: 305–316.

Dixon, J. M. S. 1960. The fate of bacteria in the small intestine. *J. Pathol. Bacteriol.* **79**: 131–140.

Donaldson, R. M. 1967. Role of enteric microorganisms in malabsorption. *Federation Proc.* **26**: 1426–1431.

Drasar, B. S., Hill, M. J., and Shiner, M. 1966. The deconjugation of bile salts by human intestinal bacteria. *Lancet* **I**: 1237–1238.

Dubos, R. 1965. *Man Adapting*. Yale Univ. Press, New Haven, Conn.

Dubos, R., Schaedler, R. W., and Costello, R. L. 1963. The effect of antibacterial drugs on the weight of mice. *J. Exptl. Med.* **117**: 245–257.

Dubos, R., Schaedler, R. W., Costello, R., and Holt, D. 1965. Indigenous, normal, and autochthonous flora of the gastrointestinal tract. *J. Exptl. Med.* **122**: 67–76.

Freter, R. 1955. The fatal enteric cholera infection of the guinea pig achieved by inhibition of normal enteric flora. *J. Infect. Diseases* **97**: 57–65.

Freter, R., and Gangarosa, E. J. 1963. Oral immunization and production of coproantibody in human volunteers. *J. Immunol.* **91**: 724–729.

Gelzayd, E. A., Kraft, S. C., and Fitch, F. W. 1967. Immunoglobulin A: localization in rectal mucosal epithelial cells. *Science* **157**: 930–931.

Gorbach, S. L., Plaut, A. G., Nahas, L., Weinstein, L., Spanknebel, G., and Levitan, R. 1967. Studies of intestinal-microflora. II. Microorganisms of the small intestine and their relationships to oral and fecal flora. *Gastroenterology* **53**: 856–867.

Lev, M. 1963. Studies on bacterial associations in germ free animals and animals with defined floras. *Symbiotic Associations, Thirteenth Symp. Soc. Gen. Microbiol.* Cambridge Univ. Press, London, pp. 325–334.

Lindenbaum, J., Kent, T. H., and Sprinz, H. 1966. Malabsorption and jejunitis in American Peace Corps volunteers in Pakistan. *Ann. Internal Med.* **65**: 1201–1209.

Meynell, G. G. 1963. Antibacterial mechanisms of the mouse gut. Role of Eh and volatile fatty acids in the normal gut. *Brit. J. Exptl. Pathol.* **44**: 209–219.

Miller, C. E., and Sack, R. B. 1968. Serotypic changes of *Vibrio cholerae* in germ-free mice. *Advan. in Germ-Free Research and Gnotobiology.* Chemical Rubber Co., Cleveland, Ohio.

Plaut, A. G., Gorbach, S. L., Nahas, L., Weinstein, L., Spanknebel, G., and Levitan, R. 1967. Studies of intestinal microflora. III. The microbial flora of human small intestinal mucosa and fluids. *Gastroenterology* **53**: 868–873.

Schaedler, R. W., Dubos, R., and Costello, R. 1965a. The development of the bacterial flora in the gastrointestinal tract of mice. *J. Exptl. Med.* **122**: 59–66.

Schaedler, R. W., Dubos, R., and Costello, R. 1965b. Association of germ free mice with bacteria isolated from normal mice. *J. Exptl. Med.* **122**: 77–82.

Smith, D. H. 1967. The current status of R factor. *Ann. Int. Med.* **67**: 1337–1341.

Smith, H. W. 1965. Observations on the flora of the alimentary tract of animals and factors affecting its composition. *J. Pathol. Bacteriol.* **89**: 95–122.

Smith, H. W. 1966. The antimicrobial activity of the stomach contents of suckling rabbits. *J. Pathol. Bacteriol.* **91**: 1–9.

Sprinz, H., Sribhibhadh, R., Gangarosa, E. J. Benyajati, C., Kundel, D., and Halstead, S. 1962. Biopsy of small bowel of Thai people. *Am. J. Clin. Pathol.* **38**: 43–51.

Thomasi, T. B. 1967. Gamma A secretory antibodies. *Hosp. Practice* **2**: 26–35.

CHAPTER 5

Comparative Demography

BRENDA K. SLADEN

Demography is the numerical aspect of population structure. The subject is introduced here because simple numerical descriptions are a necessary preliminary to discussion of biological properties of populations, including the health of human populations. The chapter will discuss various elements in population change that are measurable, what the measures are, and some of their limitations.

A living population has a size that depends on the operation of many factors, some of which can be described mathematically. These factors begin to affect the numbers of individuals under certain circumstances and may limit population growth or existence under other circumstances, or they may have more steady effects. Their action in unison may differ in different circumstances. Here they will be treated as if more simple than in reality.

There are five basic types of numerical change in populations: input—births and immigrations; output—deaths and emigrations; and changes of total size. Measurement of migrations will not be discussed.

As animals change in age their biological attributes change, as when they become sexually mature, and thus the distribution of ages in a population affects the attributes of the whole population.

The relationships of some of the methods of expressing numerical changes in populations that are considered in this chapter are given in Table 5.1.

TABLE 5.1.

Rates of Birth, Mortality, and Increase Discussed in This Chapter

Crude rates	Age-specific rates	"True" ("real" or intrinsic) rates
The age distribution in the population is ignored, and the whole population is the denominator.	Based on performance of specific age groups. Data can be arranged in life tables and fertility tables.	These assumptions are made: (1) The observed age-specific birth and death rates remain the same through all generations. (2) With condition (1), the population is considered to grow through several generations so that a stable age distribution develops.
Crude rate of increase = crude birth rate − crude death rate.	Net reproduction rate (R_0) = ratio of reproductive females from one generation to the next	*Then*: the crude rates are stable and are called true rates Intrinsic rate of natural increase (r) = true birth rate − true death rate.

47

Mathematical descriptions of these measures and their relationships are given in Birch, 1948.

Statistical descriptions are generalizations about large numbers of varying items or events but single occurrences can fall outside the described pattern. Events that are unusual in this sense can be multiplied in their effects within a single population. For instance, if a location is to be occupied by a population of one out of two very similar species, chance variations in the arrival and in the onset of reproduction in the colonizing individuals may determine which species eventually takes sole possession of the place. This multiplication of a chance effect is demonstrated in the competition experiments of Park with *Tribolium* beetles (Chapter 6).

The collection of data to test models of population growth tends to lag behind the theory because of many difficulties. Data can be collected from simplified population conditions as described in Chapter 6 or from complex natural situations as described in Chapters 7 and 8. For models of populations see Watt (1968).

CENSUS

The purposes of censusing are principally to estimate population size, the relative frequencies of the sexes and of the age groups, the extent of recent mortality, and the current amounts of reproduction. Census involves total counts, which seldom are possible, or extraction of samples. The population under study will be changing its numbers, often moving in space, may be partially or wholly out of sight of the observer, and may be very large. Problems provided by the biology of the species give great variation in application of the methods of sampling.

Population Density

Population density is a value which is expressed as numbers per area or volume. The space involved may be important in terms of food, cover, territory, or general habitat, or may be chosen by the investigator only for ease of measurement. Much estimation of population size is from evidence of activities—tracks, spoor, nests, marks on vegetation, reopening of burrows closed by the investigator, noises made by territorial animals such as birds and monkeys, and damage to the substratum as is caused by viruses in tissue culture. Another census method is to mix known quantities of recognizable units into the population and to measure by further sampling their dilution among the objects to be measured. For instance, virus population size can be estimated by addition of known numbers of small particles whose ratio to virus particles can be seen by electron microscopy. The simple formula that relates this ratio, the number initially added, and the total population size is known as the Lincoln index and an example of its use is given in Chapter 1. The most widely used application of the index is in the capture–recapture method, which involves replacement into their population of animals that have been trapped and marked with bands, tags, paint, dyes, and radioactive materials.

Trapping and other extractions from soil and water are widely used for population estimation, as are also records from hunters and from commercial fisheries. Many of the methods carry the disadvantage of biased reporting, or bias in the ability of animals to be trapped. Some animals previously attracted to bait in traps become "trap-happy." Some individuals are permanently trap-happy, or by their behavior or habitat choice are not exposed to the same risk of being trapped as are other members of the population. For instance, the age groups and the sexes often differ in their behavior and in their ranges of movement. Moreover, complete removal of samples is a form of mortality, artificially imposed upon the population under study.

Biomass

Biomass is the sum of the weights of all the individuals of a group or population. An approximate value can be obtained by multiplying a representative individual weight by the number of individuals. Biomass is sometimes preferred as a measure of population size because numbers and individual size may compensate for each other. For instance, when numbers in an age class of a fish population are unusually high, individual size is often very small; and when numbers are small in fish populations, individual size is often great.

Identifications of Sex and Age

Identifications of sex and age may be difficult or impossible at some life stages. Often sex cannot be determined from morphology without dissection, although behavior may help. Age identification has many problems, from clustering in even numbered-ages in human census data, to the impossibility of getting the information from many animals outside captivity. Age can sometimes be recognized from the following kinds of clues: rings in tooth enamel correlated with seasonal growth variations (seals), bone rings (in the otoliths of whales), and wood rings in trees; numbers of regressed corpora lutea in the ovary (female whales), numbers of empty ovarioles (female mosquitoes, Chapter 28); size in given environmental conditions (rat skulls, snail shells, fish body length, stalk length in the invertebrate rotarian *Floscularia*, based on the growth of marked individuals). Age can also be recognized from the numbers on the tags of recaptured individuals (mammals, fish, birds, turtles).

Mortality

The extent of mortality in a population is particularly difficult to assess. Sometimes animal remains are useful (skulls of Dall sheep and deer, parasite remains in dead pupae of insects inhabiting galls). Mortality is easier to measure when reproduction and migrations can be ignored and the population is repeatedly sampled for estimation of decreasing size. This is possible with animals, particularly insects, which have an annual life cycle and a reproductive period of short duration and a small range of movement, and with some small mammals in winter. If a particular age group can be recognized for some time, its decreasing numerical size may indicate the mortality it has experienced. This method has been used to determine the length of life of populations of snails that are intermediate hosts to schistosome parasites (Chapter 27).

Sex Ratios

The human male death rate is higher than that of the female and there is compensation for this by higher initial proportions of males. The ratio of males to females at birth is 106:100. The ratio at conception, the primary ratio, has not yet been determined. Natural abortions in humans are in the range of 160–200:100, but in this ratio a higher death rate of male embryos cannot be separated from the proportion of male embryos at risk. Estimates of primary sex ratio are in the range of 110–160:100. There have been attempts to ascertain primary ratios in other mammals by flushing out zygotes before implantation, with sex determined by chromosome analysis. The results indicate high inequality of numbers of the sexes, and other mammals besides man show the same pattern of unequal death rates. The higher death rate of males is at least partially attributed to alleles on the X chromosome with recessive deleterious effects (Geiser, 1924). In the male, recessive characters controlled from the X chromosome cannot be masked because the Y chromosome carries little genetic information.

In mammals the female is the homogametic sex and in birds, some reptiles, butterflies, and moths the male is the homogametic sex. It remains to be seen whether the patterns of mortality and early sex ratios are mirror images of each other in these two groups although some bird and butterfly population are known to have higher female death rates. Some ornithologists have attributed this to the hazards of motherhood, while for the human male death rate some demographers blame the hazards of working for a living. Of course, death rates also reflect differences in the biology and ecology of the sexes, for instance, female woodlice, *Porcellio*, have higher death rates due in part to their more frequent molts during which the animal is more vulnerable to environmental hazards.

Fertility and Fecundity

Some animal species are oviparous and some viviparous. What gets measured as fertility and what gets measured as early mortality depends on the particular circumstances, while the change from parental fertility to early mortality is a matter of continuity for the organism concerned. Generally the number of eggs or sperm produced is called fecundity and the viability of the eggs and sperm is not always taken into account in this measure. Fertility is the number of live young hatched or born. For example, in these terms, sperm production by the human male is male fecundity and the live births are female fertility.

Spatial Distribution

The problem of investigating and understanding spatial distributions is important. Do certain animals have a random scatter within their areas of occurrence or are they aggregated, and what forces determine aggregations? Many animals have home ranges which are particularly concerned with the nesting or roosting site, with cover, and with the distribution of food. The effect of one population upon another may depend partly upon the searching range of one of them—for instance, the area that an arthropod parasite carrying a human virus will search for its human host, or the area that an insect parasite will search for the host whose numbers it is thought to regulate (Nicholson and Bailey, 1935). Many animals migrate for long distances as seasons change—birds, butterflies, bats and other mammals, and fish. Much of this migration may be very significant in the dissemination of disease agents. But specific examples of animal migrations causing distant transmission of infection are difficult to prove.

Mosby (1960) gives the methods and literature of animal census. Table 5.2

TABLE 5.2

Census of Rats in Ricks of Unthreshed Corn[a]

Weight group (grams)	No. of males	No. of females	No. mature	No. pregnant	No. embryos	Males (%)	Females (%)
44 or less	106	93	0	0	—	48.66	44.48
45– 94	21	32	0	0	—		
95–144	39	38	11	0	—	21.84	19.57
145–194	18	17	17	1	7		
195–294	31	42	42	18	149	11.88	14.95
295–394	35	53	53	26	226	13.41	18.86
395–494	10	6	6	1	10	3.83	2.14
495 or over	1	0	—	—	—	0.38	—
	261	281	129	46	392	100.00	100.00

[a]Six corn ricks were threshed in March, and before they were dismantled all rodents were killed by gassing. There were estimated to be 603 rats (*Rattus norvegicus*) consisting of 431 "actives" and 172 nestlings, and 643 mice. Ninety percent of the rats were examined for sex, maturity (in females by corpora lutae in ovaries), pregnancy, and embryo numbers. Weight group can be translated to approximate age group, and these data are sufficient to give various estimations of rates of increase. (Leslie *et al.*, 1952.)

shows data obtained from an almost complete count of a wild animal population.

RATES

The subject of population dynamics is concerned with the change of numbers in time, the input and output of populations in terms of births and deaths, immigration, and emigration. Individuals of a population have potential mean life spans, possible maximal ones, and chances of dying at any time. Some of them for certain periods of time have certain responsibilities for the births. Rates of birth and death are the numbers of such occurrences in defined time, among the number of individuals exposed to the risk of such occurrences.

$$\text{Rate per given time} = \frac{\text{Number of occurrences in that time}}{\text{Population at risk}}$$

There is some latitude in choice of the denominator, for instance, the size of the population at risk will usually be changing because of mortality. It may be necessary to choose among at least three measures of its size—the initial size and the arithmetic and the geometric means through the time interval. Also there are varying degrees of involvement in the risk, and the demographer may be general or very specific in his choice. At the beginning of the time interval the probability of an occurrence happening throughout the time interval in an individual at risk is the same as a rate, expressed as a decimal of 1.0. For instance, if a death rate is 10 percent per week, an individual at the beginning of the week has a probability of 0.1 of dying within the week. For the total risk involved when subject to several different probabilities and for further statistical methods, see texts such as Moroney (1956.).

Crude Rates of Birth, Death, and Increase

Crude rates of birth and death are obtained by division of the number of occurrences in unit time by the population during that time or at the beginning of it.

$$\text{Crude rate} = \frac{\text{Number of births or deaths in unit time}}{\substack{\text{Population during that time} \\ \text{(or at the beginning of it)}}}$$

In Sweden the crude human birth rate per year was approximately 83 per thousand in 1750 and by 1934 it had changed to 13 (Lotka, 1936). What caused the decline? Did the birth rate per female of reproductive age decline, did a change in the number of reproductive females change the total births, or did the number of births each year remain the same while total population size increased greatly because of greater survival of older age groups? Or did all these changes occur?

The crude rate of increase is the crude birth rate minus the crude death rate. Thus the crude rate of increase also reflects change in the age distribution, which is not defined in crude rates.

These measures are valuable descriptions of events for the described time but without further data they cannot be used for prediction of future population size and increase.

Crude rates are in units of calendar time, as opposed to generation time.

TABLE 5.3

Some of the Components of Life Tables

l_o = Original group, or cohort. Several alternatives are possible:
A mean number per area or volume, which is a measure of population density (e.g., *Urophora* in Chapter 7).
A convenient number such as 100 or 1,000 which is actual size or conversion from another number, but is not a measure of population density.
A real cohort which is censused at actual intervals. This gives a *cohort life table* (e.g., laboratory populations, wild insect populations with an annual life cycle).
A population in which all age classes are represented at one time and which is censused once. A life table is reconstructed with a theoretical cohort that would have produced such an age distribution. This is a *current life table*. It is used in human demography because of the extreme difficulty of getting data for the cohort life table. The most usual procedure is to use census counts and also observed deaths during a recent period. A disadvantage is that patterns of mortality change with time, so that the reconstructed population is not likely to closely represent real events.

x = The exact age in stated units of time. Sometimes it is convenient for the units of time to be stages in the life history, although the actual time intervals may be unequal. Pivotal age is the age half-way through the time interval

$$= \frac{x + (x+n)}{2}$$

n = Time interval before the next exact age in the table. When n does not equal 1, n may be used with many of the symbols following in this table, such as $_nd_x$ when $n = 3$ weeks.

l_x = Number surviving to exact age x out of the original cohort (l_0). If $l_0 = 1$, then l_x is the probability of surviving to age x,

$$l_x = l_{(x-n)} - d_x$$

d_x = Number dying between ages x and $x+n$. Sometimes a column is put beside d_x which gives causes of death. Then d_x (and therefore q_x) can be split into components by cause, though many approximations may be involved.

$$d_x = l_x - l_{x+n}$$

q_x = The proportion dying, which is the number dying between ages x and $x+n$ ($= d_x$) divided by the number surviving at exact age x ($= l_x$).

$$q_x = \frac{d_x}{l_x}$$

$$q_x = 1 - P_x$$

m_x = The age-specific mortality rate as used in human demography. It is d_x divided by the mean size of the population surviving from x to $x+n$ ($= L_x$). *Note.* In animal demography q_x is called the age-specific mortality rate and the denominator is either l_x or L_x, and the symbol m_x is reserved for the age-specific fertility rate, as described in Table 5.5.

$$m_x = \frac{d_x}{L_x}$$

P_x = The probability at birth of being alive at exact age x. This is used in a method of calculating the net reproduction rate

$$P_x = \frac{l_x}{l_0}$$

L_x = Mean population size as it passes from age x to $x+n$

$$L_x = \frac{l_x + l_{(x+n)}}{2}$$

This column gives the age distribution of the life table.

(*continued*)

Table 5.3 (*continued*)

T_x = Total number of time units of life remaining to the surviving group at age x.

$$T_x = \int_x^{x+n} L_x dx$$

It is calculated by adding up all the remaining figures in the L_x column from the specified time x.

(T is also used as the symbol for generation time.)

e_x = Age-specific life expectancy, the average number of years remaining to a survivor at age x. At time 0, e_x is the mean length of life of the cohort.

$$e_x = \frac{T_x}{l_x}$$

Age-specific Rates, Life Tables, and Age-specific Mortality Rates

A specific rate is the number of occurrences in a specified class in unit time divided by the number of individuals in that class. Age classes are such specified classes and age-specific rates of births and deaths are widely used. When age-specific death rates are to be calculated from a population with a wide range of ages, or from a repeatedly censused population of the same age, data are conveniently arranged as a life table. This is a formal arrangement with a set of symbols in general use (Table 5.3), but the arrangement can be changed to suit particular purposes. The population is divided into age classes, and each age class occupies a new line down the left-hand side of the table; data and calculations for each specified age class are arranged across the line under column headings of symbols. Table 5.4 is a life table for voles kept in the laboratory.

Data in columns of the life table can be shown as graphs. Figure 5.1 uses survival data from the life tables for the spruce budworm, *Choristoneura fumiferana*, a pest of Balsam fir forests of the northern part of the North American continent. Life tables for the spruce budworm have been constructed for many places over many years and the one used in the figure is typical. The l_x column is plotted against time. The large drop in survivors during late August is due to aerial dispersal of tiny caterpillars.

Life tables are used in government demography (for methods see U.S. Bureau of the Census, 1951), and by life insurance actuaries. They are used in public health research, particularly by demographers, and by epidemiologists for age-specific mortality rates from specified causes. They are used to record the "life" of such inanimate objects as intrauterine contraceptive devices. Life tables are used in ecology of wild animals as a convenient arrangement of data so that causes of death and age-specific death rates can be compared from place to place and from year to year. Sometimes environmental contaminants such as atomic radiation are measured by their effect upon the life span of laboratory animals, survival being tabulated in life-table form. There is also much interest in the form of the end of survivorship curves in relation to human problems of old age and the general problem of senescence. Also,

TABLE 5.4

Life table for Laboratory Populations of the Vole, *Microtus agrestis*[a]

Approximate stage in life history at age x	x Age in weeks from birth	n Interval in weeks before next age	l_x Survivors at age x	d_x Number dying between ages x and x+n	q_x Age-specific mortality rate per 10^3 per week
2 weeks after conception	−1	1	10,000	2,107	210.07
Birth	0	2	7,893	1,121	71.0
Weaning	2	1	6,772	No data	Authors gave value of 0
Onset female fecundity	3	5	6,772	235	6.94
Onset male fecundity	8	8	6,537	470	8.96
	16	8	6,067	1,222	25.17
	24	8	4,845	1,223	31.55
	32	8	3,622	1,129	38.96
	40	8	2,503	1,180	58.92
	48	8	1,323	709	66.98
	56	8	614	425	86.52
	64	8	189	94	—
	72	8	95	0	—
	80	8	95	47	—
	88	8	48	0	—
	96	8	48	48	—

[a]From the ages of 3 weeks, the mortality was recorded from 98 male and 46 female voles. This was not a single cohort—data collected during 3 years were grouped. The females spent their whole lives, from the ages of 3 weeks, in single pairs with a fecund male (the same male until he died). Fetuses were detected and counted by manual palpation through the abdominal wall from about 2 weeks after conception (i.e., at 1 week before birth). Approximately 1,500 live young were born. Mortality data on the young were collected from 1 week before birth until weaning at approximately 2 weeks. No mortality data were obtained for age 2–3 weeks, and the value of 0 deaths was assigned to this period. (Ranson, 1941; Leslie and Ranson, 1940.)

life-table arrangements are used, along with fertility tables, for calculation of rates of increase of populations.

Fertility Tables: Age-specific Rates of Fecundity and Fertility

Many animals, particularly insects, have life cycles that start at approximately the same time for the whole population each year and end before the new generation is hatched the next year (annual life cycles with nonoverlapping generations). Such insects often have a short reproductive period. Then crude rates of reproduction for the female population are also age-specific rates, since the population is approximately the same age. However, many populations consist of individuals of all ages. In many, giving birth or egg laying occurs repeatedly in one female. Often age-specific reproductive rates change with age, as in man and the laboratory vole. In all such cases it

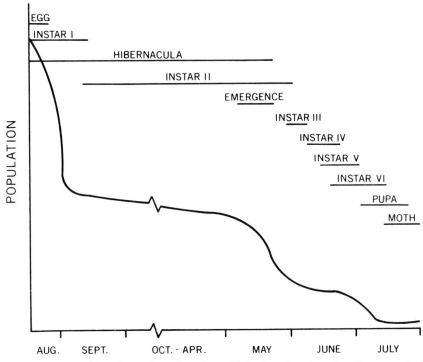

Figure 5.1. The arithmetic number of survivors (l_x) is plotted against time, from a typical life table of the spruce budworm, *Choristoneura fumiferana*, in nature. Life stages are noted against the curve. The young larvae (instars I and II) hibernate in silk tènts (hibernacula). (After Morris *et al.*, 1958.)

may be helpful to arrange reproductive data as tables, giving reproductive performances separately for each class. Age-specific rates are calculated for each age class. The symbols used are given in Table 5.5; and Table 5.6 is a fertility table for the vole in the laboratory.

The Net Reproduction Rate (R_0)

This is the ratio of live female births in two successive generations. If generation times are short the rate can be obtained by direct counts. The net reproduction rate is easiest to obtain when generations overlap very little, or are easily distinguished, and the reproduction period is short and well defined. This condition is true of many insects with annual life cycles (*Urophora*, Chapter 7). However, data on reproductive performance of two generations in some species, such as man, are not likely to be accurate or even available. The theoretical net reproduction rate is then obtained, with many assumptions, from the number of females of reproductive age, their age-specific fertility rates, and the life tables available at that time. Each newborn female is

TABLE 5.5

Components of Fertility Tables

$x=$ Exact age in stated units of time

$n=$ Time interval before next exact age in the table

$E_x=$ Mean number of females exposed to the risk of giving birth to daughters during the time interval x to $x+n$

$$E_x=f_x-1/2\,d_x$$

where f_x is the number of females alive at age x and d_x is the age-specific mortality rate of females

$m_x=$ Age-specific fertility rate.[a] Mean number of live daughters born per age interval per female alive during that age interval

$$m_x=\frac{\text{Number of live females born from } x \text{ to } x+n}{E_x}$$

---

[a]   The symbol m_x is used in human demography for the age-specific mortality rate.

TABLE 5.6

Fertility Table of a Population of the Vole, *Microtus agrestis*, in the Laboratory

x Pivotal age in weeks from birth	E_x Mean number of females exposed to the risk of giving birth to live daughters during the time interval about pivotal age x	Number of live daughters born in the time interval	m_x Age-specific fertility rate per female per 8 weeks
8	45.5	26.0	0.5714
16	42.5	112.5	2.6470
24	34.0	105.5	3.1029
32	25.5	58.0	2.2745
40	19.0	29.5	1.5526
48	11.5	12.0	1.0435
56	5.5	2.0	0.3636
64	2.0	0	0
72	1.0	1.5	1.5000

Females become fecund at 3 weeks and males at 6 weeks. In the wild, the vole requires at least 15°C and 15 hours of sunlight each day for maximum breeding, and in the laboratory breeding can be maintained throughout the year if the temperature and sunlight are kept above limiting levels. Pregnancy lasts approximately 21 days and fertilization can occur again on the day of parturition and during lactation. In the laboratory a female was paired all her reproductive life to a fecund male, each pair in a cage of its own. Eighteen percent of the animals were always bad breeders from conception to weaning of their offspring, with a high rate of resorption of embryos and a high death rate of young partly caused by abnormal parental behavior. These "rogue voles" were defined as those that consistently lost 50 percent or more of their young. This behaviour was shown all their life and influenced a sexual partner so that it shared the bad record, regardless of previous success with another partner. The origin of the behaviour was not known. Data from these pregnancies were not included in the fertility table. (Leslie and Ranson, 1940.)

TABLE 5.7

Calculations of the Net Reproduction Rate, R_0, for a Laboratory Population of the Vole, *Microtus agrestis*[a]

x Pivotal age in weeks from birth	m_x Age-specific fertility rate per female per 8 weeks	P_x Probability at birth of being alive at age x	$m_x P_x$
8	0.5714	0.81923	0.4681
16	2.6470	0.76029	2.0125
24	3.1029	0.60706	1.8836
32	2.2745	0.45382	1.0322
40	1.5526	0.31237	0.4850
48	1.0435	0.16502	0.1722
56	0.3636	0.07662	0.0279
64	0	0.02357	—
72	1.5000	0.01179	0.0177

$$R_0 = \int_0^\infty m_x P_x \, dx = 6.0992$$

[a]Leslie and Ranson (1940).

assumed to be going to experience the life-table mortality. On this basis, the probable survival of all newborn females to the same ages as their mothers is added together. R_0 for the vole is calculated in Table 5.7, and for the given conditions, it is 6.0992. This means that if the distribution of mortality and fertility as given in the life table and the fertility table is true from one generation to the next, for every female present in one generation there would be 6.1 in the next. The time used in this rate is generation time.

If a population is maintaining its numerical status over a long period, a mean value for R_0 over many generations is 1.0.

The Intrinsic Rate of Natural Increase (r)

This is a measure of rate of increase that is expressed in units of absolute time and that overcomes the influences of changing age distributions that limit the use of crude rates of increase. The measure depends partly on the proof that if a given life table and fertility table are considered to be unchanging through several generations, then the age distribution becomes stable (Sharpe and Lotka, 1911; Dublin and Lotka, 1936). Figure 5.2 shows the age distribution taken from the life table given for the vole, and also the stable age distribution calculated by Leslie and Ranson (1940) after Lotka. It follows that if age distribution and age-specific birth and death rates are all constant, crude birth and death rates are constant also. Under these theoretical circumstances the crude rates are now called the true birth rates and true death rates. The rate of increase for such a population is called the intrinsic rate of natural increase. It is the true birth rate minus the true death rate, and it is free from a changing age distribution.

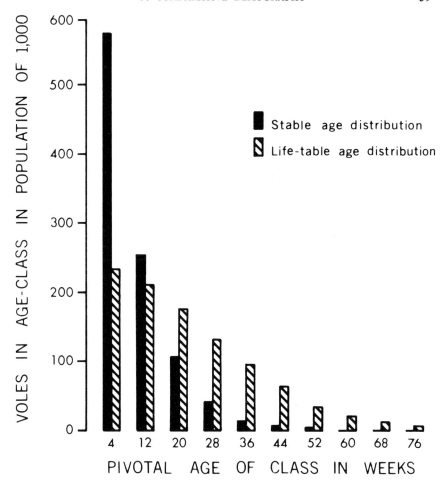

Figure 5.2. The proportion of the population that is in each age class, shown for two populations of the vole, *Microtus agrestis*. The life-table age distribution represents actual numbers of an L_x column calculated from the life-table data of Table 5.4. The stable age distribution would appear after several generations and then persist, if the age-specific birth and death rates of Tables 5.6 and 5.4 were to be constant.

Intrinsic rate of natural increase (r)=true birth rate (b)—true death rate (d).

Birch (1948) shows how r can be calculated from R_0 if the generation time is known, and also from data from life tables and fertility tables. Using some approximations, Leslie and Ranson got a value for r for *Microtus agrestis* in the laboratory of 0.0877 per vole per week. This value of r will be used in the following section to show a curve of exponential growth for a *Microtus* population.

The intrinsic rate of natural increase should not be considered to be the maximal potential of the species. Life-table and fertility-table data for calculation of r can be collected in any environmental circumstances. Calculation of r depends on the assumption that these population attributes remain the same through many generations which is very unlikely.

EXPONENTIAL INCREASE

Population size is a compound of the numbers of individuals, their rate of increase, and time. The simplest expression for this was given for bacteria in Chapter 2. Using the intrinsic rate of increase, r, exponential growth can be expressed as

$$\frac{dN}{dt} = rN$$

or

$$= (b - d) N$$

where dt is infinitesimal value of time t; dN is the infinitesimal value of population size at time t; N is population size; and b and d the true, also called infinitesimal, birth and death rates. The equation can be solved to give:

$$N_t = N_0 e^{rt}$$

where N_t is the population size at time t, N_0 is population size at time 0, and e is the base of Naperian logarithms 2.718 ... In Figure 5.3 the population size of a theoretical colony of voles with a stable age distribution is plotted against time. N_t has been calculated with t at every week from 0 to 24, N_0 is 1,000 voles, and r is 0.0877 per vole per week, as calculated previously.

On the other hand, a real population can be described by a graph in which actual numbers are plotted against time. Then the slope of any part of the growth curve can be measured, and can be expressed as dN/dt. From this and the exponential equation, a value for r at that time can be obtained.

The following observations can be made from the exponential curve of Figure 5.3:

The time that the theoretical population takes to double its size is constant. It is almost 8 weeks for the vole population.

The finite rate of increase is the number of times a population multiplies in unit time. It is 1.091 per week for the vole population.

During the first week the population increased by 92 individuals and during the last week by 724 individuals; these were *actual increases*. Since growth depends on a compound of numbers and rates, when the numbers are high actual increase can be very great.

An intrinsic rate of natural increase for tsetse flies, *Glossina* spp., has been calculated from data on age-specific survival and reproductive rates from several of the 22 species in an assortment of environmental conditions (Glasgow, 1963). Tsetse flies transmit the trypanosomes that cause sleeping sickness in man and nagana in cattle. At approximately 8 adult flies per acre in their habitat, tsetse flies are relatively rare for insects, although they

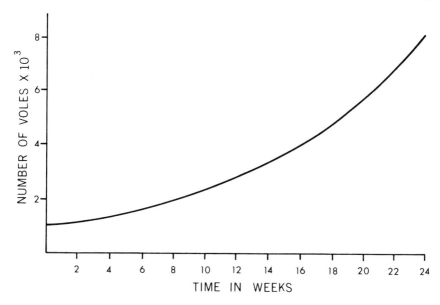

Figure 5.3. Exponential growth of a theoretical population of voles, *Microtus agrestis*, starting from a size of 1,000 (N_0) with an intrinsic rate of natural increase (r) of 0.0877 per vole per week. Population size (N_t) was determined at weekly intervals (t=time in weeks) from the equation $N_t=N_0e^{rt}$. Also note that if the numbers had been plotted as logarithms, the line would be straight—there is no lag phase in this calculated curve of population growth.

dominate the economy of large parts of the African continent. Females produce one live larva about every 14 days. Figure 5.4 gives the growth curve based on the value of r obtained of 0.0201 per fly per month. The time the population takes to double is almost 4 months. With an initial population of 1,000 adults, by $13\frac{1}{2}$ months there are 16,600 adults. Some of the biological and physical factors that are known to oppose in the field this potential for increase are discussed in Chapter 7.

POPULATION BALANCE

Populations always have a high potential for increase because individuals are endowed with high maximum fecundity and fertility and because growth rates are exponential. Populations could not survive without a high potential for increase enabling them to recover quickly from catastrophic declines. Figure 5.5 is a survivorship and reproduction curve representing many insects with annual life cycles. Changes in the population density at any one point in the life cycle can be compared from year to year. The inset curve has taken adult population density at its peaks as the point for comparison and the net reproduction rate is another such comparison.

Over many generations if the ratio from generation to generation is less

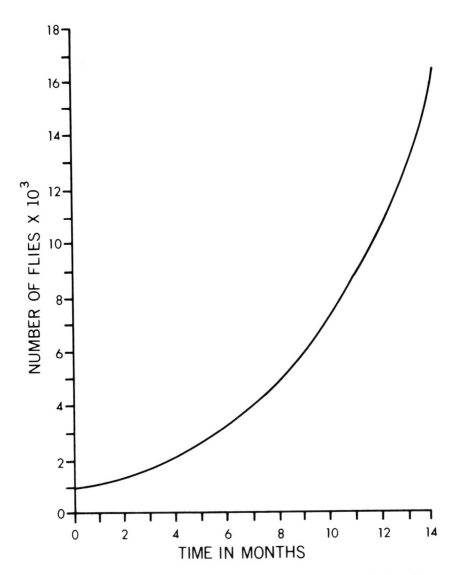

Figure 5.4. Exponential growth of a theoretical tsetse fly, *Glossina* sp., population with an intrinsic rate of natural increase of 0.0201 per fly per month. Population sizes were determined at times t according to the formula $N_t = N_0 e^{rt}$ with N_0 as 1,000.

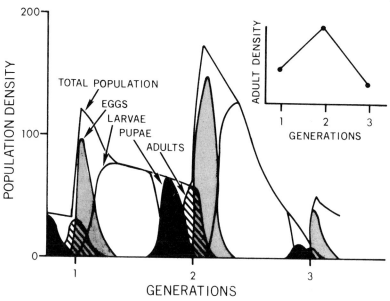

Figure 5.5. Population curves typical for an insect with an annual life cycle, a short reproductive time, and nonoverlapping generations. The total population curve (upper line) is the sum of the partial population curves for eggs, larvae, pupae, and adults. The inset shows a generation curve for the adults, in which the total number reaching the adult stage in a generation is plotted against generation number. (After Varley, in Le Cren and Holdgate, 1962).

than 1 the population is decreasing and if it continues so it will become extinct in that place. If the ratio is more than 1 the population is increasing and cannot continue so indefinitely. Maintenance of an average ratio of 1:1 over many generations is called population balance, with recognition that the population density at which balance occurs can change, or itself could have long-term fluctuations.

So populations that survive can be said to be regulated. What are the characters of the regulating mechanisms? The physical environment could not directly be a regulating mechanism because it varies very much with time for reasons that are independent of the populations present. A controlling mechanism must affect demographic rates and be sensitive to changes in population density, so sensitive that rates are changed enough to return numbers toward the level of balance. Rates showing change with population density are given in Figure 5.6 and Table 5.8. Other rates studied in laboratory animals are mentioned in Chapter 6, such as the survival rate of young *Tribolium* that is sensitive to the population density of the older cannibalistic forms.

It is unlikely that the same demographic rate would always be the controlling factor in all populations of a species, or at all times in one population, or

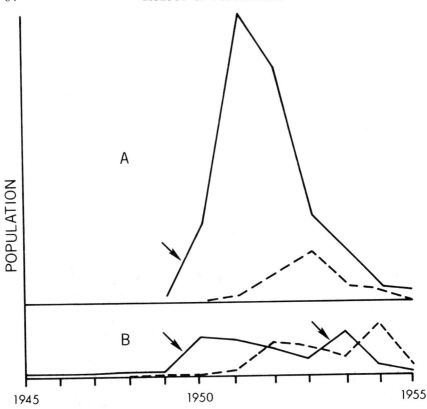

Figure 5.6. Annual population density of spruce budworm larvae (———) and the adult parasites that attack them (— — —) in two study plots in Canada. Arrows = immigration of budworm larvae. A. Budworm epidemic stopped by defoliation of trees. B. Non-epidemic area. Numbers of hosts and parasites are not on the same scale; they are plotted so that the dotted line exceeds the solid line at the levels at which the parasite, when added to other mortality factors, is capable of controlling the host. In area B the food supply does not allow outbreaks of epidemic size. These conclusions are based on analyses of very extensive data covering about 20 years. (After Morris *et al.*, 1958.)

that it would always operate alone. Also, a regulating mechanism might have more obvious features, such as being a pattern of behaviour. For instance, some species apparently regulate their density by the fact that the breeding population holds territory covering a total area that is limited and some potential breeders are excluded from the breeding territories. This mechanism operates as a regulator by changing reproductive rates with population density.

KEY FACTORS
It may also be possible to say of a population that a particular factor varies in

TABLE 5.8

Key factors of Mortality Determined from Life-Table Studies of Twelve Canadian Insect Pests[a]

Species	Generations studied	Critical age interval(s)	Key factor(s)[a]	Investigator(s)[b]
Fruit-tree leaf roller, *Archips argyrospilus*	6	Pupa/adult	Parasitism*/ migration*	Paradis and LeRoux
Diamondback moth, *Plutella maculipennis*	18	Adult	Weather	Harcourt
Imported cabbageworm, *Pieris rapae*	18	Larva	Disease*	Harcourt
Pistol casebearer, *Coleophora malivorella*	7	Larva	Parasitism*	LeRoux, Paradis, and Hudon
Eye-spotted bud moth, *Spilonota ocellana*	7	Larva	Weather	LeRoux, Paradis, and Hudon
European corn borer, *Ostrinia nubilalis*	5	Adult	Migration*	LeRoux, Paradis, and Hudon
Colorado potato beetle, *Leptinotarsa decemlineata*	6	Larva	Food supply*	Harcourt
Spruce budworm, *Choristoneura fumiferana*	15	Larva	Weather	Morris *et al.*
Winter moth, *Operophtera brumata*	8	Pupa	Parasitism*	Embree
Oystershell scale, *Lepidosaphes ulmi*	3	Egg/adult	Predation*/ parasitism*	Samarasinghe and LeRoux
Apple leaf miner, *Lithocolletis blancardilla*	3	Pupa/adult	Predation*/ weather	Pottinger and LeRoux
Birch leaf miner, *Fenusa pusilla*	9	Larva/adult	Predation*/ migration	Cheng

[a] Those factors marked with an asterisk were shown to be density dependent.
[b] References to the publications are given by the compilers, Harcourt and LeRoux (1967).

its strength from time to time, and that the variation is strongly reflected in population density. The factor may be, for instance, winter weather, so that after severe winters population densities tend to be low for some time and after mild winters numbers are higher.

The ecologist concerned with economic pests or disease vectors wishes to identify key factors to facilitate prediction or prevention of high densities. Such key factors are not necessarily regulating population balance because, although they are strongly influencing numbers, their rate of operation does not necessarily depend on population density—it may vary independently. Figure 5.7 is a series of graphs of annual variation in mortality rates from six causes and total mortality rates for a British population of the winter moth, *Operophtera brumata*, which can be considered an infection of oak trees. In this case winter mortality was a key factor in total mortality, which in turn profoundly affected population density. Table 5.8 gives key factors of mortality that have been found in twelve studies of insect pests in Canada and the life

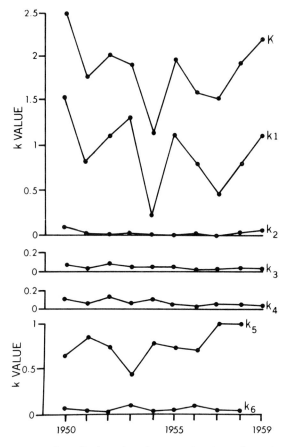

Figure 5.7. Mortality affecting the winter moth, *Operophtera brumata*, 1950–1959 at Wytham, Berkshire, on logarithmic scale.

Mortality (*K* or *k*) is measured as numbers lost during the operation of the causal factor (s).

K, total mortality $= k_1 + k_2 \ldots k_6$

k_1, winter disappearance

k_2, Tachinid parasite of larva

k_3, other larval parasites

k_4, disease (Microsporidian)

k_5, pupal predators

k_6, pupal parasite

k_1 is a *key factor* in producing annual variation in population density of this insect.

Also note that pupal mortality from predators (k_5) is a *compensating factor*. When pupal population density is low, predators find smaller proportions of the pupae and the mortality rate from this cause declines. Since at the levels of winter moth densities implied this rate changes with density and moves the population toward the level of balance, it can be concluded that it does help to regulate density. However, it is not the most important factor.

(After Varley and Gradwell, 1960.)

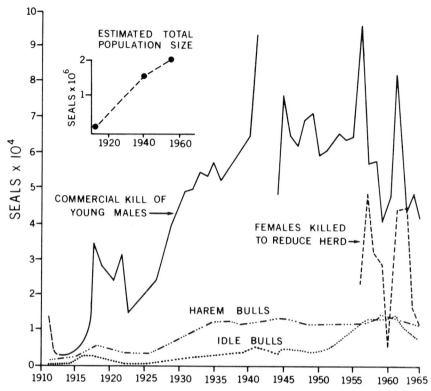

Figure 5.8. Northern fur seals, *Callorhinus ursinus*, of the Pribilof Islands, Alaska. Males aged 3 and 4 years are annually harvested. (There was some interruption in the 1940's.) There has been wider fluctuation in the harvest since 1955, because of fluctuations in the size of the age classes. Harem bulls are aged 10 and more years, and have up to 100 females in the harem. (Data from the North Pacific Fur Seal Commission, 1962, and Roppel *et al.*, 1963, 1965a, 1965b.)

stages in which they operated. Some factors are noted as being density dependent and thus capable of regulating density.

CONCLUSIONS: APPLICATIONS OF DEMOGRAPHY TO SOME HUMAN PROBLEMS

It is important at the present time that there should be widespread understanding that population growth is exponential. With large human populations and current rates of increase, within short time periods actual increase can become too great for resources and space. So rates of increase are bound to decline, one way or another.

Management of wild animal populations for food is also an important topical subject because the world food supply even now is not meeting demand. (See Le Cren and Holdgate, 1962, for papers by authors in this paragraph.) Pearsall has summarized the case for increasing the management of wild mammals in Africa as a food source, because their productivity in the African environment is greater than that of imported livestock. Man has tended to use wild populations too heavily so that yield has declined and sometimes species have been lost. Marine animals, particularly, are under-exploited or overexploited. Watt asked: "For a given population and conditions extrinsic to the population, what age distribution and rate of exploitation will yield a maximum harvest and still leave behind enough reproductive individuals so that a maximum rate of biomass yield can be sustained?" Beverton discussed some of the problems of obtaining demographic data from fish populations in order to construct harvesting models. Interesting conclusions about harvesting came from laboratory populations manipulated by Slobodkin (1960) and are also found in Le Cren and Holdgate (1962). The Pribilof Islands herd of the northern fur seal, *Callorhinus ursinus*, is an example of a managed wild resource (Figure 5.8). The population was almost killed by man by the turn of the century, but it recovered while being harvested by the United States Fish and Wildlife Service and its predecessor. At the present time it appears that population growth may be limited by intraspecific competition for food among young seals at sea. With a heavier kill man should be able to keep the population back in the phase of exponential growth and take a steadier yield, and the Fish and Wildlife Service is attempting to do this.

Demography is also used in analysis of epidemics, whether of pests of forest plantations or of human disease, and it is used in attempts to understand epidemics of human disease by constructing mathematical models (Chapter 25).

REFERENCES

Birch, L. C. 1948. The intrinsic rate of natural increase of an insect population. *J. Animal Ecol.* **17**: 15–26.

Chapman, D. G. 1964. A critical study of Pribilof fur seal population estimates. *Fishery Bull.* **63**: 657–669.

Dublin, L. I., and Lotka, A. J. 1936. *Length of Life: A Study of the Lifetable.* Ronald Press, New York.

Geiser, S. W. 1924. The differential death-rate of the sexes among animals, with a suggested explanation. *Wash. Univ. Studies* **12**: 73–96.

Glasgow, J. P. 1963. *The Distribution and Abundance of Tsetse.* Pergamon Press, Macmillan, New York.

Harcourt, D. G., and Le Roux, E. J. 1967. Population regulation in insects and man. *Am. Scientist* **55**: 400–415.

Le Cren, E. D., and Holdgate, M. W. (eds.) 1962. *The Exploitation of Natural Animal Populations.* Blackwell, Oxford.

Leslie, P. H., and Ranson, R. M. 1940. The mortality, fertility and rate of natural increase of the vole (*Microtus agrestis*) as observed in the laboratory. *J. Animal Ecol.* **9**: 27–52.

Leslie, P. H., Venables, U. M., and Venables, L. S. V. 1952. The fertility and population structure of the brown rat (*Rattus norvegicus*) in corn ricks and some other habitats. *Proc. Zool. Soc. Lond.* **122**: 187–238.

Lotka, A. J. 1936. Modern trends in the birth rate. *Ann. Am. Acad. Polit. and Soc. Sc.* 1936, Nov., pp. 1–13.

Moroney, M. J. 1956. *Facts from Figures.* Pelican Books A236, Penguin Books, Baltimore, Md.

Morris, R. F., Miller, C. A., Greenbank, O., and Mott, D. G. 1958. The population dynamics of the spruce budworm in Eastern Canada. *Proc. Intern. Congr. Entomol. 10th, Montreal*, 1956. **4**: 137–149.

Mosby, H. S. 1960. *Manual of Game Investigational Technique*. Wildlife Society, Washington, D.C.

Nicholson, A. J., and Bailey, V. A. 1935. The balance of animal populations. *J. Animal Ecol.* **2**: 132–178.

North Pacific Fur Seal Commission. 1962. North Pacific Fur Seal Commission Report on Investigations from 1958–1961. Manuscript Report, Washington, D.C.

Ranson, R. M. 1941. Prenatal and infant mortality in a laboratory population of voles (*Microtus agrestis*). *Proc. Zool. Soc. Ser. A* **111**: 45–57.

Roppel, A. Y., Johnson, A. M., Bauer, R. D., Chapman, D. G., and Wilke, F. 1963. Fur seal investigations, Pribilof Islands, Alaska, 1962. *U.S. Fish Wildlife Serv., Spec. Sci. Rep., Fisheries*, **454**.

Roppel, A. Y., Johnson, A. M., and Chapman, D. G. 1965a. Fur Seal Investigations, Pribilof Islands, Alaska, 1963. *U.S. Fish Wildlife Serv., Spec. Sci. Rep., Fisheries*, **497**.

Roppel, A. Y., Johnson, A. M., Anas, R. E., and Chapman, D. G. 1965b. Fur Seal Investigations, Pribilof Islands, Alaska, 1964. *U.S. Fish and Wildlife Serv. Spec. Sci. Rep., Fisheries*, **502**.

Sharpe, F. R., and Lotka, A. J. 1911. A problem in age distribution. *Phil. Mag.* **21**: 435.

Slobodkin, L. B. 1960. Ecological energy relationships at the population level. *Am. Naturalist* **94**: 213–236.

U.S. Bureau of the Census. 1951. *Handbook of Statistical Methods for Demographers*.

Varley, G. C., and Gradwell, G. R. 1960. Key factors in population studies. *J. Animal Ecol.* **29**: 399–401.

Watt, K. E. F. 1968. *Ecology and Resource Management*. McGraw-Hill Book Co., New York.

CHAPTER 6

Experimental Demography with Laboratory Invertebrates

BRENDA K. SLADEN

Populations have
    sizes, and spatial arrangements, and biomass
    age distributions, including age distributions of reproductive individuals
    sex ratios
    birth rates
    death rates
    emigration and immigration rates
    gene pools, gene distributions, and phenotypic variation
    social organizations
    rates of food consumption
    rates of waste production
    histories, etc.

One can begin to describe these features, and the changes in them, with laboratory populations for which some of the variables can be simplified. Eight experiments on such populations will be described in order to discuss their biological properties and to prepare for a discussion of the ecology of wild populations.

In most of these experiments there is a fixed volume within which the population can grow, and movement in and out of the population is not allowed or is completely controlled. Food is added regularly and waste is regularly removed. Age of individuals can usually not be recognized beyond broad categories of egg, larvae of various sizes, pupa, and adult. Age distribution of reproductive activity can be studied in individuals, although this may not represent performance in the population. Sex ratio can be found from sampling and it is usually near to unity. Social organization in these animals is usually not known. The genetics of the population is usually not studied, though in some cases genetic changes have been shown to have great effects upon population events. Population size and changes of size are usually measured results and the experiments are replicated in large numbers.

SINGLE POPULATIONS

Survival Directly Varying with the Physical Environment in Fleas, *Xenopsylla*
Physical conditions limit population existence and have profound effects upon population growth. Bacot and Martin (1924) took populations of rat fleas, *Xenopsylla cheopis*, to show some relationships between survival rates of the adults and the physical environment. At the onset of hot, dry weather in the

70

northern plains of India, when plague was common, there was a rapid fall in the numbers of human cases. Rat fleas can transmit the bacillus *Pasteurella pestis* to man when they leave a dead rat host and search for another host, which may be a rat or a man. It was shown that desiccation can cause death in wandering rat fleas and that infected fleas are particularly sensitive to a dry atmosphere. Survival rates were measured of groups of 100 adult rat fleas in various conditions of relative humidity and temperature. In one experiment (Figure 6.1) temperature was constant at 32°C, and at 89 percent relative

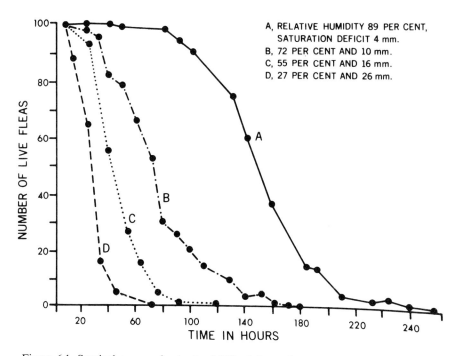

Figure 6.1. Survival curves of cohorts of 100 adult rat fleas, *Xenopsylla choepis*, in four conditions of humidity and temperature. (After Bacot and Martin, 1924.)

humidity fleas were still alive after nearly 240 hours. At 27 percent relative humidity, the whole population was gone in 60 hours. There were two intermediate curves of survival at intermediate relative humidities. The authors concluded:

A variation in saturation deficiency from 5 mm. to 35 mm. such as occurs in the plains of Northern India at different seasons would, accordingly, shorten the average duration of life of wandering rat fleas in the proportion of 15 to 1. As a rise in mean temperature may amount to a difference of 20°C. between January and June this would reduce the length of life of wandering fleas to about one-third. The effect of saturation deficiency and increased temperature will be additive and would go a long way to explain some of the climatological features of the epidemics.

Uvarov (1931) commented:

I would suggest that the existence of bubonic plague in the Caspian Steppes with their exceedingly severe winters and hot dry summers may be correlated with the fact that plague is there transmitted not by fleas of the domestic rat, but by those parasitic on wild rodents (susliks, *Spermophilus* spp.) hibernating in burrows where the fleas are perfectly protected from both winter cold and summer drought.

The Logistic Pattern of Growth in the Fruit Fly, *Drosophila*

Pearl started the field of experimental demography with experiments at Johns Hopkins University School of Hygiene and Public Health. Experiments with *Drosophila melanogaster* (Pearl, 1925) were done to obtain a pattern of population growth to be described mathematically. Adult *Drosophila* were put in half-pint milk bottles with banana-agar jelly and growing yeast for food. Adults flew above the mixture and laid eggs in the top 2 mm. and larvae burrowed within. Generation time was about 10 days and the mean length of life of the adults was 12 to 40 days, depending on population density. There soon were overlapping generations in the bottle. The food medium deteriorated and could not be renewed, and the experiments were terminated at 38 days. Only adults were censused.

The growth pattern of the adult population is shown in Figure 6.2. The upper level was at about 200 flies. Pearl saw that as population density increased, adult mortality increased and the fecundity rate rapidly declined. He thought that increased disturbance of laying females caused the decline in fecundity but it was later shown to be due to decreased food supply per female. Pearl also showed that a population of flies homozygous for a certain five mutants had a growth curve of the same form, but with an upper level at two-thirds that of the wild-type population.

Pearl used the logistic equation of Verhulst, which is

$$\frac{dN}{dt} = rN \frac{(K-N)}{K}.$$

The first part, $dN/dt = rN$, is the exponential equation for growth. K is a constant concerned with realization of the potential, and N is the number of individuals in the population. When numbers are very low, $(K-N)/K$ represents maximal realization of the potential. As numbers increase, the realization of the potential, as represented in the formula, decreases instantaneously. For the *Drosophila* the $(K-N)/K$ part of the equation represents at least changes in longevity, and changes in fecundity that depend on food supply. Thus the Pearl–Verhulst equation can be a considerable simplification of several somewhat independent forces. In this case, and in many bacterial populations, increased numbers and decreased food supply for the individual affect population increase very rapidly. In many populations, however, there is more delay in the operation of the forces that retard population increase, and this simple formula is less useful. Figure 6.2 gives calculated values of population size for *Drosophila* and census figures of the growing population. Stiven's adaptation of the logistic equation to the growth of an experimental epidemic is discussed in Chapter 25.

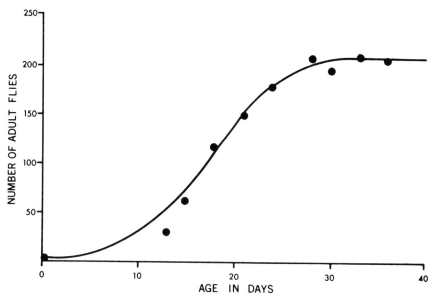

Figure 6.2. The logistic pattern of growth of a population of *Drosophila melanogaster*, only adults being shown. The line is calculated according to the Pearl–Verhulst logistic equation, and the dots represent actual censuses. (After Pearl, 1925.)

Relatively Steady Level of Population Density—The Flour Beetle, *Tribolium*

Tribolium are small beetles that have been inhabiting stored grain and grain products for probably at least 45 centuries and they are widely used in demography experiments. The natural environment is rather closely matched by a laboratory environment of sifted unadulterated flour and dried yeast in small glass vials. At 34°C and relative humidity of 70 percent the life cycle of *Tribolium confusum* is as follows. Eggs hatch at about the fourth day and there are 5 or 6 larval stages, lasting in all about 17 days. The pupal stage lasts for about 5 days, adult males have mean longevity of 178 days and females 198 days. Eggs are laid 5 to 8 days after the female emerges as an adult. Fecundity and fertility remain high for a long time, about 10 eggs being laid per female per day, and the rate varies with age and with population density. Since generation time is much shorter than a lifetime, several generations of females can be laying eggs in the same population. In the laboratory there is no migration and no hibernation. Chapman (1928) used a series of habitat sizes of 4, 8, 16, 32, 64, and 128 grams of flour at 27°C, added one pair of adult beetles to each and let them develop into populations in regularly renewed flour. After increase the populations all settled to rather steady levels of 25 beetles of all stages except eggs per gram of flour, this density being independent of the total size of the habitat. This sort of result was true for many variations of the experiment. Park kept *Tribolium* populations for 70 generations with regular renewal of the medium and showed that there are changes in population size, but that these are neither great nor regular (Park *et al.*, 1964) (Figure 6.3). Excretory products "condition" flour and con-

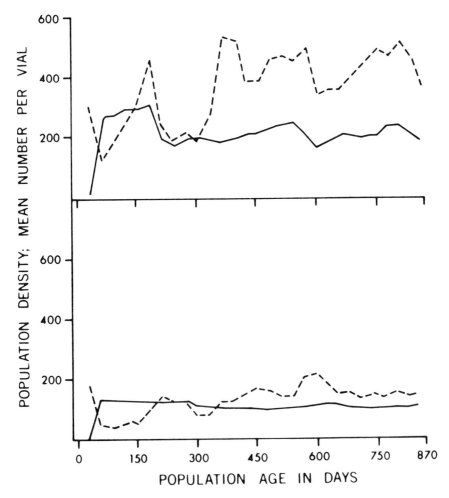

Figure 6.3. Population growth of two strains of *Tribolium confusum*, numbers of adults being shown by solid lines and pupae and larvae by broken lines. (After Park *et al.*, 1964.)

ditioned flour decreases egg-laying rates. Egg-laying rates also decrease with increase of population density when conditioning is not allowed to occur because of flour renewal. Mortality rates of adults, in the range of densities of these experiments, are not sensitive to population density. However large larvae and adults eat inactive stages, particularly eggs, females taking up to seven times more eggs than males; young adults are eaten also. As adult population density increases, cannibalism rates per individual decrease but the survival rate of eggs declines rapidly as the density of eggs increases. The total effect of increased density, which is accompanied by decreased fecundity, is such a rapid decline of survival rate of the young stages because of cannibalism, that recruitment of adults declines to very low levels. The effects of the population upon itself are summarized in Table 6.1. Chapman (1928) summarized: "A high potential is ever present, and when the environmental resistance is lowered by the death of adults, eggs hatch and produce new adults to take the place of whose which have died."

Regularly Fluctuating Population Density in Blowflies, *Lucilia cuprina*

In the wild the blowfly adult flies around and the female deposits eggs in large numbers on carcasses and the larvae eat the flesh. In experiments by Nicholson (1957), adults were kept apart from larvae but could lay their eggs on the larval food of ground liver. Food supply of the adults and food supply of the larvae were controlled independently. In the experiment reported here, adult food was unlimited, but larval food was added regularly in small, fixed amounts. Larvae compete for food in such a way that they share what is available and at some point none of them gets enough for its survival. Nicholson called this a scramble for food. Adults and larvae were censused, and the results are given in Figure 6.4. Adult population size is shown by the black line, which has pronounced regular fluctuations. Numbers of eggs that developed through to adulthood are shown by bars drawn for the day on which those eggs were laid. The figure shows that when the adult population was low, the largest numbers of eggs grew to adulthood. As the adult population increased, the number of eggs laid became enormously greater, until so many larvae were competing for the limited food that no eggs developed to adulthood. During the period of no recruitment to the adult population because of overpopulation of the larvae, the adult population began to decline in numbers. When the adult population had declined to a certain level, the input of eggs was again so low that larvae could develop through to adulthood; and thus the cycle repeated itself. Population increase was limited by larval competition for food in such a way that large regular population fluctuations occurred.

In one set of experiments with the same arrangement as that described above, Nicholson imitated a control procedure that might be used in a public health program. He removed 99 percent of all emerging adults and the result was a sixfold increase in emerging adults. By decreasing the adult population size he had relieved much of the larval competition for food, and this allowed

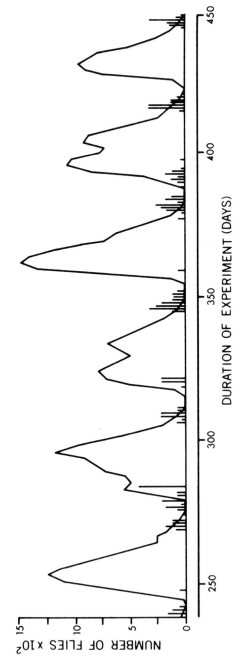

Figure 6.4. Part of the growth of an experimental population of the blowfly, *Lucilia cuprina*, with food for the larvae being restricted. Numbers of adults are shown by the solid line. A vertical bar represents the number of eggs laid that day that eventually became adults. (After Nicholson, 1957.)

TABLE 6.1

Major Components of the *Tribolium* Model[a]

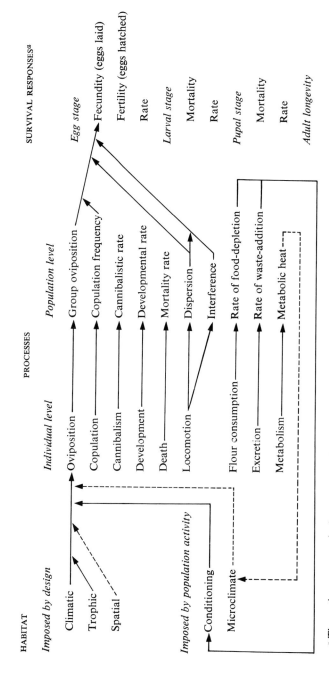

[a] The authors use survival to mean the interrelation between mortality and natality; "that is, what actually happens to the population." (Neyman *et al.*, 1956.)

greater larval survival, demonstrating that interference with a population can produce unexpected effects.

In another experiment, Nicholson (1957) furnished unlimited larval food and restricted adult food. Changes in fecundity, the numbers of eggs laid, due to food shortage, produced population fluctuations. However, the pattern of population change itself changed very markedly in the middle of the experiment. Out of the original population, adults had been selected that could lay eggs on a restricted protein diet. A genetic change had produced a change in the pattern of population growth.

TWO POPULATIONS OF DIFFERENT SPECIES

Fluctuating Host and Parasite Population Densities with the Bean Weevil, *Callosobruchus,* and its Hymenopterous Parasite

Utida (1957) used two populations of different species, a host, the bean weevil *Callosobruchus chinensis,* and its parasitoid, the wasp *Heterospilus prosopidis.* The parasite was free living as an adult, hence the term parasitoid. This was experimental epidemiology, with a parasite population and a host population changing in response to each other. The bean weevil used azuki beans for oviposition and for food, and new beans were added at regular intervals and old ones removed. Thus the weevil's food supply was probably unlimited and waste accumulation probably had little influence on the weevil population. The parasite was efficient at finding hosts and had a rather low fecundity. One parasite could lay eggs in each of about 40 hosts, and the host died when the parasite emerged, but when more than one egg was laid in a single host the host died early and no parasites emerged. Heavy parasitism brought about decline of the host population and a slightly later decline of the parasite adult population, followed by increase again of the host population. The adult censuses of both populations for 115 generations covering $1\frac{1}{2}$ years are shown in Figure 6.5. The fluctuations were not always regular and were rather diminished at times. In some of the replicates of this experiment the bean weevil population became extinct at times of very heavy parasitism.

Competition between Two Species with Identical Requirements Leading to the Extinction of one; the Flour Beetles, *Tribolium*

Ecological competition among populations is interference among themselves, leading to changed reproduction, mortality, or migration rates, or to two or all of these, so that at least one population is lower in numbers or biomass than if it were present alone. Competition does not include predator-prey relationships or parasitism. It probably plays a large role in the control of population distribution and abundance, although it tends to be difficult to recognize in nature. Park (1948 and Park *et al.,* 1964) used two closely related species of *Tribolium* that had small morphological differences and some differences in their population attributes. The conditions of culture were the same as for single populations of *Tribolium* but the two species were inoculated into the same vials in equal numbers. When environmental conditions

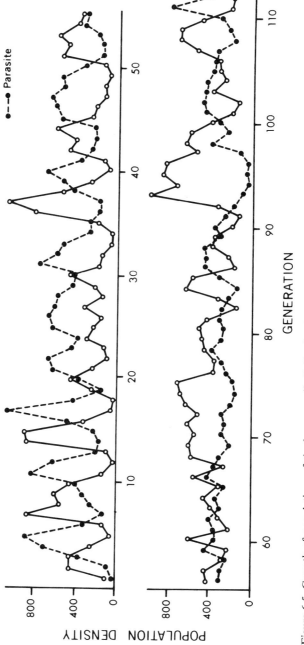

Figure 6.5. Growth of populations of the bean weevil. *Callosobruchus chinensis*, and its hymenopteran parasitoid, *Heterospilus prosopidis*. (After Utida, 1957.)

of temperature and humidity were kept steady there was an invariable result—after from 180 to 1470 days only one species remained but which species would disappear could not be predicted in any single experiment. Only a probability could be given and it varied with temperature, relative humidity, and presence of the sporozoan gut parasite *Adelina tribolii*. Results are shown in Figure 6.6.

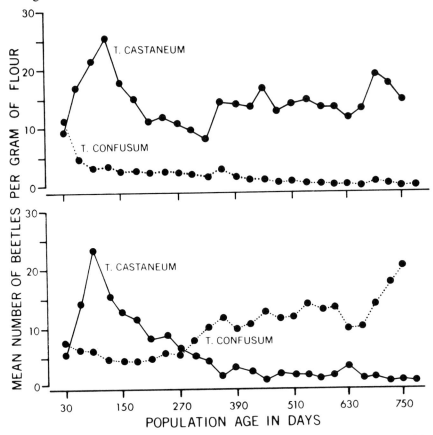

Figure 6.6. Populations of *Tribolium confusum* and *T. castaneum* growing in single vials. *Tribolium confusum* became extinct in the experiment represented by the upper graph, and *T. castaneum* became extinct in the experiment represented by lower graph. The numbers refer only to adult beetles. (After Park, 1948.)

The differences between the two species were investigated but not exhaustively and not in all conditions of population density. Cannibalism played a large role in limiting numbers of the mixed populations and there were differences in the voracity of the species toward each other, depending on strains of beetles used. However, the active and the passive stages were not of exactly the same duration in the two species. The pressure of different fecundity rates was also an important factor in the elimination of one of the species, and the different frequencies of elimination of each species at different environmental conditions was thought to be in part due to different responses of fecundity rates to different conditions. When the parasite was present its different effects on the two hosts probably played a large part in the species elimination. In many of the replicates *Tribolium castaneum* first had the lead in numbers but was depressed by 300 days. It was not depressed when cultured alone and unparasitized, but was so when alone with the parasite. During the depression of *T. castaneum*, *T. confusum* was able to increase and probably further affected *T. castaneum* by being, for it, a source of further infection.

The development of the two species populations may be seen as unfolding in a typical, particular way for a given set of circumstances, to lead to the extinction of one species. But in exceptional cases the development might proceed along other pathways and lead to extinction of the other species. Park said of these experiments;

> ... for some treatments, the dependence of the survival of a species on the characteristics of the environment appears to have the nature of a chance event (not necessarily a 50–50 even). From this point of view some generality might be gained by considering the survival of a given species always as a chance event with variable probability P of the survival, where, on occasion P may be unity or zero.

The way of life of a species is called its niche. This includes the components of the broad habitat in which it lives, its requirements of the environment, and its "role" in the community (Chapter 7). It is a debated matter in ecology whether two species of the same niche can maintain populations in the same place at the same time. Park's two species of the same niche under the conditions of his experiments could not maintain themselves together. The next experiment carried the analysis of this matter a little further.

Two Competing Species with Almost Identical Requirements, surviving because of a Habit Variation: Flour Beetles *Oryzaephilus* and *Tribolium*

Tribolium and *Oryzaephilus* (Crombie, 1946) have the same habitat of stored grain. Pure cultures of each species will survive for many years in the laboratory in regularly renewed medium. Each species is cannibalistic and when put together in flour they ate each other as well as their own species, but the more rapacious *Tribolium* took a greater toll of *Oryzaephilus*. The unvarying result was that after about 250 days *Oryzaephilus* had disappeared from the mixed populations and *Tribolium* continued alone. This so far was like the previous experiment, except that the winner could be predicted. Having established this result Crombie introduced into the flour lots of small pieces of glass tubing. The bore of the tubes was large enough to contain an *Oryzaephilus* larva of any size and its pupa but the larger *Tribolium* larvae and the adults could not enter. With this small habitat variation, the *Oryzaephilus* species was sufficiently protected to maintain itself as a continuing population along with the *Tribolium* (Figure 6.7).

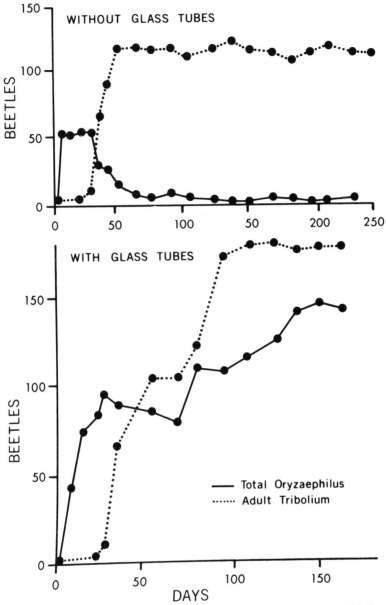

Figure 6.7. Growth of populations of beetles of two closely related genera *Tribolium* and *Oryzaephilus* cultured in single vials. In the experiment represented by the upper graph the medium was plain flour, and *Oryzaephilus* always became extinct. When small pieces of glass tubing were added, both populations survived, as shown in the lower graph. (After Crombie, 1946.)

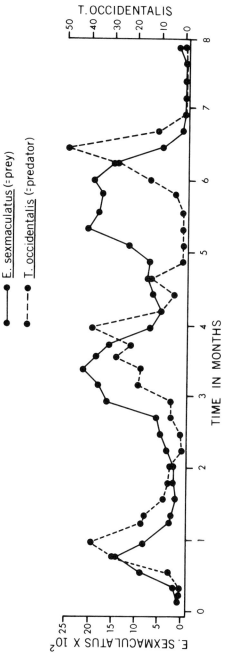

Figure 6.8. Population growth of the orange-feeding mite, *Eotetranychus sexmaculatus*, cultured in a complex environment with its predator, the mite *Typhlodromus occidentalis*. (After Huffaker, 1958.)

Predator Population and Prey Population Surviving Together because of Habitat Variation—The 6-Spotted Mite and Its Mite Predator

Experimenters with populations of predator and prey had always experienced rapid destruction of the system by overpredation. Huffaker (1958) sought to maintain such populations for long periods and found that for his species some complexity of the environment was required. He used the 6-spotted mite *Eotetranychus sexmaculatus* which lives and feeds on oranges, and the mite *Typhlodromus occidentalis* which feeds on *Eotetranychus*. In the wild, *Typhlodromus* will develop in great numbers on oranges infested with the 6-spotted mite, destroy the prey population, and then die itself in large numbers. Laboratory colonies were kept in the dark with temperature at 83° and relative humidity not below 55 percent. The size of the feeding areas was controlled by having parts of the orange surface covered with paper sealed with wax and the orange surface was covered with lint. To create complexity oranges were mixed with other oranges completely covered in wax and all were arranged in areas with limited access created by barriers of petroleum jelly. Posts made of sticks were included. It was thought that there was no food shortage for the prey. One-quarter of the oranges was replaced every eleventh day, with precautions that this did not involve removal of large parts of the populations. The mites had short generation times and the system carried 10,000 to 12,000 mites when the populations were most dense. When a population was dense there was much migration to other food sources and some mites became isolated on posts. Not only did the prey have to find another place to feed and multiply, but the predator had to reach the same place too in order to survive. When Huffaker first set up the experiment the system was not complex enough. The prey and predator would multiply and both populations would crash and become extinct. The prey needed to be able to survive in small pockets of the environment. By making this a game of hide-and-seek enough of both populations could survive from periods of low density and be able to start a cycle again. The populations continued through at least three big cycles of abundance, lasting in all about a year (Figure 6.8).

CONCLUSIONS

Concerning the logistic pattern of growth, Lotka said, "The logistic expresses what may be regarded as a fundamental law of population, that a population cannot increase indefinitely in a constant geometric progression." Gause (1934) said the same thing that the logistic equation says—population growth = the potential rate of increase, x, the degree to which the potential is realized. To requote Chapman (1928) about his *Tribolium* experiment: "A high potential is ever present, and when the environmental resistance is lowered by the death of adults, eggs hatch and produce new adults to take the place of those which have died." Inserting environmental resistance into Gause's definition above gives another statement of Chapman's—population growth rate = the potential rate of increase *minus* environmental resistance.

Types of environmental resistance that were important in these experiments were as follows:

Fleas: the physical factors of the environment—heat and humidity.

Drosophila: food supply, as it affected fecundity.

Tribolium: other members of the same population were cannibals; presence of other members of the same populations affected fecundity; sometimes spoilage of the medium by waste created by other members of the population.

Blowflies: larval food supply and competition for it by other members of the population.

Bean weevil and parasite: population density of the parasite, for the host; population density of the host, for the parasite.

Tribolium species and *Oryzaephilus*: competing species.

Mite and predator: population density of a predator, for the prey; population density of the prey, for the predator.

The experiments demonstrated that environmental resistance, as used above, does not consist only of factors of the physical environment. Other members of the same population, and members of other populations, exert resistance which prevents the realization of the potential for increase.

This conclusion leads to a discussion of ecology in general. Ecology is particularly concerned with mixtures of populations in natural conditions. It is concerned with the normal microbial populations of the human intestine or those of the nose and throat; with the bacteria, worms, and insects concerned in sewage breakdown; with disease vectors; with domestic pests and their parasites and predators; with the very complicated mixtures of species in soil or in woodland. Ecology is concerned with how these populations and the rest of their environment react upon each other. It is concerned with the organization within groups of populations, with the distribution and cycling of materials and energy, and with limitation of numbers.

An interesting central question in ecology is this: if any one species present in nature has a high potential for increase, why does it not overrun all the other species and take over as a pure culture? Part of an answer can be found in what has just been discussed about environmental resistance; part also in the experiments of Crombie and of Huffaker concerning the protection afforded by complex environments. Under natural conditions the physical components of the environment are highly variable in climatic and diurnal changes and in the form of the substratum. Consider what hiding places from other species there are in a woodland tree trunk with its cracks and fissures, water holes, pieces of loose bark, wood for boring into, and so on. It is partly because of the great complexity of the environment that such complexity of species can coexist.

In the next chapter a wild insect population will be discussed and the components of the environmental resistance that hold in check the potential for increase of that species.

REFERENCES

Allee, W. C., Park, O., Emerson, A. E., Park, T., and Schmidt, K. P. 1949. *Principles of Animal Ecology*. Saunders, Philadelphia, Penn.

Bacot, A., and Martin, C. J. 1924. The respective influences of temperature and moisture upon the survival of the rat flea (*Xenopsylla cheopis*) away from its host. *J. Hyg.* **23**: 98–105.

Chapman, R. S. 1928. The quantitative analysis of environmental factors. *Ecology* **9**: 111–122.

Crombie, A. C. 1946. Further experiments on insect competition. *Proc. Roy. Soc.* (London), **Ser. B 133**: 76–109.

Gause, G. F. 1934. *The Struggle for Existence*. Williams and Wilkins, Baltimore, Md.

Huffaker, C. B. 1958. Experimental studies on predation: dispersion factors and predator-prey oscillation. *Hillgardia* **27**: 343–383.

Neyman, J., Park, T., and Scott, E. L. 1956. Struggle for existence. The *Tribolium* model: biological and statistical aspects. In *Proc. Third Berkeley Symp. on Mathematical Statistics and Probability*. Vol. 4. Univ. of California Press, Berkeley, Calif.

Nicholson, A. J. 1957. The self-adjustment of population to change. In *Population Studies: Animal Ecology and Demography. Cold Spring Harbor Symp. Quant. Biol.* **22**: 153–172.

Park, T. 1948. Experimental studies of inter-species competition. 1. Competition between populations of the flour beetles, *Tribolium confusum* Duval and *Tribolium castaneum* Herbst. *Ecol. Monographs* **18**: 265–308.

Park, T., Leslie, P. H., and Merz, D. B. 1964. Genetic strains and competition in populations of *Tribolium. Physiol. Zool.* **37**: 97–162.

Pearl, R. 1925. *The Biology of Population Growth*. Knopf, New York.

Utida, S. 1957. Population fluctuation, an experimental and theoretical approach. In *Population Studies: Animal Ecology and Demography. Cold Spring Harbor Symp. Quant. Biol.* **22**: 139–150.

Uvarov, B. P. 1931. Insects and climate. *Trans. Entomol. Soc. London*, **79**: 1–247.

CHAPTER 7

The Ecology of Animal Communities

BRENDA K. SLADEN

ECOLOGICAL COMMUNITIES*

A place in nature, such as a cluster of flowers, bottom of a pond, nest of a burrowing rodent, is occupied by populations of several or many species. This aggregation of populations depends in large part on the concentration of particular kinds of food there or on conditions for the production of food for plant populations, and then on the distribution of the food from population to population by eating and in turn by being eaten. Species are specialized in their requirements, and many factors help determine the presence of particular species. Some populations use a place for mating or resting, and there can become prey or host to other populations. Thus there is much interaction among the populations in terms of predation, parasitism, scavenging, and competition. This gives the aggregations properties of their own; they are ecological communities.

Ecological community includes the plant, animal, bacterial, and other microorganismal components, or is limited to some or one of those parts. An animal community lives in squirrel carcasses in woodland; the botanist may talk of the plant communities of alpine meadows.

All communities share some individuals with other communities. For instance, many species are specialized for living and feeding as young stages in cow dung which is scattered over pasture fields, in woodland, and at the edges of rivers. Some specialized feeders will be present in each individual dung dropping and with them there will be associated and dependent species. But all the species will be found at some time away from the cow dung, if only when they are in the process of moving to new droppings. Some predators that feed mostly in other places will occasionally visit the dung for food. In most insects the young and the adults have rather different requirements and adaptations and thus tend to belong to different communities for part of their lifetime. Thus the boundaries of communities are not complete barriers. Communities, then, are groups of populations that are more or less aggregated at centers of action. The centers of action may be small in size, or very large, like soil litter in an extensive woodland of few species of trees.

ECOSYSTEMS

Ecological communities use materials from the nonliving environment and return materials to it. Such a community or group of communities and the

*An *ecological community* is defined here as a group of populations of different species living together and having certain interactions with each other.

nonliving environment, that together have some long-time stability of their own, can be considered to be working as a single system, or ecosystem. A tank of fish, algae, weeds, and other organisms, with enough light and air can be self-supporting and can be considered an ecosystem. A lake, a sea, a desert are also ecosystems. An ecosystem may be large or small but has some stability.

THE ANIMAL COMMUNITY ASSOCIATED WITH BLACK KNAPWEED

To introduce some of the principles of community ecology, Varley's (1947) study of the black knapweed gallfly, *Urophora jaceana*, will be used. The knapweed gallfly is about 5 mm long, and is an adult in July. All stages except the adult live in the flower head and fruit of the black knapweed, *Centaurea nemoralis* (Figure 7.1). The adults in Varley's study did not fly more than

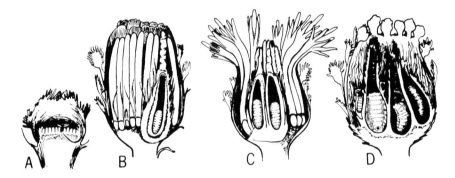

Figure 7.1. Some events in the life history of the knapweed gallfly, *Urophora jaceana*. A. Knapweed head of buds containing four gallfly eggs and showing the track of the ovipositor. B. Most of the florets are almost ready to bloom. A gall, surmounted by the remains of the pappus of the fruit from which it has been formed, contains a second-instar larva. The dark woody layer of the gall is beginning to form. C. The gall contains third-instar larvae which have already consumed a large part of the fleshy gall tissue. D. Knapweed flower head after flowering, in September. The fruits have dropped out. The gall is hard and woody, and contains (left) a fully fed gallfly larva, (center) a larva of a chalcid parasite, and on the remains of the host can be seen three of its egg shells. On the right is a brown gallfly puparium containing a larva of the chalcid parasite *Eurytoma curta*. To the extreme left is a slightly swollen fruit containing a larva of the small gallfly *Urophora quadrifasciata*. (Redrawn from Varley, 1947.)

66 feet from their place of emergence. Many factors influencing the survival of this species leave their traces within the flower head and can be identified and quantified later. The figure shows the sequence of events from oviposition (remains of the egg persist so that fecundity and fertility can be estimated) through larval stages and gall formation by the plant around the feeding larva, through pupation and emergence of the adult. Remains of the seed

head and galls often fall to the ground in winter. Parasitism and the demography of some of the parasites can also be studied in the flower heads.

NICHE

An animal's niche is its way of life. *Urophora jaceana*'s niche is to be free flying as an adult for a short time of the year and to develop in a flower head, feeding on plant materials. Niche implies a "type of work done," or "role" in the total series of biochemical processes of the ecosystem. Elton (1927) commented that one had the same sort of understanding of role when saying, "There goes the vicar." In Varley's study there were 8 other species with the same niche in this same place. There were 3 other gallflies, 2 gall midges, and 3 moths. The moths were also predators of the other species. The moth larva moved in a flower head as it ate, and if it came across *Urophora* it ate that as well.

HABITAT

Habitat is a place in which to live, in terms of physical topography and/or organic material, living or dead. It is usually in broader terms than the place implied by niche. The habitat of *Urophora* was "Field-type" mixed with "Scrub" on the "Edge of wood". This simple terminology is from the classification of habitats by Elton and Miller (Elton, 1966). Their classification was designed for use on punch-cards, with many levels of subdivision. It covers all habitat types—for instance, aquatic, tropical, mixtures and edges, dead and decaying organic material.

Human interference with habitat can rapidly remove some species that are troublesome to man. This was well illustrated by attempts to control the populations of nesting albatrosses, *Diomedea nigripes* and *D. immutabilis* on Midway Island, Hawaii (Kenyon *et al.*, 1958). Nesting albatrosses had to be removed from the edges of airstrips because the birds used the airstrips for their own takeoff and landing. When nesting birds were killed, nest sites were reoccupied by other individuals. The recruitment was complicated by the fact that it takes these species 4 years to reach sexual maturity. Therefore for 4 years groups of birds that were attempting to breed for the first time, and had been reared at those sites, might be particularly strongly attracted to the cleared area for nesting territory. Eventually the part of the population adjacent to the airstrips was removed only when sand dunes habitat was replaced by tarmac. Another example is discussed later, that tsetse flies can be removed from an area if scrub and sparsely scattered trees are cut down or are allowed to grow into dense forest.

Changing habitats can bring other changes that create problems. For instance, schistosomiasis is spreading on the African continent along with the creation of new irrigation channels and other bodies of water. Many agricultural pest problems are in part due to the presence of large monocultures of plants and of few areas of other habitats with more variety of plant species and animal communities. Thus the animal communities living on agricultural crops consist of comparatively few species, which have a tendency to build into large numbers. Elton (1958) advocates conservation of variety of habitats and their communities for helping to overcome agricultural and other environmental problems.

ECOLOGICAL SUCCESSION

The *Urophora* habitat was in an unstable state because the scrub was periodic-ally cut to leave only the field type. The scrub would grow up again to elimin-ate the areas of field type, and if left uncut it would change to woodland. Succession of plant, animal, and microbial populations is a common process. Some species are successful invaders of certain habitats and they change the character of the habitat by their presence. The changed environment favors the establishment of other species, which themselves may further change the nature of the environment. Eventually the original invaders do not survive because the environment is no longer suitable for them. Succession of species may be stopped by various conditions: by exhaustion of resources, as in the breakdown and decay of dead organic material such as milk and carcasses; by animal activities such as the effects of rabbits on plant succession on English chalk hills*; by human activity such as mowing of lawns; by climate as on the Arctic tundra; and when the complex had reached limits inherent in the living forms themselves, as in climax forests.

IN ECOSYSTEMS THERE IS CYCLING OF MATERIALS
AND ENERGY FLOW

Sunlight is the source of energy necessary for life, except that chemical energy from some inorganic compounds can be used by some microorganisms. The only organisms that are able to transform solar energy to chemical energy necessary for life are those that have chlorophyll. Other living forms must directly or indirectly use photosynthesizing organisms as their energy source. A consumer will use the energy available in various ways, particularly in metabolic processes that are reflected in respiration, and in heat loss, which will be much greater in some species than in others. By body growth and waste deposition much energy will become available to other species—predators, parasites, and scavengers. Passage of this available energy through another organism and the dissipation of much of it thereby, again reduces the energy remaining available to further consumers. Thus the number of con-sumers through which the energy originally stored up by green plants can be passed is limited to four. The dependence of living forms upon others for energy, and the limits to the transfer of energy, give structure and organization to communities.

The number of species in the world is vast. The analysis of many communi-

* The rabbit populations were reduced to one-tenth by myxomatosis (Chapter 25) and succession is going toward climax forest again on the chalk hills.

ties in terms of the biology and demography of the species involved is a huge research task. Slobodkin (1960) and others believe that further unifying concepts in community and ecosystem ecology will first come in terms of energy relationships because models for energy transfer will apply to many kinds of communities, regardless of the species involved. These will undoubtedly be of great practical value as man has to take over and maintain more of the world for his own support.

FOOD WEBS AND THE PYRAMID OF NUMBERS

The community of species living in knapweed flower heads was relatively small and simple. Varley was concerned with nine plant-eaters of the same niche. They and the predators and parasites of *Urophora jaceana* are shown in Figure 7.2, where they are arranged to show the web of feeding relationships. There was a predatory fly whose larva burrowed in the flower head and ate the gall contents. In winter, mice and voles removed animal contents of the flower

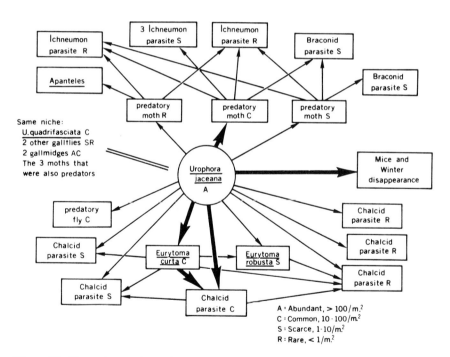

Figure 7.2. The main members of the ecological community around the knapweed gallfly in the study area, including members of the same niche, predators, and parasites. Chalcids (including *Eurytoma*), braconids, and ichneumonids belong to the wasp order. Arrows indicate feeding relationships, the heavier ones representing the greater losses to the prey or host. (After Varley, 1947.)

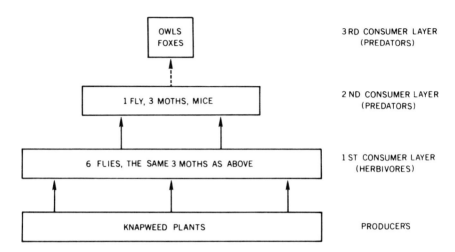

Figure 7.3. Producers, herbivores, and predators of the knapweed community arranged as a pyramid. This is so small a group of species that it does not fully illustrate the properties of such a pyramid, as discussed in the text.

heads and left opened remains in piles. The chief parasites were *Eurytoma curta* and another chalcid wasp.

Species in a community can also be arranged in a diagram as a pyramid of numbers (Figure 7.3). The knapweed plants were the producers, and nine animal species were primary consumers eating the knapweed plant. Some species were predators of the primary consumers and further species were predators of the predators.

Some generalizations can be made about such pyramids involving producers, primary consumers, and predators. From the first consumer layer, the herbivores (and, in some characteristics, from the producers), to the top layer (the ultimate predators) there tends to be:

Decrease in the number of species present, or having the potential of being present, that is that could fill that niche in the geographic area.

Decrease in the standing crop. Standing crop is the amount of plant, animal, or microorganism present at one time, measured as numbers, mass, energy, or other content.

Decrease in the rate of natural increase per unit time. Involved with this is increase in generation time and decrease in productivity. Productivity is the amount of new plant, animal, or microorganism produced per unit time, measured as numbers, mass, energy, or other content.

Increase in the size of individuals.

Increase in the range of movement of individuals.

The number of consumer layers in an ecological pyramid seldom exceeds four. This limit is imposed by the limits of animal size and, as mentioned previously, the dissipation of energy during its use.

The productivity of the producers, the photosynthetic plants, must be greater than that of the primary consumers. In some aquatic communities the producers are microscopic algae and these plants may have very high rates of birth and death, so that though the productivity be high the standing crop may be comparatively small.

Parasitism involves inverted pyramids of numbers. Primary parasites tend to be larger, fewer in numbers and species, greater in range of movement, than their own parasites and the parasites of these. Rates of increase of numbers per unit time increase from primary parasite to the third parasite consumer layer. Such a series would be seen in a mammal, its arthropod ecto-parasites, their intestinal bacterial flora, and the bacterial viruses.

There is another pathway of feeding relationships, that of the scavengers. These include feeders on excreta and carcasses and their breakdown products. They also have consumer layers of predators and parasites.

Competition is usually considered to occur only within a consumer layer, predator-prey and host-parasite relationships being excluded. Thus during a locust plague a locust can be a conspicuous competitor of the cow; they occupy the same consumer layer, indeed almost the same niche.

EVOLUTION ACTS UPON THE INTERRELATIONSHIPS OF POPULATIONS

Lotka (1944) said: "Survival may be assured either by a relatively high protection against foes combined with a relatively low fertility, as is the case with man. Or, *vice versa*, a species may survive in spite of very high vulnerability and by being highly prolific as is the case with certain fishes. Now we might ask which of these two types of organisms will be selected by preference for survival. The question is badly put. They must both survive side by side, for the former are dependent on the latter for their food supply. Such organisms survive together as interdependent couples or systems. It is such systems as a whole that have the requisite qualities for survival."

A predator that eliminates its prey, or a parasite that eliminates its host, also eliminates its own food supply. Thus natural selection tends to favor relationships where populations do not overexploit each other, with the result that there is evolution of tolerance. This is particularly seen in the evolution of disease (Chapter 24). It can sometimes be seen that a predator can be "beneficial" to its prey population in that it tends to remove the old and the diseased members (Errington, 1967). Slobodkin's (1960) investigations with laboratory populations of the water flea, *Daphnia pulex*, showed that removal of old animals was the mechanism of harvesting most efficient in his carefully defined situations of steady maximal harvesting yield in relation to caloric consumption by the prey. He wrote "To remove animals that are growing slowly, have lived most of their time and have low reproductive value is the epitome of prudent predation and therefore has a high population efficiency." Moreover, many species to some extent avoid shortages in the populations upon which they depend by being catholic in their tastes, and thus some parasites have several alternate hosts, and predators and herbivores may be able to eat more than one species.

 There is a growing acceptance among biologists that ability of a population to regulate its density before overexploitation of its environment, has long-term survival value. At least among higher organisms, behavior can play a large role in the regulation of density.

DEMOGRAPHY OF THE BLACK KNAPWEED GALLFLY

In Varley's study demographic data on the most common species were collected from individuals brought into the laboratory and in the field. In a greenhouse the following numbers of eggs were laid by *Urophora jaceana* females: 216, 277, 29, 78, 125, and other similar figures were obtained, and all the females contained mature eggs when they died. In 1935 the fecundity per female in the field was 70 ± 19 and in 1936 44.8 ± 8.5. So the potential fecundity rate was higher than the actual; and weather seemed to have much influence on performance. One year 7 percent of the field egg batches were infertile and four intraspecific matings had been seen, and the *U. jaceana* population density was low. So the cause of raised infertility may have been interspecific matings resulting from low population density of adults. The males give an appearance of maintaining territory in that they tend to stay separated in the same places and are aggressive to each other.

Life Tables

Varley took cohorts of eggs, which were actual numbers per square meter in the field, each year and constructed life tables for several species. One covering $1\frac{1}{4}$ generations of *U. jaceana* is in Table 7.1 along with measures of fecundity and fertility rates, and sex ratios. Mortality was subdivided by cause as far as it could be determined.

Natural Control of the Gallfly Population

At no time during Varley's observations were food plants fully occupied. It can be assumed that at that time availability of flower heads for oviposition did not limit the population. In 1935–1936 0.5 percent of the eggs survived to produce a mature female. That is, survival from the egg to female adulthood was 1/200. Winter disappearance, which was largely attributed to mice, was the greatest cause of death, accounting for 22 percent, perhaps 43 percent, of the total kill. This mortality factor gave no indication of changing in a regular fashion with change of population density. Parasites sometimes accounted for much of the total kill. Varley suspected that parasitism by *Eurytoma curta* was most likely to have a density-dependent rate and be a controlling factor of the *Urophora jaceana* population, although the observations did not con-

TABLE 7.1

Life Tables and Fecundity, Fertility, and Sex Ratios of the Knapweed Gallfly, *Urophora jaceana*, Using Actual Densities per Square Meter

Time interval[b]	Causes of mortality[c] and other data	No. alive per sq. meter at beginning of interval l_x	No. dying or not hatching during interval d_x	Percent d_x/l_x q_x
1935				
July	Emerged per square meter, 42.5% were females	6.9		
	Mean number of eggs laid: 70 per female			
	Infertile eggs	203	18.3	9
	Larvae died before forming galls	184.7	37.1	20
	Larvae died in galls due to unknown cause	147.6	3	2
	Parasitized successfully by *Eurytoma curta*	144.6	65.8	45.5
Aug.–Sept.	Miscellaneous causes:	78.8		37
	Habrocytus trypetae parasitism		5.6	
	Eurytoma robusta parasitism		4.1	
	Torymus cyanimus parasitism		3.7	
	Tetrastichus sp. parasitism		0.6	
	*Destroyed by caterpillars		14.8	
1936				
Winter	*Winter disappearance	50.0	30.8	61.5
	*Destroyed by mice	19.2	12.2	64
	Larvae died due to unknown causes	7.0	1.8	26
	Miscellaneous causes:	5.2		31
	*Birds		0.4	
May–June	*Habrocytus trypetae* parasitism		0.7	
	Macroneura vesicularis parasitism		0.25	
	Tetrastichus sp. parasitism		0.25	
July	*Drowned in floods	3.6	1.57	44
July	Emerged per square meter, 42.5% were females	2		
	Mean number of eggs laid: 52 per female			
	Infertile eggs	44.8	6.9	15.3
	Larvae died before forming galls	37.9	9.9	26.2
	Larvae died in galls due to unknown causes	28.0	1.2	4.3
	Parasitized by *Eurytoma curta*	26.8	7.2	27
Aug.–Sept.	Miscellaneous causes:	19.6		36
	Habrocytus trypetae parasitism		0.2	
	Eurytoma robusta parasitism		1.6	
	Torymus cyanimus parasitism		1.5	
	Tetrastichus sp. parasitism		0.2	
	Killed by predatory fly		0.2	
	*Destroyed by caterpillars		3.3	
Winter		12.6		

[a] Modified from Varley (1947).

[b] Time intervals are irregular, being based on life stage, activity, and ease of sampling.

[c] Mortality affecting the gallfly and its parasites indiscriminately is marked with an asterisk. Causes of mortality are analyzed as far as possible.

tinue long enough for proof. For instance, from 1934 to 1936 the numbers of host larvae per square meter and the percentage of parasitism by *E. curta* were 43 and 15, 144.6 and 45, and 26.5 and 27.

Summary of the Gallfly Study

The gallfly had a fairly high potential for increase—in some cases more than 270 eggs per female were laid in the greenhouse, and from 45 to 70 was a common range of field performance per female. As the animals developed many were lost to other populations in the community. The population was a part of bigger systems, the community and beyond that, the ecosystem. Ecosystems are complex mechanisms in which energy and materials are incorporated, used, and lost. The gallfly population served as a supply of prey, hosts, and waste materials, having obtained its own food from a producer, the knapweed plant. There is no doubt that one or some of these factors had the effect of regulating the density of the gallfly population.

FACTORS OPPOSING INCREASE OF TSETSE FLY POPULATIONS

The previous ideas will now be more directly related to public health problems, first by considering tsetse flies *Glossina* spp. which are vectors of trypanosomes causing sleeping sickness of man and nagana of cattle. In Chapter 5, a rate of increase was given for tsetse flies compiled from data about several of the 14 species. Under some circumstances they have the potential to at least double the population size every 4 months. The following conditions in nature are known to oppose the potential for increase (summarized in Glasgow, 1963).

Habitat

"Any species of *Glossina* can be destroyed by destroying the woodland and replacing it with grass. I know of no proved exception to this providing the grass is not higher than a man's waist" (Glasgow, 1963). For instance, in Northern Rhodesia in certain valleys which were very favorable habitats, 50 female flies could be caught in a 10,000 yard transect, and all had mated. Then the trees were felled and during the following 5 years numbers collected in identical circumstances were 35, 15, 1.7, 0.1, and 0. One year 1 uninseminated female was found, the next year 12 of 136 were uninseminated, and the next year 29 percent were so. At some point, the decline in numbers associated with change of habitat began to be accelerated by failure of mating, probably because of low population density.

Weather

Dry seasons have been shown to markedly increase the death rate and there is no reproduction outside the temperatures of 22–30°C.

Food

Food shortage affects fertility by causing abortions. In certain places large numbers of leopards have sometimes reduced the ungulate populations on which tsetses feed, and tsetse numbers have become markedly reduced. Four rinderpest epidemics have disastrously reduced numbers of the mammal hosts of the tsetse, and these epidemics were followed by great reduction of tsetse populations. *Glossina mortisans* disappeared after one of these epidemics from the northern Transvaal, large areas of Southern Rhodesia, and from other places.

Parasitism and Predation

Heavy parasitism of tsetses has sometimes been found in the field; in one study, 33 percent were found parasitized by a Calliphorid fly. They have predators—*Glossina* are absent from a particular island in Lake Victoria, and this is attributed to the great abundance there of certain predatory spiders.

Competition

Four of the *Glossina* species that overlap in their ranges at only very narrow belts are known to cross-mate very freely where the ranges of the species meet. A tsetse female stores sperm from a single insemination to use all her reproductive life. The cross-matings produce sterile eggs. A single male could sterilize many females in this way and eliminate them from an area by stopping their effective reproduction. This is interspecies competition by sterilization. It is like Knipling's (1959) method of eliminating the screw-worm fly, *Callitroga hominivorax*, an economically damaging parasite of livestock, from southeastern United States. Fifty million male screw-worm flies sterilized by X-ray irradiation, but not impaired thereby in their mating behavior, were released every week over an area of 70,000 square miles, which was the condensed winter range of the species in the East. At all times this part of the population was fairly well separated from other geographical areas in the range of this species. Like the tsetse, female screw-worm flies are inseminated only once. After 6 months the natural population was eliminated by failure to reproduce successfully. However, man cannot apply this control method to tsetses, because they cannot be reared artificially in large numbers.

Tsetse are thought to compete among themselves and with Tabanid flies for vertebrate blood. The attack of one bloodsucking fly does not cause an ungulate host to change its behaviour, but when several flies attack at once the host changes its movements so that none can settle. Calculations of the amount of blood lost in a single tsetse bite show that the host could not tolerate many such losses in a short time.

Thus it is seen that many of the limits to activity, survival, and reproduction of tsetse flies are due to other animals of the same species, to other species of the same genus, to predators, parasites, hosts, and other competitors. They are the consequences of the fact that tsetse flies are parts of ecological communities.

UPSETTING THE RELATIONSHIPS IN ANIMAL COMMUNITIES

The balance of populations within a community can be destroyed when insecticides are used. One example is from use of dieldrin in outdoor primitive lavatories in Georgia (Kilpatrick and Schoof, 1956). For some years it was noted that houseflies were emerging in pest proportions from human excrement in which they were breeding. This was occurring only if the lavatories had been treated with dieldrin (Table 7.2).

TABLE 7.2

Average Number of Houseflies Emerging per Lavatory per Month[a]

Treatment	May	June	July	August	Sept.
30 Treated in late March to early April with dieldrin	193	254	1374	1204	1946
10 Untreated	3	13	5	16	10

[a] Kilpatrick and Schoof (1956)

The excrement supported an animal community potentially consisting of 123 species. They fed on waste materials, bacteria, other animal species, or were parasites. Larvae of soldier flies were abundant and they were competitors of housefly larvae in that they rendered the excrement into a liquid mess which inhibited growth of housefly larvae. The soldier fly numbers were very reduced by dieldrin, as were most other members of the community. For some reason, perhaps by later arrival or by greater resistance, the housefly larvae were not so inhibited by the insecticide and had the habitat mostly to themselves, and so they greatly increased in numbers.

In the antimalaria campaign in 1961 in Sarawak and Borneo, spraying DDT on the roofs thatched with attap leaves caused the roofs to rot (Cheng, 1962). It was shown that the roofs were being eaten by larvae of *Herculia nigrivitta*. This pyralid moth was parasitized by a tiny wasp that was free-flying as an adult and was very susceptible to DDT. The moth was also susceptible to DDT but very strongly avoided it and could always find plenty of unsprayed leaves on a treated roof. So spraying increased numbers of the moth by releasing it from its parasite and this was causing the extensive roof damage.

RADIOACTIVE WASTE AND OTHER DANGEROUS MATERIALS CAN BE CONCENTRATED IN SOME PARTS OF ECOLOGICAL COMMUNITIES

A last example of practical significance of community interrelationships comes from accumulation of radioactive substances in an aquatic animal community. The basic part of the relationships among communities is food supply and its distribution through all the members of the community. A Canadian lake (Ophel, 1963) contained radionucleotides from seepage from a liquid disposal area of an atomic reactor plant. There was a fairly constant input of radionucleotides, and lake water was regularly depleted in nucleotide level in the warmer months because of biological activity. Strontium-90 in bones of perch reached a constant after 5 years. In shorter-lived animals an equilibrium of content was reached in 2 years or less. Figure 7.4 shows how the waste product passed from species to species in the community and accumulated in some species more than in others. Strontium-90 was in perch and muskrat bones at levels 3,000 and 3,500 times the concentration in water. Thus it is important to remember that, partly because of the network of feeding relationships, species can enormously concentrate dangerous materials that are added to the nonliving environment.

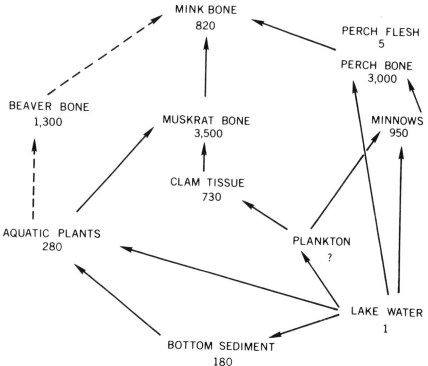

Figure 7.4. Average concentration factors for strontium-90 in Perch Lake food web. Strontium-90 level in lake water is taken as 1. The content of strontium-90 in all other measured material is valued as

$$\frac{\text{amount per gram of wet weight}}{\text{amount per ml. of water, taken as 1.}}$$ (After Ophel, 1963.)

CONCLUSIONS

Man is part of ecological communities, and as yet his food and most other physiological requirements depend upon the other plant, animal, and microbial populations. Also, in the public health field it is valuable to comprehend his diseases in their true ecological setting. Moreover, with the current population increase it becomes necessary to study and speculate about environmental damage. For it is typical in populations that there is a time lag before all the consequences of environmental change reflect back upon the populations.

REFERENCES

Cheng, F. Y. 1962. Deterioration of thatch roofs by the larva of a moth after house-spraying in a malaria eradication project in North Borneo. WHO/Mal/368, unpublished document.

Elton, C. S. 1927. *Animal Ecology*. Sidgwick and Jackson, London.

Elton, C. S. 1958. *The Ecology of Invasions by Animals and Plants*. Methuen, London.

Elton, C. S. 1966. *The Pattern of Animal Communities*. Wiley, New York.

Errington, P. L. 1967. *Of Predation and Life*. Iowa State Univ. Press, Ames, Iowa.

Glasgow, J. P. 1963. *The Distribution and Abundance of Tsetse*. Macmillan, New York.

Kenyon, K. W., Rice, D. W., Robbins, C. S., Aldrich, J. W. 1958. Birds and aircraft on Midway Islands, November 1956–June 1957 investigations. *U.S. Fish Wildlife Serv., Spec. Sci. Rep., Wildlife*, **38**. Washington, D.C.

Kilpatrick, J. W., and Schoof, H. F. 1956. Fly production in treated and untreated privies. *Public Health Rept. (U.S.)* **71**: 787–796.

Knipling, E. F. 1959. Sterile-male method of population control. *Science* **130**: 902–904.

Lotka, A. J. 1944. Evolution and thermodynamics. *Science & Society*, **8**: 161–171.

Odum, E. P. 1959. *Fundamentals of Ecology*. 2nd ed. Saunders, Philadelphia, Penn.

Ophel, I. L. 1963. The fate of radiostrontium in a freshwater community. In *Radioecology*. V. Schulz and A. W. Klement (eds.). Reinhold, New York, and A.I.B.S., Washington, D.C.

Slobodkin, L. B. 1960. Ecological energy relationships at the population level. *Am. Naturalist*, **94**: 213–236.

Varley, G. C. 1947. The natural control of population balance in the knapweed gall-fly (*Urophora jaceana*). *J. Animal Ecol.* **16**: 139–187.

CHAPTER 8

Fluctuations of Vertebrate Populations

CHARLES H. SOUTHWICK

The study of vertebrate populations is an important topic in public health for two primary reasons: population fluctuations of vertebrate animals may determine the pattern and extent of the transmission of zoonoses to human populations; and vertebrate populations often illustrate essential aspects of population biology and can be useful for the analysis of specific population problems. This chapter will develop both of these topics with emphasis upon the latter.

The study of vertebrate populations is complicated at the outset by difficulties of obtaining data on population sizes, densities, sex and age structures, reproductive rates, and mortality rates. The reasons for these difficulties are fairly obvious. Many vertebrates are nocturnal or are almost impossible to observe. Animals are unevenly distributed, rare in some places, abundant in others, and many are very mobile. Thus the study of vertebrate populations is fraught with problems of field techniques, sampling procedures, and statistical analysis. In virtually every study of vertebrate populations the available data are less than those desired. Reference to procedures for sampling vertebrate populations is in Chapter 5.

THE DENSITIES OF SOME VERTEBRATE POPULATIONS

The abundance of certain vertebrate populations is interesting and sometimes startling. For example, a survey of rat populations of Baltimore during World War II indicated a population of approximately 400,000 Norway rats in Baltimore City (Davis, 1953). They were estimated to consume 20 tons of food per day. A study of a rodent outbreak in California estimated that house mouse populations increased within 6 months from less than 5 mice per acre to 300 mice per acre in one area (Pearson, 1963). Surveys of rodent populations in stacks of unthreshed grain on English farm lands indicated that wheat stacks of 100 cubic yards sometimes contained up to 2,000 mice (Southwick, 1958). Damage in such cases was estimated to be 50 to 75 percent of wheat harvest.

A survey of rhesus monkey populations in northern India in 1960 indicated that the province of Uttar Pradesh had a population of approximately 1,000,000 rhesus monkeys. Birds may also have large populations. In Maryland, Pennsylvania, and Virginia in winter, starlings and blackbirds frequently number 1,000,000 to 5,000,000 in a single roosting site. Numerous other examples of vertebrate populations could be cited, not only for their great abundance, but also for extreme rarity, in which cases the interesting problems concern population maintenance rather than limitation.

VERTEBRATE POPULATIONS IN ZOONOSES

Vertebrate populations have direct bearing on public health and biomedical problems. There are more than 150 zoonoses, diseases which occur naturally in both animal and human populations (Chapter 29). For example, yellow fever virus in Central and South America is maintained in monkey populations involving more than 10 species of primates. Under certain conditions the virus may be transferred by *Haemagogus* or *Aedes* mosquitoes into a human population. Because of the numbers, distribution, and habitat of the monkeys it would be virtually impossible to eradicate yellow fever. Rocky Mountain spotted fever is a rickettsial infection which in Maryland and Virginia has a reservoir in various small animals, including rabbits, wood mice, field mice, opossums, and raccoons. An infected opossum can show a high rickettsemia with no overt illness. It usually carries many *Dermacentor variabilis*, the tick vector of spotted fever. Modern suburban areas have an interspersion of housing, woodlots, vacant fields, and golf courses that bring people, wild animals and domestic pets into close and frequent contact which permits the exchange of ticks.

A major characteristic of vertebrate populations is their tendency to fluctuate widely. It is not uncommon for fivefold fluctuations in population size to occur within and between years. Probably these fluctuations in population can determine the role that vertebrate populations play in disease transmission. In periods of high density, vertebrate populations may disperse, with individuals moving into new habitats, creating new ecological contacts, and exposing different members of the biotic community to vectors of infectious agents. For example, Audy (1958) showed in Malaya that the presence of trombiculid mites, vectors of scrub typhus, indicated that individual rats had been moving beyond the village into cultivated fields and surrounding grasslands or forests.

THE INFLUENCE OF MAN ON VERTEBRATE POPULATIONS

Vertebrate population fluctuations can be classified as those which are predominantly determined by human influences and those which are not. This may seem artificial but is emphasized because man's influence has been so great.

Elton (1958) discussed the fact that the present mingling of thousands of different kinds of organisms from different parts of the world by means of human trade routes, is setting up huge dislocations in nature. Drastic changes in the natural populations of the world have occurred in the last few hundred years. Slobodkin (1962) emphasized the same phenomenon: "Not only is man the most shocking innovation since the first appearance of the terrestrial vertebrates, but his activities are proceeding at an accelerating rate... The importance of man continues to increase and the possibility of the biologic world ever being as stable as it was in prehuman times becomes more and more remote."

Mankind has dominated and altered most of the major ecosystems. Even Antarctic penguins contain DDT although hundreds of miles from direct contamination (Sladen *et al.*, 1966). There is no better example of mortality caused directly by man than in the American bison. In 1800 *Bison bison* covered more than half of North America and a conservative estimate of the population of bison in 1800 was 60 million individuals. Severe hunting pressure came with the frontier period in the American West, particularly from 1820 to 1870, and sometimes millions of animals were killed in a year. In 1889 a careful estimate of the total bison population of North America was 540. Several other factors were involved in the decline, including agricultural development of the Great Plains. Since the 1890's the population has rebounded to the tens of thousands and there is no longer threat of extinction. But some other species became extinct. The passenger pigeon, Arizona elk, great auk, Labrador duck, heath hen, Carolina parakeet, and Eskimo curlew were all residents of North America, and many of them became extinct through shooting and other human influences on mortality.

Man has also caused increases and several species have been transplanted and have done well. The ring-necked pheasant was imported into the United States in about 1880 and is now one of North America's finest game birds. The Hungarian partridge was introduced into the Northern Plains in 1908. The brown trout is a European species successfully introduced into the United States. The striped bass of the Atlantic Coast was introduced into the Pacific Coast and is now an important game fish in the West. The wild turkey was exterminated in Ohio but now it has been successfully restocked there; the mountain goat has been successfully restocked in a number of states. The prong-horned antelope was eliminated from Texas, New Mexico, and Montana and now all of these states have fine restocked populations.

Along with desirable propagation there have been undesirable introductions. The English sparrow and the starling were deliberately introduced into North America in the 1800's. The carp was introduced by the United States Fish Commission and it now has become a detrimental ecological force in many lakes.

Probably the best examples of unwise and disastrous transplants can be found in New Zealand. The species of native land vertebrates are limited to two bats, a few flightless birds, and a few reptiles and amphibia. Colonists, starting in 1770, made over 200 introductions of vertebrates. Many were useful to man in their native countries, but introduced into "vacuum niches" in New Zealand became so numerous that they are pests. Red deer, which are a valued species in Scotland, have become a serious forest and agricultural pest although more than 100,000 are shot annually. Japanese, white-tailed, and Himalayan deer and the California quail are also pests. Durwood Allen (1954) said of New Zealand, "A herbivorous opossum from Australia is eating the forest and orchard from the top down, while deer, goats, and rabbits from Europe and North America are denuding the land from the ground up."

There have also been many indirect human influences by habitat modification—grassland ploughing, marsh drainage, and forest cutting. Since 1915 there has been a notable upsurge in white-tailed deer populations, *Odocoileus virginianus*, in northeastern United States. This has been attributed to forest cutting and replantation, producing lush secondary growth suitable for deer browsing, also to predator control, and buck laws which prevent the killing of does. The events in Pennsylvania have been typical; throughout the latter part of the nineteenth century there was heavy cropping of deer so that by 1906 the populations were severely depleted. Then a thousand white-tailed deer were imported. After 1910 ideal forest conditions were present, also there were no wolves and cougars, and buck laws operated. By 1930 the deer had increased to about 800,000 and by 1946 to 1,000,000, 1 deer to 26 acres of total land. The carrying capacity of the 13,000,000 acres of forest in Pennsylvania is thought to be about 250,000 deer, approximately 1 deer to 52 acres. Dense populations severely overbrowse coniferous and deciduous forests, and this leads to starvation, particularly in winter when deer tend to congregate. Starvation is especially likely when there is heavy snow.

Protection of the Kaibab forest of Arizona was also an example of poor ecological management. In 1906 Kaibab became a national game preserve. From 1906 to 1930 there was extensive predator control and all deer hunting was restricted. Large numbers of mountain lions, coyotes, bobcats, and wolves were exterminated. The original herd of the mule deer, *Odocoileus hemionus*, was estimated by Leopold (1943) to have 4,000 deer, but at one time it was increasing by 20,000 per year. By 1924 there were probably 100,000 deer. Leopold estimated that in many areas of the Kaibab 80 to 90 percent of the forage and browse had gone. In the particularly severe winter of 1924 about 60,000 deer died of starvation. This was a tragic demonstration of the necessity of keeping predators.

Occasionally excessive predator control has not only resulted in explosive growth of the prey species, but it has been accompanied by a deterioration of the prey species as well. Seventy percent of the bighorn sheep, *Ovis canadensis*, of Yellowstone Park died in an epizootic disease in the mid 1930's following excessive predator control. Since then the herds have not thrived. The animals are prone to lungworm, pneumonia, and hemorrhagic septicemia. This disease susceptibility is not well understood, but it is probably related to poor range condition and lack of removal of weak individuals by predators.

SEASONAL AND ANNUAL FLUCTUATIONS OF NUMBERS

Most animal populations fluctuate seasonally with declining populations in winter, low points in spring, increase in the summer, and high points in fall. Figure 8.1 of a California quail population, *Lophortyx californicus*, shows this decline in winter and spring and a sudden upsurge in June when broods appear (Emlen, 1940). The same phenomenon is shown for bobwhite quail populations, *Colinus virginianus*, in Figure 8.2 (Errington, 1945). In every year of a 15-year study of a 4,500 acre reserve the spring population was lower than the autumn population. Significant variation from year to year also occurred. From 1932 to 1935 quite high populations were present; then in 1935 a severe

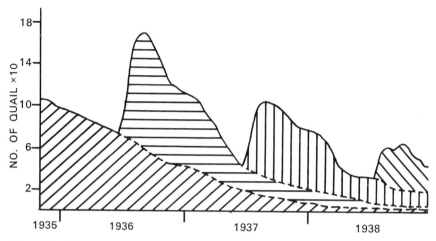

Figure 8.1. Changes in a population of the California quail, *Lophortyx californicus*. Birds present in late 1935 and produced in 1936, 1937, and 1938 are represented by differently shaded blocks; thus age distribution can be followed. (After Emlen, 1940.)

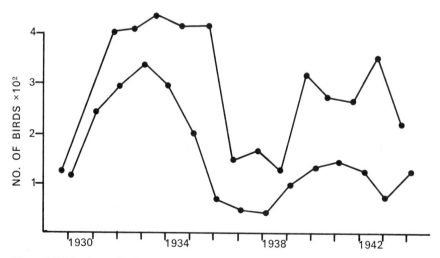

Figure 8.2. Numbers of bobwhite quail, *Colinus virginianus*, on 4500 acres of farmland. The upper curve shows the numbers attained at the end of each summer, the lower curve the numbers in spring; comparisons of the upper and lower curves indicate extent of deaths in winter and births in summer. (After Errington, 1945.)

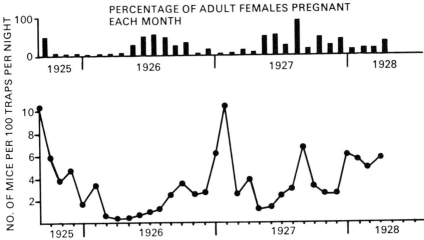

Figure 8.3. Number of wood mice, *Apodemus sylvaticus*, and percentage of adult females pregnant by months, against time. (After Elton, 1942).

winter reduced the population to 100 birds. After a few years of low population in the late 1930's, numbers rose again, but were not as high as formerly.

Figure 8.3 shows annual fluctuations and seasonal fluctuations in populations of *Apodemus sylvaticus* the English field mouse (Elton, 1942). However, this figure shows the errors that can enter population study by assuming that trap results represent population density. The lowest numbers of mice trapped per 100 traps per night were in April, May, and June, with highest numbers in January and February. This figure is a graph not of population size but of trapability. Food supplies presumably are lower in the winter and the animals are more prone then to use the food and shelter provided in a trap.

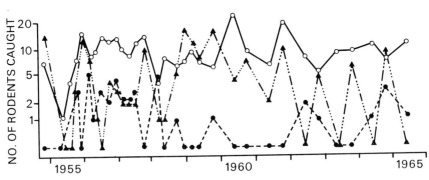

Figure 8.4. Variations in numbers caught from October 1954 until May 1965 of the field mouse *Apodemus sylvaticus* ▲— · · · —▲, and the voles *Microtus agrestis* ●— — —● and *Clethrionomys glareolus* O———O. (After Ashby, 1967.)

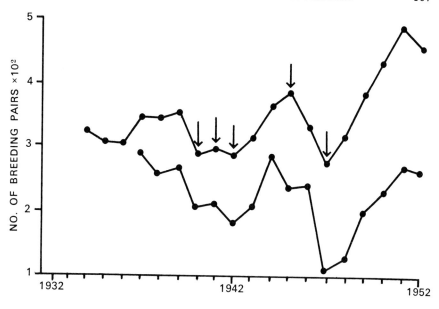

Figure 8.5. Numbers of breeding pairs of the heron, *Ardea cinerea*, in two parts of England. The arrows indicate hard winters. (After Lack, 1954.)

Ashby (1967) showed a similar phenomenon in *Apodemus sylvaticus* populations (Figure 8.4). Numbers varied considerably over a 12-year period, but in most years maximum catches occurred in the winter and minimum catches in the summer. Ashby felt on the basis of the statistical distribution of trapping results that these were true population patterns. Populations of the voles *Microtus* and *Clethrionomys* fluctuated independently of *Apodemus*. *Clethrionomys* showed maximum numbers in the summers, and in the years 1956 and 1957, 1960 and 1961, and 1963 to 1965. Figure 8.5 shows fluctuations of the numbers of breeding herons, *Ardea cinerea*, in two parts of England (Lack, 1954) and declines, which coincided with severe winters.

LONG-TERM CYCLICAL FLUCTUATIONS OF NUMBERS

The most dramatic animal fluctuations are those which are cyclical. Some animal populations, such as those of the snowshoe hare, *Lepus americanus*, in North America have a periodicity of abundance of 9 to 10 years. Figure 8.6

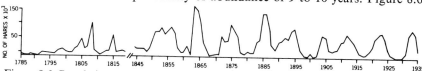

Figure 8.6. Population trends of the snowshoe hare, *Lepus americanus*, in the Hudson Bay watershed. (After MacLulich, 1937.)

shows snowshoe hare population estimates from Hudson Bay fur returns and from a variety of other sources for nearly 150 years. Starting about 1865, the beginning of good data, peak populations occurred every 9 to 10 years. Animals which tend to show these 9- to 10-year cycles are the snowshoe rabbit, the ruffed grouse, and some of the predators of these species, for example, lynx and red fox. To a lesser extent marten, fisher, mink, and muskrat may show 10-year cycles, but not so clearly. Figure 8.7 shows a 10-year cyclical

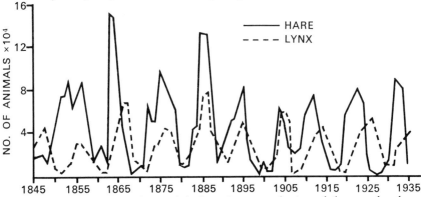

Figure 8.7. Changes in abundance of the lynx, *Lynx canadensis*, and the snowshoe hare, *Lepus americanus*, as indicated by the number of pelts received by the Hudson Bay Company. (After MacLulich, 1937.)

pattern of the snowshoe hare and of the lynx, *Lynx canadensis*, a prominent predator of the showshoe hare. The predator cycle follows the prey cycle, lagging slightly behind it in most cases. Ten-year cycles have been reviewed by Keith (1962).

Other animal populations have a cyclical periodicity of 3 to 4 years. These include the lemming of Scandinavia and North America, the field mouse, *Microtus agrestis*, and some of the predators of these two, notably the Labrador coloured fox, *Vulpes fulva* (Figure 8.8) and the snowy owl, *Nyctea scandiaca*. Not all populations of these animals always fluctuate cyclically. When they appear to do so, sophisticated statistics may be required for proof.

EXTRINSIC CAUSES OF POPULATION CYCLES

Some extrinsic factors have been considered to be causes of cycles. They are cycles of meteorological and cosmic changes, interspecific competition, predator-prey relationships, disease, parasitism, and food supply.

Climatic and meteorological changes have not been satisfactorily correlated with regular long-term cycles of population change. At one time sunspot cycles were considered to be correlated. Sunspot maxima have occurred every 10 to 11 years and hare population cycles are sometimes synchronized with these. Figure 8.9 indicates, however, that they often go out of phase.

Changes of populations of predators such as the lynx, fox, and snowy owl

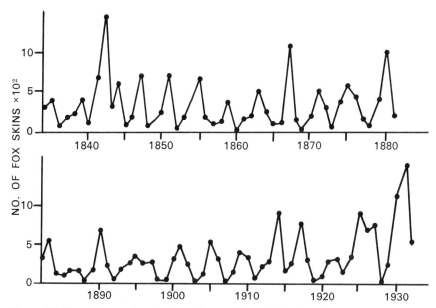

Figure 8.8. Population cycles of colored foxes, *Vulpes ?fulva*, in Labrador computed from pelt returns. (After Elton, 1942.)

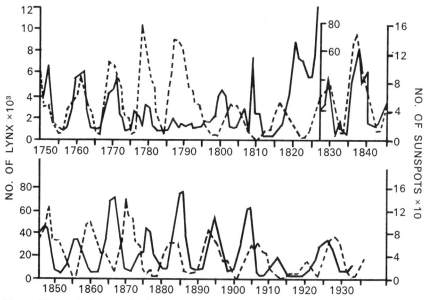

Figure 8.9. Population trends of the lynx, *Lynx canadensis* (solid line), and sunspot numbers (dotted lines). The change in scale in the upper graph applies to the numbers of lynx. (After MacLulich, 1937.)

are clearly correlated with cycles of their prey. But they cannot be shown to cause the prey cycles, for prey cycles often occur in areas devoid of predators. For example, snowshoe hare populations have cycles of change in areas where no lynx are found.

There has been some attention given to disease and parasitism as the cause of long-term cycles. Twenty years ago there was interest in hypoglycemic shock in snowshoe hares. Animals at peak periods of abundance seemed to have low blood sugar levels and go into states of shock and death. This was interpreted as the alarm reaction of Selye's general adaptation syndrome, which is discussed later.

In general, no infectious agent has been found to be consistently associated with rodent declines, but there has not been adequate study of this. Most ecologists have not been trained in pathology, bacteriology, and virology. The field has many opportunities.

Food supply has often been considered important in the decline of animal populations. Lack (1954) proposed that food is the main limiting factor in bird populations as it is for deer populations in North America.

Most field studies of rodents have demonstrated that even at peak periods of abundance ample food is present. Chitty studied grass during and after high vole plagues and found ample food. There are exceptions; for instance, during lemming peaks food becomes severely depleted and this is one of the factors that is thought to start lemming migrations (Clough, 1965). Pitelka (1958) feels that insufficient attention has been paid to qualitative study of food. The estrogens and steroids in growing plants, particularly grasses and legumes, are now being studied. It has been observed in the Orient, for example, that upsurges of rat populations are often timed with the flowering and sprouting of bamboo shoots. Then there is not only increased supply, but the young growing plants are more nutritive and estrogens are increased. The estrogens have direct influence on fertility and fecundity of the animals feeding on them.

INTRINSIC CAUSES OF POPULATION CYCLES
Social Stress
In the last 20 years there has been increasing interest in the role of behavioral, social, and physiological factors in control of population size. The observation of hypoglycemic shock in dense populations of snowshoe hares led Christian to develop the social stress theory of population limitation (1950, 1963). Christian's theory states that population increase results in crowding and increased behavioral interactions, particularly fighting. Eventually these begin to operate as stressors in individuals. This starts the manifestation of Selye's general adaptation syndrome, and as a result two major changes occur in the population: fecundity is reduced and mortality is increased. The operation of Selye's general adaptation syndrome (G.A.S.) is outlined in Figure 8.10. Stress operates through the sensory pathways to affect the hypothalamus, neurohypophysis, and adenohypophysis so that the endocrine production of

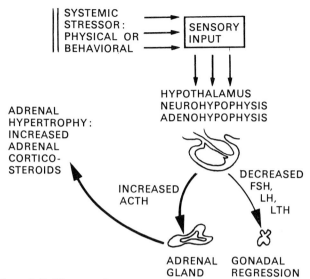

Figure 8.10. The operation of Selye's General Adaptation Syndrome.

the anterior pituitary gland is changed. There is decreased production of the gonadotropic hormones FSH (follicle-stimulating hormone), LH (luteinizing hormone), and LTH (luteotropic hormone), and this decrease is responsible for decline of fecundity. There is increased production of ACTH (adrenocorticotropic hormone), which stimulates the adrenal glands to increase production of the adrenocorticosteroids. This enables the animal to maintain homeostasis under stress. Blood glucose levels and sodium and potassium balances are kept within normal limits and other physiological demands are met. Stress continued for a long time leads to Selye's diseases of adaptation, then adrenal exhaustion and death. Some of these features can be demonstrated in laboratory animals. A rat or a mouse exposed to a simple measurable stressor, such as cold temperature, trauma, or surgical injury, will show various stages of the general adaptation syndrome (G.A.S.). Figure 8.11 shows development of the G.A.S. under prolonged stress, and some of the typical symptoms which may develop in experimental animals.

Christian thinks that the phenomenon operates in wild mammal populations and starts population declines. This is based on his work with laboratory populations of mice in crowded conditions. The mice show adrenal hypertrophy and gonadal regression. Chitty (1958, 1960) and Louch (1958) have shown these conditions in some dense field and laboratory populations of *Microtus*, and Christian and Davis (1964) in Norway rat populations in Baltimore City. However, most studies on crowded populations in nature have failed to find such changes in the adrenals and gonads. These studies include the work of McKeever (1959) on vole populations in the United

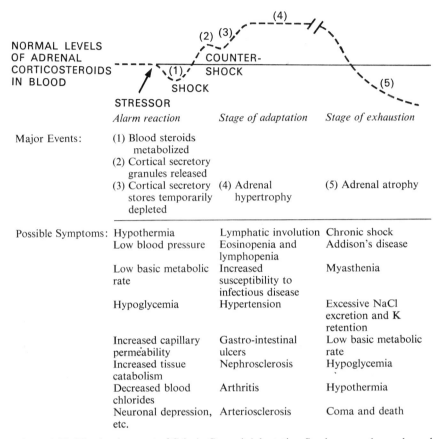

Figure 8.11. The development of Selye's General Adaptation Syndrome under prolonged stress, and symptoms in experimental animals.

States, Chitty (1961) on wild voles in Canada, Southwick (1958) on wild rodents in England, and Negus *et al.* (1961) on the rice rat on Cape Breton Island. Thus, field observations do not confirm that the Selye-Christian phenomenon operates commonly in natural populations. Most field studies indicate that other factors such as food supply, predation, aggression, aberrant parental behavior, and cannibalism operate before symptoms of the G.A.S. are evident. Anderson (1961), writing about rodent populations, concluded: "The gonadal-adrenal-pituitary relationship explored by Christian would be involved only rarely in these phenomena, and its operation in the regulation of rodent population seems to be abnormal or secondary to some other primary function. It thus appears that under most commonly encountered environmental conditions social and behavioral mechanisms alone may control population density, without immediate involvement of the endocrine system."

Negus *et al.* (1961) in a field study of population ecology of the rice rat, *Oryzomys palustris*, stated: "The adreno-pituitary-hypothesis states that the growth of mammalian populations is controlled by a socio-physiological feedback system that responds to changes in population density. We find this hypothesis difficult to accept. In fact, direct response to climate and food and often other extrinsic factors offers a more plausible explanation of population behavior than does the concept of social stress."

Christian's theory must not be rejected, but it is doubtful if it occurs very often as a major mechanism in natural populations.

The Olfactory Block to Pregnancy in Mice

Parkes and Bruce (1962) elucidated another aspect of behavioral physiology which may limit population growth. They showed that the fertility rate in female house mice, *Mus musculus*, is very sensitive to the presence of strange males. When recently impregnated females are exposed to strange males, there is an 80 to 90 percent failure of implantation. This failure does not occur if the sense of smell has been blocked. Physical contact is not necessary for this olfactory block to pregnancy to occur in intact mice; only odor is required.

In natural populations female house mice tend to associate with other females and during pregnancy and nesting they become territorial, excluding males from their nest sites. In crowded populations this territorialism breaks down and there is increased contact with different mice. Hence, it seems likely that the Parkes-Bruce phenomenon may operate in crowded populations and result in a declining fertility rate. It is not uncommon to find in dense rodent populations that there are normal rates of male and female fecundity, but declining pregnancy rates. No one has yet demonstrated olfactory block phenomena in natural populations, but the theory is compatible with some field data where the Christian social stress theory does not fit.

A discussion of the role of behavior, particularly in mice, in regulation of numbers is continued in Chapter 20.

CONCLUSIONS

The study of vertebrate populations is important for understanding the epidemiology of zoonoses and basic problems of population biology.

Various factors limit vertebrate populations. They include food, predation, and probably disease. Interacting with these are intrinsic mechanisms, acting through behavioral and physiological feedback systems to exert control. In some mechanisms a change may go far enough in one direction to bring into play another group of factors. These counteract the change and return the system to its former state. Mechanisms in most populations appear to be so beset by lags that steady control is rarely achieved. As a result the populations fluctuate.

Two prominent feedback mechanisms that have been suggested are the Christian-Selye social stress theory, and the Parkes-Bruce olfactory block phenomenon. Both can be demonstrated in the laboratory; neither has as yet been well documented in the field.

REFERENCES

Allee, W. C., Emerson, A. E., Park, O., Park, T., and Schmidt, K. P. 1949. *Principles of Animal Ecology*. Saunders, Philadelphia, Penn.

Allen, D. L. 1954. *Our Wildlife Legacy*. Funk and Wagnall, New York.

Anderson, Paul K. 1961. Density, social structure, and nonsocial environment in house-mouse populations and implications for the regulation of numbers. *Trans. N.Y. Acad. Sci.*, Ser. 11, **23**: 447–451.

Andrewartha, H. G., and Birch, L. C. 1954. *The Distribution and Abundance of Animals*. Univ. of Chicago Press, Chicago, Illinois.

Ashby, K. R. 1967. Studies on the ecology of field mice and voles in Houghall Wood, Durham. *J. Zool.* (London), **152**: 389–513.

Audy, J. R. 1958. The localization of disease with special reference to the zoonoses. *Trans. Roy. Soc. Trop. Med. Hyg.* **52**: 308–328.

Chitty, D. 1958. Self-regulation of numbers through changes in viability. In *Cold Spring Harbor Symp. Quant. Biol.* **23**: 277–280.

Chitty, D. 1960. Population processes in the vole and their relevance to general theory. *Can. J. Zool.* **38**: 99–113.

Chitty, H. 1961. Variations in the weight of the adrenal glands of the field vole, *Microtus agrestis. J. Endocrinol.* **22**: 387–393.

Christian, J. J. 1950. The adreno-pituitary system and population cycles in mammals. *J. Mammal.* **31**: 247–259.

Christian, J. J. 1963. Endocrine adaptive mechanisms and the physiologic regulation of population growth. In *Physiological Mammalogy* W. V. Mayer and R. G. Van Gelder (eds.) Academic Press, New York.

Christian, J. J., and Davis, D. E. 1964. Endocrines, behavior and population. *Science* **146**: 1550–1560.

Clough, G. C. 1965. Lemmings and population problems. *Am. Scientist* **53**: 199–212.

Davis, D. E. 1953. The characteristics of rat populations. *Quart. Rev. Biol.* **28**: 373–401.

Elton, C. E. 1942. *Voles, Mice and Lemmings. Problems in Population Dynamics*. Clarendon Press, Oxford, England.

Elton, C. 1958. *The Ecology of Invasions by Animals and Plants*. Methuen, London.

Emlen, J. T. 1940. Sex and age ratios in survival of California quail. *J. Wildlife Mangt.* **4**: 92–99.

Errington, P. 1945. Some contributions of a 15-year local study of the northern bobwhite to a knowledge of population phenomena. *Ecol. Monographs* **15**: 1–34.

Keith, L. B. 1962. *Wildlife's Ten-year Cycle*. Univ. of Wisconsin Press, Madison, Wisconsin.

Lack, D. 1954. *The Natural Regulation of Animal Numbers*. Oxford Univ. Press.

Leopold, A. 1943. Deer irruptions. *Wisconsin Conserv. Bull.* August, 1943.

Louch, C. D. 1958. Adrenocortical activity in two meadow vole populations. *J. Mammal.* **39**: 109–116.

MacLulich, D. A. 1937. Fluctuations in the numbers of the varying hare (*Lepus americanus*). Univ. of Toronto Studies, *Biol. Ser.* 43.

McKeever, S. 1959. Effects of reproductive activity on the weight of adrenal glands in *Microtus montanus. Anat. Record* **135**: 1–5.

Mullen, D. A. 1960. Adrenal weight changes in *Microtus. J. Mammal.* **41**: 129–130.

Negus, N. C., Gould, E., and Chipman, R. K. 1961. Ecology of the rice rat, *Oryzomys palustris* (Harlan), on Breton Island, Gulf of Mexico, with a critique of the social stress theory. *Tulane Studies in Zool.* **8**: 95–123.

Odum, E. 1959. *Fundamentals of Ecology*. Saunders, Philadelphia, Penn.

Parkes, A. S., and Bruce, H. M. 1962. Pregnancy-block in female mice placed in boxes soiled by males. *J. Reprod. Fertility* **4**: 303–308.

Pearson, O. P. 1963. History of two local outbreaks of feral house mice. *Ecology* **44**: 540–549.

Pitelka, F. 1958. Some aspects of population structure in the short-term cycle of the brown lemming in northern Alaska. In *Cold Spring Harbor Symp. Quant. Biol.* **23**: 237–251.

Sladen, W. J. L., Menzie, C. M., and Reichel, W. L. 1966. DDT residues in adelie penguins and a crabeater seal from Antarctica. *Nature* **210**: 670–673.

Slobodkin, L. B. 1962. *Growth and Regulation of Animal Populations*. Holt, Rinehart and Winston, New York.

Southwick, C. H. 1958. Population characteristics of house mice living in English corn ricks: density relationships. *Proc. Zool. Soc.* (London) **131**: 163–175.

Welch, B. L. 1964. Psychophysiological response to the mean level of environmental integration. In *Symposium on Medical Aspects of Stress in the Military Climate*. Walter Reed Army Institute of Research, Washington, D.C., pp. 39–99.

CHAPTER 9

Man's Ecological Environment

GEORGE B. SCHALLER

Ecology is the science of the relation of living things to each other and to their environment, and it includes human ecology. Man like any other animal requires food, water, space, and protection from disease. To obtain these he must fit into an ecological system even though this system may be a modified one of his own creation. For centuries man has attempted to set himself aside from the ecological system in which he lives and has attempted to subdue nature rather than understand it. He is now only beginning to comprehend that it is not enough to improve the political system or economic system, but that he must reach a healthy relationship with his environment.

Man in one form or another has been on the earth for well over a million years. For most of this span he was a close part of the ecological community. Small groups of people scattered in Africa and the tropics of the Old World lived by hunting and fishing and by gathering fruits, nuts, and other vegetable food. Dried human feces found in caves of Mexico and dating to 7000 B.C. contained grass and pumpkin seeds, agave fruits, grasshoppers, egg shells, bird bones, lizard, deer hair, and *Peromyscus* mouse remains. When the food supply in one area was depleted, these early people apparently moved to another site. Their populations were probably fairly stable. They probably lived somewhat like the bushmen of Africa and the pygmies of the Congo forest. The population density of the Australian aborigines has been correlated with rainfall and hence food supply (Birdsell, 1957). Disease and starvation, even infanticide, kept the populations low. In crowded animal populations the same checks operate, including cannibalism. This is not to imply that earlier man did not affect the environment, even as a primitive hunter. Fire has been in man's possession for perhaps half a million years. Primitive peoples today burn forests and grasslands to drive out and kill the game, and there is no reason to suppose that it was not employed in the past, too. This may have been part of the cause of the expansion of grasslands and change of grasslands to deserts. Man has been implicated in the extermination of several Pleistocene mammals such as the woolly rhinoceros and mammoth. However, man's influence was local and usually temporary.

During the Mesolithic period, about 10,000 B.C., a far-reaching revolution occurred in Western Asia, the transition from hunting and gathering to farming. The birth of farming occurred somewhere in Palestine, Syria, Northern Iraq, and Iran. It was a time when the last glaciers began their retreat from Europe. Western Asia was at that time clothed with forests resembling those of Europe and to the south these gave way to grasslands. In this area the prototype of wheat and barley flourished. For the first time man planted and reaped his crops. Wild sheep, wild pigs, and wild oxen, the proto-types of the domestic ones, lived in the forests and hills. The dog was already domesticated. Sheep and goats were domesticated by 8000 B.C., cattle by 5000 B.C. Man in the Near East began to wield the plow, the ax, and fire ruthlessly. He did not know that if he removed the plant cover, the soil could erode away and leave a desert. As the glaciers moved northward the country became drier. This, and careless farming, turned what in the Bible was considered the Garden of Eden into a desert. Usually the end of the Babylon-ian Empire is written in terms of war with the Persians. No weight is given to the fact that the city of Ur, once a great metropolis, now lies abandoned 150 miles in a desert. The wealth of all nations lies in their soil.

As man developed increasingly complex societies and technology, he increased the gaps between individuals and the natural processes upon which human populations still depend. The world of nature may appear as a world apart from man, and there is resistance to the idea that ecological principles apply seriously to him. Some of this attitude comes from scientists who are more aware of existing reserves than of the existence of limiting factors (Sears in Thomas, 1956). Herein lies an all-pervasive danger, the curious develop-ment of a false sense of detachment from the environment.

DEPLETION OF RESOURCES AND DESTRUCTION OF HABITAT

Man exists by virtue of an environment which is also highly evolved, having come to its present state during more than two billion years. Civilized man is still dependent on plants for his nutrition and energy, which are derived either directly from foodplants or indirectly when they are cycled through domestic livestock. Food chains always start with green plants because of their unique ability to capture solar energy and to synthesize nutrients from raw materials such as carbon, nitrogen, and phosphorus. Each ecological system tends to evolve toward a climax state. The climax is the most complex biological community an environment will sustain. In other words, a community evolves toward a maximum flow of energy.

The rain forest is a good example of a climax. It is the most complex one on earth, but paradoxically also one of the most delicate and easily damaged. There is a perfect balance between growth and decomposition, for many bacteria have their optimum activity at 90–100°F. If man removes the tree cover, the humus is burned by the sun and heavy rains leach the minerals from the soil. The soil cover is washed away. If domestic crops are planted the soil is exhausted within 3 to 4 years and must then be left alone for at least 20 years to regenerate. As a result, man has to shift his fields continuously. Even today the only pattern of cultivation in the rain forest areas of the Congo, South America, and Southeast Asia is the shifting one. If an attempt is made to shorten the cycle of cultivation, the forest reverts to scrub and grass and finally to desert. In Africa today, because of misuse of the forest cover, the Sahara desert is advancing southward at the rate of a half mile a year.

Perhaps the largest area of misused land in the world is the tropical savannah of Africa. This area is of low productivity to begin with, and yet this marginal land is plowed. The topsoil is blown and washed away. The soil structure changes from granular, which provides aeration and lets the water seep in, to very dense. This causes the rains to rush off, which in turn causes erosion. The uncultivated grasslands are burned yearly and fire reduces many of the nutrients to inorganic ash. As a result much of the African savannah is becoming more arid. Slowly it is becoming desert.

North America too is a good example of a land in decline. The Europeans that arrived only 350 years ago were the most destructive group of humans this earth has ever seen. Half of the United States was covered with forests. There were only about 1,000,000 Indians. The immigrants abandoned land as rapidly as it was cleared; the watersheds were denuded. Today an estimated six-sevenths of the range land in the American West is in degraded condition. Within the past 150 years the United States has lost one-third of its topsoil, over half of its high-grade timber, and an unknown amount of fresh water.

Even the fragile climax communities of the subarctic have suffered. Prospectors set fire to the spruce forests to expose the bedrock. This burned away the lichens, which are primitive plants. These lichens are the main winter food of caribou, an Arctic deer. Lichens grow slowly, requiring 120 years to reach maturity. With the lichens burned off, the caribou declined from an estimated 30,000,000 during the past century to 300,000 in 1955. With the decline of the caribou, Eskimos who depended on them starved. Recently it was found that lichens accumulate strontium-90, the radioactive material in the air as a result of atomic explosions, much more than other plants. The caribou eat lichens, the Eskimos eat caribou and end up with a very high radioactive count. This is a good example of the interrelationships of organisms in a community.

Approximately half of the arable land in the world has in the past 10,000 or so years been ruined or has had its productivity lowered drastically. Man has advanced materially simply by breaking into the stored energy of the world's ecological climaxes. It has been estimated that 2.5 acres of arable land per person are needed for an adequate standard of living. There are about 3 billion acres of land adapted for food production in the world, about 1 acre per person. China has less than $\frac{1}{2}$ acre. Europe has 1 acre. One may argue that Europe for the most part has an adequate standard of living. This is true but only because Europe feeds on the lands of other people. Even in 1900 England imported 60 percent of its food.

Livestock Problems

When man domesticated the cow, sheep, and goat he released a terrible scourge, since these animals have denuded large areas of the earth. Cattle eat the grass down, the sheep follow, and the goats even climb the trees to get at the foliage. In northern Nigeria, the Sudan, Near East, and Mexico the land is being overgrazed by livestock. Wherever man sees grass in almost unlimited quantities he assumes that it should be eaten by his livestock. It is particularly so in the savannahs of Africa. Much of the area is unsuitable for farming because of periodic droughts. Many areas are infected with nagana, a form of sleeping sickness caused by *Trypanosoma*, which is fatal to domestic cattle and is spread by the testse fly. However, the wild animals are immune to the disease and they act as reservoirs. In order to use cattle in these areas the tsetse fly has to be eliminated. This can be done in two ways. The tsetse fly likes brush and shade, therefore the brush was removed. This certainly eliminated the fly, but vast tracts were made into a desert of no value to either man or beast because the soil eroded away when the bush was removed. Then it was decided to get rid of the reservoir of the disease, the wild game. In Rhodesia alone 700,000 head of game were killed. This had little effect on the fly, because it merely used small game as a reservoir. This is a typical illustration of trying to solve a problem without studying it first. In recent years the problem has been approached more sensibly. Since the wildlife was obviously adapted to the area and survived well without ruining the habitat, why not raise wildlife instead of cattle for meat? Why not harvest wild game on a sustained yield basis by shooting the annual surplus each year (Pearsall, 1962)?

Some interesting facts were discovered when this idea was put into operation. It was found that the total weight of game on land of low quality is greater than that of livestock on good quality land. Here the concept of carrying capacity is needed. Each plot of earth can support only so many pounds of animal life on a permanent basis. This biomass can be all in mice or rats, or it can be one deer or one elephant, but usually it is a combination. The concept of carrying capacity is readily seen when, for example, only 1 acre is needed to support a cow in Maryland but in Texas 30 to 50 acres may be needed. The reason why an acre can support a greater biomass of wildlife than domestic cattle is again because natural selection tends to maximize the energy flow in a community. Evolution does this by diversifying animal and plant types. For example, in the Serengeti plains there are over 15 species of hoofed animals. Each has its own food habits and preferences, its own niche. There are those that browse on treetops (elephant, giraffe), on various levels of shrub (several antelopes, rhino), on long grass, on short green grass (zebra, gnu), on roots (wart-hog). There are species confined to plains, to woodlands, to thickets, to river banks. As a result the vegetation is used to the maximum extent. But cattle eat only certain types of grasses; they do not use the habitat fully. Cattle also need water daily, so they concentrate around the water sources. Many wild species, on the other hand, have a water-saving mechanism and they do not need to drink. In dry areas many of the waterholes have harmful amounts of fluorine (25 parts per million). At this high concentration cattle get soft bones, but wildlife is not harmed. Wildlife is able to use the eaten food more fully. Wild buffalo, for example, thrive on grass that will not maintain weight in livestock. Table 9.1 illustrates the difference in carrying capacity. The habitat is savannah with scattered *Acacia* trees. The figures are the number of pounds which a square mile has been able to support on a permanent basis.

In one study in Southern Rhodesia it was found that if the wildlife was

TABLE 9.1

The Carrying Capacity of Various Places in Africa in Terms of Wild Species and Cattle

Place	No. wild species in area	Biomass (lbs/sq. mile)
Southern Rhodesia	16	18,700
Serengeti Park	15	28,000
Nairobi Park	18	76,000
Queen Elizabeth Park	11	107,000
Albert Park	11	32,000–135,000
Domestic cattle (East Africa)		11,200–16,000

harvested on a sustained yield basis, that is, only the annual surplus was removed, the rancher obtained 4 pounds of meat per acre per year. But if he raised cattle he obtained only 3 pounds. This is just a small illustration of how with some ecological research the wildlife can be saved, the soil better utilized, and the land can be economically more productive.

FOOD PRODUCTION

Even though man has abused the soil terribly and half of the world population is underfed, man could, it he applied himself, restore much of the land and feed the present world population effectively. He could improve the strains of crops used and find crops that furnish a higher yield of protein. He could see that there is less waste. Two hundred million dollars worth of damage is done yearly in the United States to crops by rodents. He could also control insect pests. But such control should not be by indiscriminately spraying poisons over the land. Spraying may have serious effects by disturbing the balance of the ecosystem. Many of the poisons directly affect man adversely and many are cumulative.

The tobacco which is smoked now still contains arsenic which was sprayed on the crops 30 and more years ago. It remains in the soil and cycles around in the ecosystem. Control measures should be specific for a certain species. The screw-worm was eradicated in Florida by releasing infertile males; *Tribolium* beetles are known to die from high frequency sounds; some species of insects when fed certain substances fail to mature.

Man can also find new sources of food. He can obtain protein from leaves. A substitute for milk can be made from leaves without the wasteful method of having a cow do it. Moreover, man has not learned to effectively harvest the resources of the sea.

POLLUTION OF THE ENVIRONMENT

"Probably the greatest problem in public health in the near future is not the control of specific diseases but the control of pollution of the general environment (water, air, soil), which can bring with it ill-health of all kinds" (Odum, 1959).

TABLE 9.2

Primary Emitters of Environmental Pollution in the United States[a]

Environment	Primary emitters	Primary pollutants
Air	Motor vehicles	Carbon monoxide
		Hydrocarbons
		Oxides of nitrogen
	Public power utilities	Hydrocarbons
		Oxides of nitrogen
		Sulfur dioxide
Water	Municipalities and	Biological oxygen demand
	general industry	Phosphates and nitrates
	Land runoff	Phosphates and nitrates
		Silt
	Public electric utilities	Heat
Land	Homes	Refuse
Additional pollution from:	Nuclear testing	Carbon-14
		Strontium-90
		Cesium-137
	Agriculture	Pesticides

[a] National Academy of Sciences (1966).

Environmental pollution, in the biological sense, is the introduction into the cycles of the ecosystems of substances that are able to stop, control, or change biological activities. The primary pollutants in the United States and their sources, in general, are given in Table 9.2.

The Air

Emission of the principal pollutants to the atmosphere in the United States totals about 125,000,000 tons per year (Table 9.3). Even if it is assumed that 50 percent of the nation's electric generating capacity will use nuclear energy by the year 2000, if present population trends continue, pollutants from fossil fuel generation will double by 1980 and redouble by 2000 (National Academy of Sciences, 1966). The portion of the earth's atmosphere available to dilute

TABLE 9.3

The Amounts of the Principal Pollutants Emitted into the Atmosphere in the United States[a]

Pollutant	Million tons per year
Carbon monoxide	65
Oxides of sulfur	23
Hydrocarbons	15
Particulate matter	12
Oxides of nitrogen	8
Other gases and vapors	2
	125

[a] National Academy of Sciences (1966).

air pollution is about 6 miles thick. It serves as a temporary reservoir and pollutants are transformed in a variety of ways, by oxidation, electrostatic aggregation, and cloud formation, into forms that are deposited by gravity or swept away by other forces. However, much of man's pollution is localized. A single fossil-fuel power-generating plant may emit several hundred tons of sulfur dioxide daily, locally overburdening the air.

Air pollution damages vegetation, livestock, structures, and human health. Serious damage has occurred in cities to pine, tobacco, grape, vegetable field and flower crops, and ornamental plants. A crop can be made unmarketable by a single exposure of only a few minutes, and this can occur as a result of breakup of a temperature inversion of the air. Livestock damage is usually chronic. The ingestion of large quantities of fluorides, derived from ore-processing, causes loss of teeth and weakening of the skeleton. Metals corrode and tarnish, fabrics weaken and fade, leather weakens and becomes brittle, rubber cracks and loses elasticity, paint discolors, glass is etched, paper becomes brittle, building stone discolors and erodes, and electrical circuit elements behave erratically. Exposure to extreme pollution can kill people, and chronic effects are especially prevalent among persons with respiratory disorders (National Academy of Sciences, 1966).

Water

The average annual stream flow which discharges into the oceans from the continental United States is 1,100 billion gallons a day. In 1954 some 300 billion gallons were removed daily, 110 billion were depleted, and 190 billion carrying wastes were returned. It is predicted that by the year 2000, the return will be 889 million gallons. Thus, whereas in 1954 withdrawals amounted to less than one-third of the available total and waste-ridden returns less than one-fifth, in the year 2000 withdrawals will be over four-fifths of the available water and polluted returns two-thirds of the entire United States stream flow. Sewage returns are expected to double by the year 2000, and 95 percent of the 280 million people will average 132 gallons per day per person. A sevenfold increase is expected in purely industrial wastes produced by the large water-using industries. Major withdrawal will be for the generation of steam-electric power.

An idea of the enormity of the problem can be gained by the consideration of only pollution by domestic sewage and other oxygen-demanding wastes. These wastes are reduced to stable compounds through the action of aerobic bacteria that require and obtain oxygen from the water. The resultant oxygen reduction can have serious impact on life in the water. It is estimated that by 1980 the oxygen demand of treated effluents will be great enough to consume the entire oxygen content of a volume of water equal to the dry-weather flow of all the United States' 22 river basins. Other pollutants include infectious agents, plant nutrients, organic chemicals (insecticides, pesticides, detergents), sediments, other minerals and chemicals, radioactive substances, and heated water (National Academy of Sciences, 1966).

The Soil

The principal categories of the land-pollution problem are disposal of solid wastes and maintenance of agricultural soils in proper condition for efficient crop production. Per capita production of refuse has grown from 2.75 pounds per day in 1920 to 4.5 pounds per day in 1965. Total refuse production is increasing at about 5 percent per annum. The storage, collection, and disposal of solid wastes in metropolitan areas have public health implications and land for deposition grows scarcer. In the soil itself two types of pollutants are of particular interest, radioactive elements and pesticides. Present evidence is that pollution of soil by fallout has been small, with increases in levels being less than natural variability between soils of different areas. Neither strontium-90 nor cesium-137 seem at present to be a hazard to plants or animals. Levels of radiation from fission products in United States soils are generally about one order of magnitude lower than natural radiation from the soil.

The chemicals designed to kill either plants or insects reach the soil through direct application or through application to plants. Many of these resist biological destruction and present current and potential problems (National Academy of Sciences, 1966). The beneficial effects of pesticides in agriculture and public health must be weighed against the harmful effects many of these chemicals have in the ecosystem. Excessive applications of DDT can seriously reduce populations of other animals. DDT and dieldrin are examples of chlorinated hydrocarbons which are not degradable. Thus, toxic residues occur in soils and accumulate in other parts of the ecosystem.

Example of an Urban Concentration, the Lower Delaware River Area

(Report of the Environmental Pollution Panel, 1965)

In 1960 the lower Delaware River area which is composed of adjacent parts of Pennsylvania, New Jersey, and Delaware contained 377 municipalities, 5 million people, and 100 large industrial plants. Giant urban concentrations such as this dispense huge quantities of diverse materials into the air, soil, and water. Of the 3.5 billion gallons of water available, 2.9 billion are now being used. The forecast is for a threefold increase in use of water in the next 30 to 40 years. Water pollution, in terms of biological oxygen demand, is 225 million pounds a year, and a 50 percent increase is estimated for 1975. In 1959 in Philadelphia 830 tons of sulfur dioxide, 300 tons of nitrous oxide, 1,350 tons of hydrocarbons, and 470 tons of particulates were released daily. Philadelphia has temperature inversion of the air on approximately 200 days of the year, and severe pollution of the lower layer of the air exists on 100 of those days. Within a 15-mile radius of the city, particulate fallout is 7,000 tons per month. In southern New Jersey at least 36 species of plants have been shown to be injured by pollutants from the Philadelphia area. About 2 million tons of solid waste have to be disposed of annually, a figure that will probably increase by a factor of four by the year 2000. In fact, the rates of pollution are increasing more rapidly than the population.

Pollution of the Great Lakes (Sperry, 1967)

Lake Erie has 761 miles of shoreline, and within its drainage area are five states and one Canadian province. It is valued as a water supply, for sewage disposal, economic transportation, recreation, and for commercial and sport fishing. Unfortunately it is becoming severely changed by eutrophication. Eutrophic lakes contain substantial amounts of nutrients so that they have great primary productivity—greater amounts of littoral vegetation and plankton and algal "blooms"—and summer stagnation. The algal blooms deplete oxygen from the bottom layers of lake water. When the algae decay they foul beaches and may cause odor and discoloration of drinking water. Lake Erie is naturally evolving toward the eutrophic state but this is being accelerated by the daily addition of 150,000 pounds of phosphates, of which 100,000 are retained. The sources of the phosphates (measured by pounds delivered daily) are Lake Huron (20,000), urban land runoff (6,000), rural land runoff (2,000), detergents (70,000), human excreta (30,000), and direct industrial discharge (6,000). Eighty percent of the phosphates are first treated by municipal sewer plants, and of those 66 percent originate in detergents. Only 2 years before the report the two billion dollar-a-year detergent industry completed a changeover to the manufacture of biodegradable detergents. The change cost them 150 million dollars. They changed because they had been accused of turning many lakes, rivers, and water supplies into giant bubble baths.

Detroit, Cleveland, and other cities could increase their efforts to remove phosphates from their disposed wastes. Storm sewers are combined with sanitary sewers and thus carry organic wastes in times of heavy rainfall. Five hundred thousand people in the drainage area use unregulated septic tanks, and there is no control of waste from ships, and no control of agricultural chemicals.

The results of this pollution on Lake Erie are very extensive. The beaches are sometimes fouled with from 6 inches to 3 feet of decaying algae, and drinking water has been discolored and odoriferous. The major commercial species of fish have nearly disappeared while less desirable fish have increased, and the bottom fauna has also extensively changed.

THE POPULATION INCREASE

One threatening aspect of environmental damage is its rate of occurrence; and this rate can only increase because of the exponential increase of people.

It seems unlikely that the population of the world before the Neolithic revolution was more than 10 million. It was not until about 1650 that the population reached half a billion. Then came the great explorations and industrial technology. The world population grew until by 1900 there were about 1.5 billion people. In 1965 the world population size was about $3\frac{1}{3}$ billion. At present crude rates of increase it will double about every 40 years. However, the crude rates of increase are themselves increasing. This is because many populations have or will have larger proportions of reproductive age groups mainly as a result of the current rapid declines in death rates of infants and children.

Modern rates of increase are due to declining death rates of all ages except the old, without much decline in age-specific birth rates. Mean length of life is changing from 30 or fewer years to as much as 70 years. Figure 9.1 shows

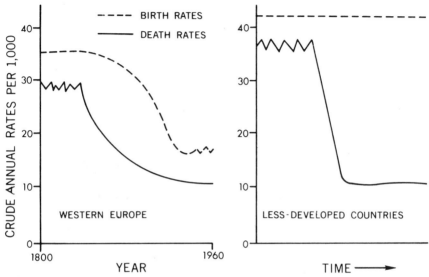

Figure 9.1. Schematic presentation of birth and death rates in Western Europe, 1800 to 1966, and in less-developed countries in the mid-twentieth century. (After National Academy of Sciences, 1963.)

that, in general, birth rates have not declined at all in less-developed countries; in the rest of the world they have declined, but not enough to balance deaths. The present crude rate of increase in the United States is 1.6 per thousand per year (currently an actual increase of about 3 million people per year) and in Africa it is about 2.3 per thousand and in South America 2.7.

The life span has lengthened largely because of the control of disease and famine that has accompanied technological change. For instance, when malaria was eradicated in Ceylon by use of insecticides the death rate fell within one year from 20 to 14 per thousand. It is difficult to realize how effectively famine and disease have acted in the past as checks on population growth. At least one-fourth of the European population died around the middle of the fourteenth century when the Black Death swept through. In 1832, 1849, and 1854 epidemics of cholera raised the death rate in New York City to more than 45 per thousand per year. From 108 B.C. to A.D. 1911 there were an estimated 1,828 famines in China, almost one per year. Between 1200 and 1600 England suffered a famine every 15 years.

The carrying capacity of the world for human populations is not known. In the recent past great population increase was absorbed by the opening of new land. For example, between 1800 and 1950 the population of Europe increased from 10 million to 55 million. But during that same time a further 60 million emigrants left Europe to settle in other places. No longer is so much new land available. If present rates of increase continue, there could be 500 billion people in the year 2200. If uniformly distributed over the world, covering the Antarctic and the deserts, the population density would be as it is today in Washington, D.C. Even assuming that the necessary food could be obtained, the lack of space would clearly be intolerable.

This increase has vast implications in relation to the economy of every country. Many nations of Asia and Africa do not have the capital and skill to feed and clothe their rapidly growing populations and at the same time raise the standard of living. Some of the nations like India have put their faith in industrialization, which they feel will raise the standard of life. But it is impossible to industrialize an underdeveloped country without financial capital and human skill. For instance, Coale and Hoover (1958) concluded that if India does not cut its birth rate by 50 percent in 40 years, or does not increase its death rate, it will never be able to develop into an industrial nation and the standard of life will go down instead of up.

If the rate of human increase is not reduced, all possibility of future improvement is jeopardized. Birth control has long ceased to be the problem of any one country and is now a problem of the whole world. There are two ways to control the rapid increase of numbers, by high mortality or low fertility. The latter is certainly the more desirable alternative, but the time for making a free choice may soon pass.

There must be conscious effort to apply ecological principles if man is to live harmoniously within his limited environment.

REFERENCES

Birdsell, J. B. 1957. Some population problems involving Pleistocene man. *Cold Spring Harbor Symp. Quant. Biol.* **22**: 47–69.

Coale, A., and Hoover, E. M. 1958. *Population growth and Economic Development in Low Income Countries: A Case Study of India's Prospects.* Princeton Univ. Press, Princeton, New Jersey.

National Academy of Sciences. 1963. *The Growth of World Population.* Publ. 1091, *Nat. Acad. Sci.*, Washington, D.C.

National Academy of Sciences, Nat. Res. Council. 1966. Waste Management and Control. Publ. 1400, *Nat. Acad. Sci.*, Washington, D.C.

Odum, E. P. 1959. *Fundamentals of Ecology.* Saunders, Philadelphia, Penn.

Osborn, F. (ed.) 1963. *Our Crowded Planet.* Doubleday, New York.

Pearsall, W. H. 1962. The conservation of African plains game as a form of land use. In *The Exploitation of Natural Animal Populations*, E. D. Le Cren, and M. Holdgate (eds.) Wiley, New York.

Report of the Environmental Pollution Panel. President's Science Advisory Comm. 1965. *Restoring the Quality of Our Environment.* The White House, Washington, D.C.

Sperry, K. 1967. The Battle of Lake Erie: Eutrophication and Political Fragmentation. *Science* **158**: 351–355.

Thomas, W. L. (ed.) 1956. *Man's Role in Changing the Face of the Earth.* Univ. of Chicago Press, Chicago, Illinois.

II

POPULATION GENETICS

Variation in a Herd of *Bos africanus* (Lyre Shaped Horns) and *Bos brachyceros* (Thick Horns Projecting Forward in this Picture). This is part of a copy of a rock painting by Neolithic "Bovidian" pastoralists of the Tassili mountains of southeast Algeria. Neolithic people dwelt in these extensively furrowed mountains in what is now the heart of the Sahara desert. The climate then was damp, and the flora and fauna were rich and tropical. Rock paintings and engravings show early Neolithic negroids who hunted buffalo. These were succeeded by sixteen to thirty types of people, identified by differences in the subject matter and style of their paintings, and showing influence and invasion from the Upper Nile. The Bovidian period followed, provisionally dated as starting in 3,500 B.C., and continued for several thousand years. Bovidian art included warlike frescoes, pictures of weapons, theft, and protection of herds. Two further periods, falling within historic times, are recorded in Tassili: the Protohistoric or Equine period of pastoralists with chariots and cavalry, and the Camel period. (Reproduced from *The Search for the Tassili Frescoes* by H. Lhote, translation from the French version published in 1959 by E. P. Dutton, New York.)

SECTION II

POPULATION GENETICS

The preceding section described the continuous opposition of the forces for reproduction and increase, against what may be generally called the resistance of the environment; and in any population these opposing forces are composites of many factors. All species have high potential for increase, negated by a multiplicity of factors which kill individuals before reproduction and prevent reproduction itself, and those individuals that do reproduce have varying degrees of success. The differential survival and reproduction causes variation from generation to generation of the types in a population; the change is the measure of natural selection. Both genetic continuity and variability are encoded in the DNA molecules of individuals, and "nature" by selecting the more satisfactory types, changes the genetics of populations. Thus by loss of individuals and by replacement, populations can discard old genetically based adaptations and expand new ones. This is clearly seen when man attempts to permanently reduce or remove certain microbial or arthropod populations with lethal chemicals, and then sees the rapid return of many of them now resistant to certain doses of the chemicals and thus readapted to the changed environment.

In order to begin to understand patterns of expression of genetically based properties within populations, and the nature of genetic change in populations, there must be some recognition of the intermediate processes: the assembly of biochemical information and its variants within the nucleus; expression of variation in individuals; and the roles of reproduction in maintaining variation and change.

Rates of change of the coded information and its arrangement in the nucleus are increased by many chemicals and by radiations of atomic nuclear energy and allied radiations. Chapter 16 speculates about the effects of such increased rates of change upon human populations and human health.

CHAPTER 10

Nucleic Acids and Hereditary Mechanisms in Microorganisms

ROGER M. HERRIOTT

The advances in knowledge of heredity have recently been so rapid that the magnitude of the change may not be appreciated by those not working in this field. These developments will be reviewed in the sequence of steps of the advance. Other areas of understanding important to public health await comparable expansion. Areas such as mental physiology and embryonic differentiation, where knowledge of the mechanisms of action is very thin, are ready for rapid expansion. These two fields stand now where heredity stood before World War II. Much can be said about them of a descriptive nature, but there is virtually nothing known about the means by which information is stored in the brain or how it is recalled, or what determines the differentiated units in organ formation. However, there is reason to believe that startling advances in at least the latter field may be near.

The rapid developments in the knowledge of hereditary processes can be largely attributed to studies of viruses and of bacterial genetic transformation. In the 1920's knowledge of the site of hereditary processes was known to be in the cell's nucleus where, in higher organisms, microscopic bodies called chromosomes could be seen. Chromosomes were composed of nucleic acid, histones, and some basic proteins, and no correlation of biological function with chemical substance was possible.

In the 1930's Stanley, then of the Rockefeller Institute, and one of the leading biochemical virologists of that period, pointed out on numerous occasions that viruses had many of the properties of genes, for they multiplied and underwent mutation. During this period, especially following Stanley's isolation of tobacco mosaic virus as a crystalline material, many viruses were isolated, purified, and analyzed chemically. They all contained protein and nucleic acid. Since these components were very complex and their functions were not understood, and because there were two kinds of nucleic acids in greatly varied amounts in viruses, no definitive answers came from this quarter. But the search for hereditary material had been reduced to only two substances, nucleic acid and protein.

BACTERIAL GENETIC TRANSFORMATION

The first clear evidence of the chemical nature of the hereditary material came from the analysis by Avery, MacLeod, and McCarty (1944) of a phenomenon described in 1928 by Griffith. Griffith was a health officer in London, and he noticed that the type of pneumococcus producing deaths in Britain was gradually changing. He experimented with mice and various strains of pneumococcus. His critical experiment was the inoculation into the same mice of heat-killed cultures of virulent encapsulated organisms of one type followed by an inoculum of viable but avirulent, unencapsulated, rough pneumococci of a different type. This combination produced death of the mice, and virulent organisms similar to those which had been heat-killed were isolated from the mice. Control experiments in which the heat-killed preparations or the avirulent organisms were inoculated each into different mice, produced no comparable illness and no virulent organisms were isolated. The explanation of Griffith's results involved either revival of heat-killed cells or a transfer of hereditary characteristics. At that time such conclusions were either not considered or there was some reticence to suggest such bold possibilities. Neither Griffith nor several workers who followed made any mention of the implications this held for hereditary processes. Not until 1936 was the implication stated clearly, by Landsteiner, the brilliant discoverer of blood group substances, in a book on serology.

Avery, MacLeod, and McCarty in 1944 showed that the active factor in Griffith's heat-killed culture which determined the pneumococcal types and virulence, was deoxyribonucleic acid (DNA). This was the first demonstration of the function of nucleic acids, and of the chemical nature of the carrier of hereditary information.

Genetic transformation, as Griffith's phenomenon is called, involves the uptake by bacteria of extracellular DNA derived from a related strain, and the incorporation of this DNA into the genetic equipment of the receptor cell. After this, the phenotypic property of the donor cell develops. Theoretically, any property of the donor cell can be transferred in this way, but unless the investigator has some means of identifying or separating the cells with a particular property, it is not possible to recognize the transformants. The genes conferring resistance to antibiotics are nearly ideal markers, for all cells not carrying such markers are inhibited or killed by the antibiotic added to the agar plate. The cells transformed with resistance markers form colonies and are readily counted. Nutritional markers permitting transformed cells to utilize fewer nutrients than the receptor cells are also readily assayed.

Since the work of Avery *et al.* (1944), genetic transformations have been demonstrated for ten to twenty species of bacteria and in human cell lines, for instance by Szybalska and Szybalski (1962). As usually grown, most transformable strains of bacteria are not able to take up DNA. A special physiological state known as competence must be developed before they will take up extracellular DNA. Retardation of growth, while allowing protein synthesis to proceed, favors the development of competence. This and other work leads to the following tentative interpretations of competence. It is associated with the appearance of a special protein in the cell membrane which can interact with the extracellular DNA, or it is associated with the removal of a cell membrane component, through unbalanced metabolism, which allows the DNA to be drawn into the cell by some unknown mechanism.

Some recent experiments suggest that the DNA of a bacterial cell is all in one very long fiber. The usual procedures for isolation of DNA result in its being broken down into many, perhaps a hundred, pieces of about 15,000,000 to 30,000,000 molecular weight. Each of these pieces probably carries different information. It is quite reasonable then that while only one molecule or piece of DNA in every 100 carries a particular genetic marker, this is equivalent to suggesting that there is one particular marker per donor cell.

Each of the above pieces of DNA is large enough to carry perhaps as many as 25 separate genetic markers. When only a dozen or so markers for the whole cell are recognized, it is surprising that one should find two of these few markers on the same piece of DNA. In our laboratory three different and independent antibiotic resistance markers have been found on the same piece of DNA. The evidence for this is that DNA isolated from cells resistant to these three antibiotics transforms sensitive receptor cells to resistance to all three antibiotics. This would not occur with any appreciable frequency if the genetic units were not on the same segment of DNA. There is no cross-resistance among these antibiotics. These markers are therefore described as being linked. Those which do not cotransfer are described as unlinked markers.

Transformation occurs most easily when the DNA is from the same species, but it is known to occur between different organisms, for instance, streptococci and pneumococci and between species of *Neisseria*. More detailed information on transformation is given by Ravin (1961) and Schaeffer (1964).

BACTERIAL VIRUSES

In 1945 many scientists were released from war duties and were eager to find new outlets for their energies. Some found particularly attractive the biological questions which could be studied in the rapidly developing field of bacterial virus (phage) systems, especially in the *Escherichia coli* T phages. Phages were considered to be not fundamentally different from animal viruses, yet they had many advantages such as a faster and more precise assay. The virus and host stocks could be obtained easily, and genetic uniformity could be developed in a short time. They could be grown in synthetic media in very large quantities, and in media labeled with radioisotopes. Previous studies by Burnet, Gratia, Northrop, Schlesinger, Delbrück, and Luria had opened a new world of possibilities, and scientists from various disciplines were attracted. Their sharp, incisive experiments had an immediate impact. Then in 1956 Dulbecco broke away from the phage field and developed the plaque technique for isolating and quantifying human and animal viruses grown in tissue culture. The rapid recognition, isolation, and study of mutant strains of polio virus by Sabin was a direct application of knowledge derived from bacteriophage studies. This illustrates how experience in more easily handled systems, which at first appear to be unrelated to human problems, can produce rich dividends for medicine and public health.

HYBRIDIZATION OF VIRUSES

In 1946 Delbrück and Hershey independently discovered that cells infected with two or more related, but genetically distinguishable, virus particles could yield hybrid viruses carrying characteristics of both parent viruses. This meant that new viruses or viruses with new properties could be expected when a host was infected with two related viruses. Since then, Burnet and Lind (1951) and others have seen this same result in influenza virus.

MULTIPLICITY REACTIVATION

Luria (1947) uncovered an amazing phenomenon. Bacterial cells were exposed to phage that had been killed by ultraviolet (UV) light. Those cells that received more than one killed phage became infected with and produced live phage. This suggested that each viral particle is made up of several independent subunits. If the subunits damaged by ultraviolet radiation were different in the two phage particles, then the remaining subunits of the two could complement each other and produce one fully active unit. This is an illustration of how material tested in the usual manner may be found to be inactive, but under slightly different conditions can produce infection. Similar results have since been seen with animal viruses (Gotlieb and Hirst, 1956; Galasso and Sharp, 1963).

THE ECLIPSE PHASE

In trying to follow the course of phage infection, Doermann broke open cells soon after infection was initiated and was unable to detect any infectious unit. Yet his method of breaking open cells did not destroy free phage. If free phage were inside the cell, Doermann should have been able to detect it. The phage had apparently gone into an eclipse phase. Shortly afterwards, this was more fully explained by the experiments of Hershey and Chase (1953).

THE HERSHEY-CHASE EXPERIMENT

These workers prepared two lots of the same phage type. One was grown in medium containing S^{35} as sulfate, which labels protein and not nucleic acid. The other lot was grown in P^{32}-labeled phosphate medium, which labels the nucleic acid and not the protein. With the phage labeled with P^{32} in the nucleic acid, they found that after infection the label was in the interior of the bacterial cell. The P^{32} came out when the cells were opened, showing that the nucleic acid had penetrated into the cell. With the S^{35}-labeled phage, the S^{35} component remained attached to the exterior cell wall after infection. This showed that on infection the nucleic acid and protein components of this virus separate, with the nucleic acid penetrating into the interior of the cell.

This was an important result, for up to that time virtually all thinking in the field of virology had assumed that viral agents infected cells much as do larger parasites. It was assumed that the whole virus entered and underwent division in the usual biological manner. Hershey's experiment demonstrated that there was a separation of viral components and that the nucleic acid was probably the essential material. After infection he could even shear off most of the protein coat (S^{35}-labeled) left on the outside of the cell, without modifying the course of the infection. This experiment also provided very strong evidence that in viruses, as had been found for bacteria, the nucleic acid is the essential hereditary material.

These results of Hershey and Chase explained Doermann's inability to find infectious phage inside the cell immediately after infection for there was no whole phage inside. It also resulted in a new view of the mechanism of infection, for the phage particle behaved like a biological syringe containing infectious nucleic acid which was injected into the susceptible host. This suggested to some workers that perhaps genetic transformation and viral infection had more in common than was first supposed. This similarity received further support when it was found that the nucleic acid from some animal, plant, and bacterial viruses is infectious without any protein component, even though the frequency of infection is somewhat lower.

RNA VIRUSES

Only DNA-containing viruses have been discussed so far, yet many viruses have RNA, ribonucleic acid, as their carrier of hereditary information. This suggests, among other things, that the storage of hereditary information is not restricted to one molecular system. It will also be found that whereas the DNA viruses can, under certain circumstances, integrate with the genome of the cell and undergo temporary suppression of its viral action, no RNA virus is known which undergoes a comparable change.

METABOLIC EFFECTS OF INFECTION

The intracellular metabolic changes during virus infection have been carefully examined for certain bacterial viruses. The first change to occur in T_2-phage-infected *Escherichia coli* is the synthesis of a small amount of RNA, then some protein followed by DNA, and then another protein. The second protein has the immunological properties of the virus, whereas the earlier protein may not. It has been found that much of the early protein includes enzymes necessary to synthesize certain viral components, some of which the uninfected host cell is incapable of making. The bacterial host's genome is broken down during infection by this virulent virus and all the new DNA synthesized is viral DNA. This is known because the T_2 viral DNA has a slightly different pyramidine which can be easily assayed. By some unknown mechanism, this new phage DNA is wrapped up in the protein coat and begins to appear inside the cell in the last half of the infectious cycle, which is only about 23 minutes. At the end of the cycle, the cell bursts, liberating about 100 fully infectious particles. Under special conditions of the inhibition of lysis, the average yield can be raised to 500 per cell. RNA phages have recently been discovered. Some of these are much smaller and the yield per cell of these viruses may reach 10,000 per cell.

The DNA of T_2 phage of *E. coli* is one continuous thread of 120,000,000 molecular weight. By comparison, the molecular weight of the RNA of polio virus is about 2,000,000. A more detailed discussion of bacterial viruses is given by Stent (1963) and of viruses of man and animals by Stanley and Valens (1962) and Horsfall and Tamm (1965).

LYSOGENY

The bacterial viruses considered thus far infect the host cell and produce visible cytopathogenic effects. The host is nearly always destroyed by the infection. There is another class of viruses infecting bacteria which produce a similar result in some of the cells but fail to produce this effect in the others. These are known as temperate viruses. Soon after infection by one of these viruses, a "decision" is made, determined by conditions within the cell that are not thoroughly understood, as to the direction the infection will take. If the viral genome is suppressed, the cell survives. Otherwise the virus causes a full infection with the usual lysis of the cell and the release of active infectious whole virus. Cells in which the virus reproduction is arrested are called lysogenic. Although they contain no infectious virus particles, at some later period the suppressed activity can be released (induced) to continue the infection. Inhibition of protein synthesis tends to suppress the infection and send it into the lysogenic phase. This is a close parasitic relationship. Phage DNA becomes intimately associated with the host cell's genome. Every time the cell duplicates, the viral factor, called prophage or provirus, also duplicates. So every daughter cell carries the potential for producing virus. The prophage behaves very much like a cellular gene except that under certain conditions it can revert to its virulent state and replicate as a virus. The prophage even passes along with the bacterial genome into the spore phase of growth. Experiments by deJong, confirmed by Northrop, showed that spores carrying prophage do not lose it on boiling. Since the whole virus is readily destroyed by boiling, it is clear that the prophage is a much more stable material than the whole virus.

In a large population of lysogenic cells, the prophage in a few cells may become induced to an active state. Such populations will have a little virus present most of the time. Studies of the conditions leading to activation of the prophage suggests that any damage to the host's genome interferes with the mechanism by which the prophage is suppressed. The means of inducing prophage are varied and not always unrelated. They include radiation by ultraviolet light, alkylating agents, imbalances in nutrients including salts. No one lysogenic system responds to all these methods of induction, but each of them is known to affect one type of prophage. In the case of lysogenic *Escherichia coli* K12 (λ) virtually every cell in a population is induced by a small dose of UV radiation. Some cells may carry several different proviruses concurrently.

The close parasitic relationship of the provirus DNA with the host genome also renders such a cell immune to infection by this virus and by most of the closely related viruses. This is a new kind of immunity, different from the humoral immunity seen in man and other vertebrates. It is immunity at the cellular level. This may have important significance for higher animals, but there are few data concerning this.

A case of lysogeny that will have greater interest for health officers is seen in the diphtheria organism. It was discovered by Freeman (1951) that avirulent diphtheria could be converted to virulent by the introduction of a particular temperate phage. Subsequent studies, for instance, by Groman (1955) have associated the development of toxin with the lysogenic cells. Recently Rajadhyaksha and Srinivasa (1965) have reported that the toxigenic factor is an independent cytoplasmic factor T^t transduced into the host by a defective temperate phage. They were able to destroy the infection with acriflavine without loss of toxin production. These cells are then susceptible to reinfection. Certain streptococcal and staphylococcal toxins are also reported to be associated with lysogenic infections. Possible lysogenic systems in mammalian tissues are discussed by Habel (1963).

CONJUGATION IN BACTERIA

Lederberg and Tatum (1946) mixed two strains of *Escherichia coli* that had different characteristics and after a short time they isolated bacteria with properties of both strains. Although the numbers were small, this result could not have occurred by chance. Lederberg received the Nobel Prize for this and other work.

Hayes in England discovered that there was a unidirectional exchange of genetic properties in these bacteria of Lederberg's. One line of cells lost the properties and the other gained them. This was confirmed by Skaar and Garen (1955), who labeled DNA of the donor cell with isotopes and showed that it was transferred to the recipient cells. This capacity to act as a donor was found to be a specific factor which, strangely, was found to be infectious. The factor could convert a number of recipient F^- cells into F^+ donor cells (Hayes, 1964).

The work has progressed so far that cultures of F^- and F^+ can be mixed and in a matter of moments electron microscope pictures reveal that every F^- cell is paired with an F^+ cell, identified by several morphological characteristics. A bridge develops between the two and very soon genetic material begins to pass in an ordered fashion from the F^+ to F^- cells. The union can be interrupted at any time by strong shearing forces, such as are obtained by violent stirring. As the time elapsing before stirring is increased, more and more of the genetic characteristics of the F^+ cell pass over to the F^- cell. By correlating the time of interruption with transfer of genetic properties, it has been possible to map the relative positions of many of the genes. It was very gratifying to find that the order of genes observed by this method agreed with that obtained by other procedures.

This is a very important discovery, for it shows that the mixing of genes by sexual recombination is a property of even unicellular units and is not restricted to higher organisms.

PHOTOREACTIVATION

A phenomenon of some interest in the present story and of possible importance to public health was discovered about 15 years ago. Bacteria and bacteriophage inactivated by exposure to ultraviolet light were revived when the phage-infected or uninfected bacteria were exposed to visible light. No such restoration occurred if the bacteria were kept in the dark. Since illumination of the phage alone had no reactivating action, it was clear that something in the cell was essential, and an enzyme responsible for this restoration was isolated. The enzyme and light repair the UV damage to transforming DNA, which is a simpler system than whole virus or a cell. This established that the primary target of the UV light is DNA and that the site of action of the photoreactivating enzyme is DNA (Rupert et al., 1958).

Mutations, and the induction of lysogenic cells by UV light can be reversed if the cells are exposed to visible light soon after the irradiation and before the cells have synthesized new DNA. The region of the spectrum most effective in restoring the biological properties of DNA is close to 3,600 angstroms. This subject is reviewed by Jagger (1958) and Rupert (1961).

INFECTIOUS VIRAL NUCLEIC ACIDS

In 1956 it was discovered that viral nucleic acids free of viral protein would infect host cells or tissues, but at a lower rate than whole virus. The result suggested that the protein may serve either as a protection for the nucleic acid against the insults of the environment (especially enzymatic ones) or it may have specialized functions for increasing the chances of infection such as an injection mechanism, as seen in bacteriophages. A great variety of human, animal, plant, and bacterial viral nucleic acids was soon found to be infectious. Included were poliomyelitis, encephalitis virus, and tumor viruses. The free nucleic acids have properties which make them particularly interesting to consider in connection with certain epidemiological and latency problems (Herriott, 1961, 1969). The first property is that nucleic acids are neither antigenic nor are they affected by antibodies to the whole virus. Infection by viral nucleic acid proceeds quite effectively in the presence of antibodies which would thoroughly neutralize hundreds of times the amount if it were whole virus. Thus antibody level of serum does not necessarily measure immunity to infection. Next, the broader host range of nucleic acid infectivity as compared with that of the homologous virus is of considerable interest. Rabbits or rats are susceptible to polio viral nucleic acid whereas the whole virus produces no infection at all. It should be noted that the infection produced by viral nucleic acid may stop after only one cycle of virus reproduction, since whole virus is the main product of the infection. Whole virus will not in this case carry the infection further. Nevertheless this one round of infection may have important possibilities as an intermediate reservoir, and this has not been given much consideration in the past. The viral nucleic acids are very vulnerable to the action of the enzymes, nucleases, which are present in many tissue or cellular fluids. Suppression of these nucleases through the use of high salt concentrations is often necessary in order to demonstrate infection by low levels of nucleic acid that would otherwise be destroyed before infection could occur. The infectious nucleic acids are more resistant than their parent viruses to many agents or treatments which destroy infectivity. The infectivity of some double-stranded viral DNAs remains after boiling or exposure to strong phenol, formaldehyde, or iodine. These properties suggest that some hygienic procedures for decontamination of infected areas may need reexamination.

CONCLUSIONS

In certain viruses, in lysogenic bacteria and in genetic transformation, the essential material is DNA, the carrier of hereditary information. No cell of any kind is known which does not use DNA as a coded blueprint of its hereditary structure. This is a startling fact. It hardly seems possible that there are enough ways by which the DNA can vary to permit so vast an amount of information to be stored in it. The next chapter will consider the probable manner by which this marvelous feat is accomplished.

REFERENCES

Avery, O. T., MacLeod, C. M., and McCarty, M. 1944. Studies on the chemical nature of the substance inducing transformation of pneumococcal types. *J. Exptl. Med.* **79:** 137–159.

Burnet, F. M., and Lind, P. E. 1951. A genetic approach to variation in influenza viruses. 3. Recombination of characters in influenza virus strains used in mixed infections. *J. Gen. Microbiol.* **5:** 59–60.

Freeman, V. J. 1951. Studies on the virulence of bacteriophage-infected strains of *Corynebacterium diphtheriae. J. Bacteriol.* **61:** 675–688.

Galasso, G. J., and Sharp, D. G. 1963. Homologous inhibition, toxicity and multiplicity reactivation with ultraviolet-irradiated *Vaccinia* virus. *J. Bacteriol.* **85:** 1309–1314.

Gotlieb, T., and Hirst, G. K. 1956. The experimental production of combination forms of virus. VI. Reactivation of influenza viruses after inactivation by ultraviolet light. *Virology* **2:** 235–248.

Groman, N. B. 1955. Evidence for the active role of bacteriophage in the conversion of nontoxigenic *Corynebacterium diphtheriae* to toxin production. *J. Bacteriol.* **69:** 9–15.

Habel, K. 1963. Malignant transformation by polyoma virus. *Ann. Rev. Microbiol.* **17:** 167–178.

Hayes, W. 1964. *The Genetics of Bacteria and Their Viruses*, Wiley, New York.

Herriott, R. M. 1961. Infectious nucleic acids, a new dimension in virology. *Science* **134:** 256–260.

Herriott, R. M. 1969. Implications of infectious nucleic acids in disease. In *Progress in Medical Virology*, Vol. 2. J. L. Melnick (ed.), Karger, Basel.

Hershey, A. D., and Chase, M. 1953. Independent functions of viral protein and nucleic acid in growth of bacteriophage. *J. Gen. Physiol.* **36:** 39–56.

Horsfall, J., and Tamm, I. (eds.) 1965. *Viral and Rickettsial Infections of Man.* 4th ed. Lippincott, Philadelphia, Penn.

Jagger, J. 1958. Photoreactivation. *Bacteriol. Rev.* **22:** 99–142.

Lederberg, J., and Tatum, E. L. 1946. Novel genotypes in mixed cultures of biochemical mutants of bacteria. In *Cold Spring Harbor Symp. Quant. Biol.* **11**: 113–114.

Luria, S. E. 1947. Reactivation of irradiated bacteriophage by transfer of self-reproducing units. *Proc. Natl. Acad. Sci. U.S.* **33**: 253–264.

Rajadhyaksha, A. B., and Srinivasa, R. S. 1965. The role of phage in the transduction of the toxinogenic factor in *Corynebacterium diphtheriae. J. Gen. Microbiol.* **40**: 421–429.

Ravin, A. W. 1961. The genetics of transformation. *Advan. Genet.* **10**: 61–163.

Rupert, C. S. 1961. Repair of ultraviolet damage in cellular DNA. *J. Cellular Comp. Physiol.*, Suppl. 1, **58**: 57–68.

Rupert, C. S., Goodgal, S. H., and Herriott, R. M. 1958. Photoreactivation in vitro of ultraviolet inactivated *Hemophilus influenzae* transforming factor. *J. Gen. Physiol.* **41**: 451–471.

Schaeffer, P. 1964. Transformation. In *The Bacteria*, Vol. 5. I. C. Gunsalus and R. Y. Stanier (eds.). Academic Press, New York.

Skaar, D., and Garen, A. 1955. (abstr.) Transfer of DNA accompanying genetic recombination in *E. coli* K-12. *Genetics* **40**: 596.

Stanley, W. M., and Valens, E. G. 1962. *Viruses and the Nature of Life.* Dutton, New York.

Stent, G. S. 1963. *Molecular Biology of Bacterial Viruses.* Freeman, San Francisco, Calif.

Szybalska, E. H., and Szybalski, W. 1962. Genetics of human cell lines. IV. DNA-mediated heritable transformation of a biochemical trait. *Proc. Natl. Acad. Sci. U.S.* **48**: 2026–2034.

CHAPTER 11

The Biochemistry of Some Genetic Processes

ROGER M. HERRIOTT

The material presented in the previous chapter indicated that DNA is the carrier of hereditary properties, but there was no clue as to how biological properties are derived from it. In addition, as late as 1950 there were no well conceived answers to any of the following fundamental biological problems:
the molecular structure of DNA (deoxyribonucleic acid)
the manner of storing or coding hereditary information in molecular or structural elements
the manner of faithfully duplicating DNA
the function of RNA
the mechanism of protein synthesis
    In the decade from 1953 to 1963 evidence suggesting answers to most of the above problems was described. Many important aspects are not yet totally understood, but the basic outlines are probably correct. In the material to follow, the key discoveries and the ideas leading to the present state of understanding in this field will be reviewed (Hayes, 1964; Watson, 1965).

THE COMPOSITION OF RNA AND DNA

Ribonucleic acid (RNA) is mainly but not exclusively a constituent of the cytoplasm. It is a polyribose phosphate polymer, with the nitrogenous bases adenine (A), guanine (G), cytosine (C), and uracil (U) attached to the end carbon of each ribose (Table 11.1). DNA is found predominantly in the nucleus. It is polydeoxyribose phosphate, with the bases adenine (A), guanine (G), cytosine (C), and thymine (T) attached to the end carbon of the deoxyribose sugar. One of the chemical differences between the two nucleic acids is the methyl group of thymine in DNA that is missing in uracil of RNA. Also, the sugar of DNA has a hydrogen on the second carbon, whereas in RNA there is a hydroxyl group. This hydroxyl group is responsible for the relative instability of RNA in alkaline solutions.
    Before 1946 the components, including the bases of the two types of nucleic acid, seemed to be constant and were present in very nearly equal quantities. There was, therefore, no suggestion of the manner by which biological variations were registered. The nucleic acids were known to be of high molecular weight but little else was known.

THE WATSON-CRICK MODEL OF DNA

An important discovery was made by Chargaff and his associates (1951), though its significance could not then be appreciated. Careful analyses of the

138

TABLE 11.1

Comparison of Components of RNA and DNA

	RNA	DNA	Base formula
BASES			
Purines	Adenine	Adenine	
	Guanine	Guanine	
Pyrimidines	Cytosine	Cytosine	
	Uracil	—	
	—	Thymine	
SUGAR	Ribose	2-deoxyribose	
PHOSPHATE	PO^-_3	PO_3	

bases of highly purified DNAs from different species showed consistent differences in the composition. But in all cases after hydrolysis the numbers of molecules of A were equal to the numbers of molecules of T, and also the molecules of G were equal to those of C. The A/G or T/C ratio varied with

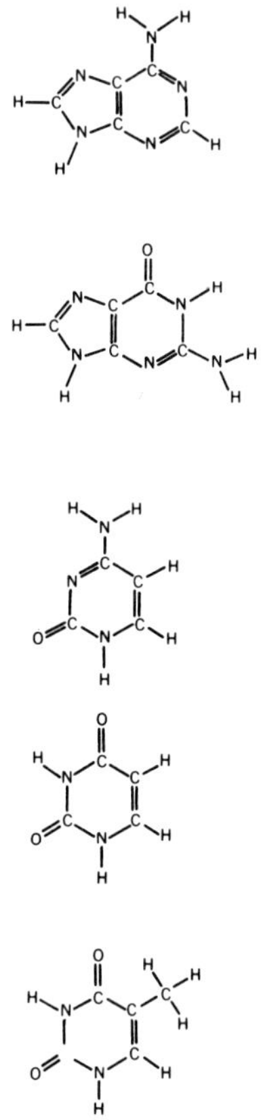

Figure 11.1. Base pairing in DNA showing complementary sequences in the two strands of the double helix.

the source of the DNA, but

$$\frac{A}{T}=\frac{G}{C}=1.$$

The next big step toward understanding the structure of DNA was Wilkins' X-ray picture of DNA (Wilkins *et al.*, 1953), which showed it to be a double-stranded helix. Crick and Watson (1953), who had been contemplating the nature of the structure, saw the relationship between Chargaff's and Wilkins's results. They saw that if A were in one strand of the helix and T were opposite in the other strand, and similarly G were opposite C, then this would allow A

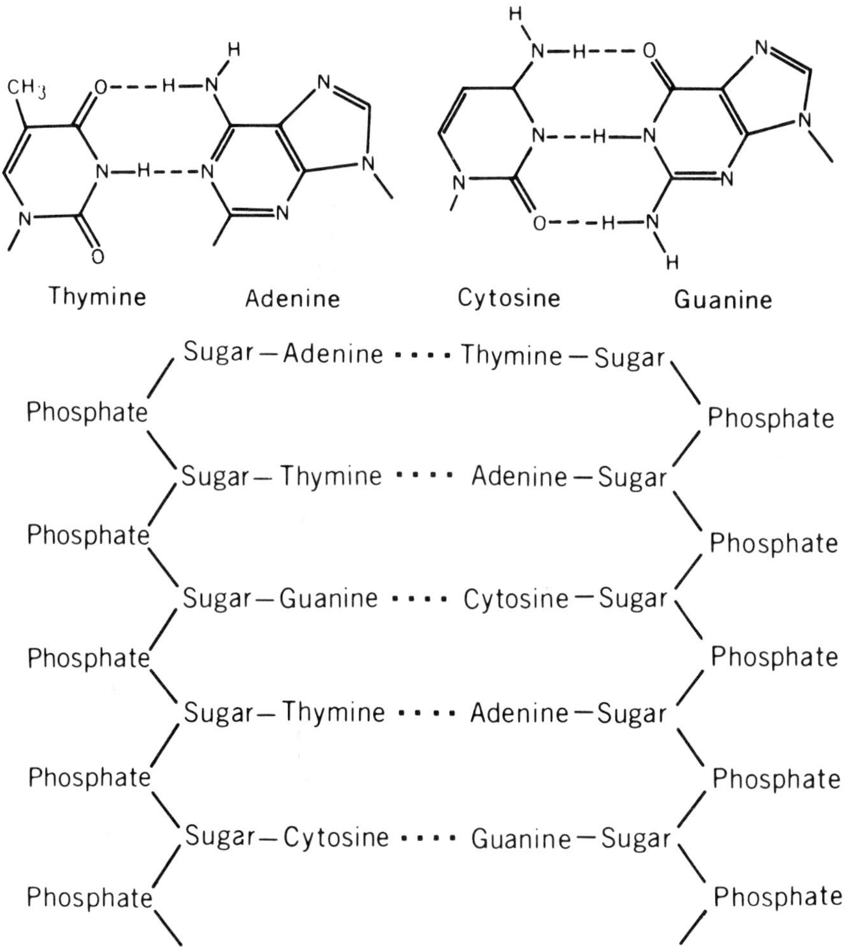

Figure 11.2. A schematic diagram of the DNA molecule showing the base pairing and the helical structure. C, G, T, A are the bases cytosine, guanine, thymine, and adenine; S is sugar; P is phosphate; – – – are hydrogen bonds. (After William D. McElroy, *Cell Physiology and Biochemistry*, 2nd. ed., ©1964. By permission of Prentice-Hall, Inc., Englewood Cliffs, N.J.)

to vary relative to G and C but not to T. In the same way G could equal C with no clear relationship to A or T.

Watson and Crick found that the bond angle geometry of the pairing structure of the four bases permitted A to pair with T and G with C (Figure 11.1). Other combinations were not excluded on this basis, but these were favored.

In this 1953 model of DNA, Watson and Crick proposed that the bases project inward at right angles to the length of the molecule and pair with a complementary base partner in the other strand (Figure 11.2). The two bases were thought to be held together by hydrogen bonds, which, although relatively weak, nonetheless could provide considerable stability to the molecule over its great length. These workers recognized that the number of bases could vary along the length of the strands and permit a coding of the genetic information.

The idea of coding all possible genetic variables by means of a 4-unit code may seem incredible. But the Morse Code is made up merely of dots, dashes, and spaces. This simple code can translate anything that can be written. Any language can be coded on such a simple 3- or 4-unit code.

In the above model of DNA the strands are complementary and not identical. Watson and Crick visualized the replication of DNA beginning by a separation of strands which involved only the breaking of the weak hydrogen bonds and then each strand serving as a template. Complementary bases were attracted into position and linked to form a new complementary strand by some biological process, probably enzymatic. Any error in base pairing would provide a basis for mutations.

In a very brief time, Watson and Crick had suggested answers to several of the problems listed earlier. Their model has withstood nearly 16 years of critical experimental testing.

Thus DNA is seen to consist of two extremely long deoxyribose phosphate polymeric strands with paired bases holding the strands together like steps in a spiral stairway. The arrangement of the bases along the strands determines the coded message. The two strands are complementary, not equal, so only one of them must carry the exact message. But either strand can generate the other, a feature which conforms to the knowledge of biological growth of bacteria by a geometric progression from 1 to 2 to 4 to 8.

DENATURATION AND RENATURATION OF DNA

The separation of strands would appear to be essential to replication, and denaturation brings about separation of strands. Renaturation is a reversal of the procedure.

Any reaction which depends on the rupturing of many chemical bonds of equal strength will exhibit a sharp temperature transition. This is seen in the melting of pure solids, such as ice. It is also true of DNA dissolved in aqueous solutions. The bonds broken are the hydrogen bonds holding the strands together, not the sugar phosphate bonds. The biological activity of viral or

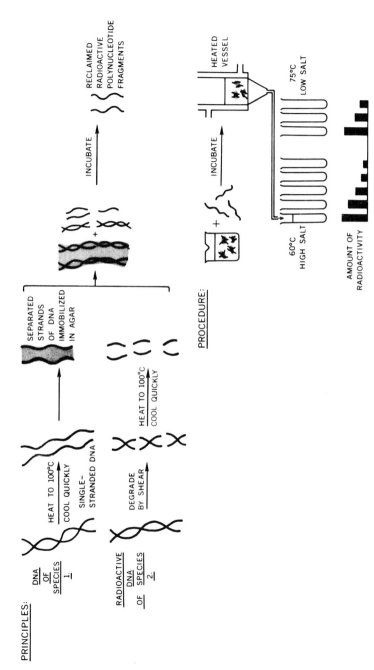

Figure 11.3. Complementary pairing in DNA molecules, used to evaluate the relatedness of species. Denatured DNA of one species is fixed in agar beads and denatured radioactive DNA of another species is passed over the beads in high salt solution at 60°C. The radio-activity of the effluent wash measures the amount of DNA that did not pair. The temperature is changed to 75°C, and a solution of lower salt content is passed through. This releases the paired, homologous or similar, radioactive DNA. The radioactivity of this effluent measures the amount of DNA that did pair. (After Hoyer et al., 1964.)

transforming DNA dissolved in dilute salt solutions is lost abruptly at specific temperatures between 60 and 95°C that rupture the bonds. The exact temperature depends on the salt concentration and the base composition of the DNA. It has been shown that, in general, after the hydrogen bonds are ruptured, the strands of the double helix untwist and separate. Marmur and Lane (1960) discovered that if the denatured inactive DNA was cooled slowly from 100°C, some of the biological activity was recovered. This functionally good DNA was described as renatured DNA. Marmur proposed that during the slow cooling the strands of DNA search one another until they find a partner with exact complementary base pairing. After this the attractive powers are greater than the diffusion forces that tend to separate them, and they twist together, forming a native, stable molecule.

Marmur *et al.* (1963), Hoyer, McCarthy, and Bolton (1964) and others have used this phenomenon of molecular recognition of complementary partners during renaturation to evaluate the relatedness of species. Denatured DNA from one species fixed in agar beads is placed in a chromatographic column. Radioactive denatured DNA from another species is passed through the column at about 70°C, where renaturation by complementary pairing is optimal. The unpaired DNA goes on through the column and the amount is measured by radioactivity (Figure 11.3). The method is not very precise. However, some strong inferences about evolution were obtained. Hoyer and his associates found that bacterial DNA was not retained by human DNA, but DNA from apes was strongly retained, rat DNA was moderately held, and fish only slightly. They also noted that there is a linear relationship between the logarithm of the percentage of DNA retained and the time in millions of years since species divergence occurred. This suggests that there is some underlying principle, probably of great importance.

SYNTHESIS OF DNA *IN VITRO*

Although synthesis *in vitro* of biologically functional DNA has not been established, the work of Kornberg and his colleagues has gone far to indicate many of the essential features of the process. Thus, they can obtain a synthesis of DNA, but for some unclear reason the new material is not functional, despite its having a base composition nearly identical to the natural active DNA. The reason may be trivial in relation to the major processes. Kornberg (1959) showed that for *in vitro* synthesis of high molecular weight material with the properties of DNA, it was necessary to include in the reaction mixture all four deoxynucleotide bases as the triphosphates, a specific enzyme which was purified from *Escherichia coli*, and also some DNA.

The triphosphates are "energy rich" just as adenosine triphosphate, ATP, found in muscle and cells is energy rich and provides the necessary energy for work or synthesis of most cellular substances. The need for added DNA in the synthesizing mixture had not been anticipated. It was discovered only when purified enzyme failed to bring about the synthesis obtained with a crude enzyme (polymerase) preparation. The DNA was needed as a template, and

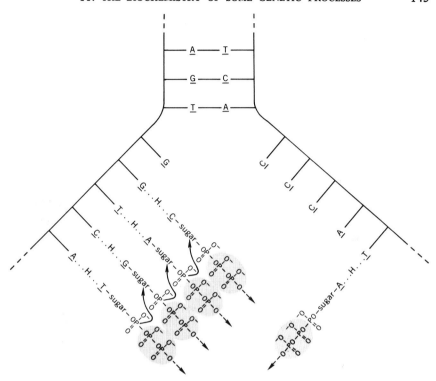

Figure 11.4. Synthesis of DNA. The DNA strands are separated at one end by some un-known mechanism. Nucleoside triphosphates

$$
\text{(Base–Sugar–O}\overset{\displaystyle O^-}{\underset{\displaystyle O}{\overset{|}{\underset{||}{P}}}}\text{–O}\overset{\displaystyle O^-}{\underset{\displaystyle O}{\overset{|}{\underset{||}{P}}}}\text{–O}\overset{\displaystyle O^-}{\underset{\displaystyle O}{\overset{|}{\underset{||}{P}}}}\text{–O–)}
$$

are then attracted and held to the complementary bases in the strand by hydrogen bond formation. Enzyme (polymerase) joins the primary phosphate of one nucleotide to the sugar of the next nucleotide and simultaneously splits out pyrophosphate

$$
\text{(O}\overset{\displaystyle O^-}{\underset{\displaystyle O}{\overset{|}{\underset{||}{P}}}}\text{–O}\overset{\displaystyle O^-}{\underset{\displaystyle O}{\overset{|}{\underset{||}{P}}}}\text{–O–).}
$$

This results in a new strand being formed that is an exact complement of the parent strand on which it was formed and an exact duplicate of the other parental strand.

the newly synthesized DNA resembled the template. This confirmed that the synthesis was in fact a copying process. The mechanism of synthesis is indi-cated in Figure 11.4. An enzyme or some force probably separates the strands and the triphosphate-nucleosides are attracted to their complementary bases in the separated strands. The paired bases in the new strand are then linked as the separation proceeds. Pyrophosphate is the by-product, and two double stranded units, identical with the original template DNA, are produced as predicted by Crick and Watson.

IN VIVO SYNTHESIS

Meselson and Stahl (1958) proved that the synthesis of DNA *in vivo* involves the separation of parental strands so that each daughter molecule has one new and one old strand. They grew cells in a medium in which the nitrogenous components contained the heavy isotope of nitrogen, N^{15}, rather than the common N^{14}. This made the materials, including both strands of the DNA, slightly heavier (more dense). The cells were then allowed to grow in N^{14} medium through just one or two cell divisions. The DNA was harvested and centrifuged in a cesium chloride gradient. By this method, heavy molecules of DNA, all containing N^{15}, would form a band at the bottom of the tube. Light molecules, all containing N^{14}, would form a band higher in the tube. Molecules with some N^{15} and some N^{14} would band in between the other two. If replication occurred without separation of the parent strands of DNA, there would be no middle layer. There would be only heavy parent molecules and light, newly formed molecules. However, most of the molecules after the first generation were found to be in the middle, that is, to have both N^{14} and N^{15}, showing that duplication of DNA was indeed based on separation of the parent strands (Figure 11.5).

RNA AND PROTEIN SYNTHESIS

As yet there has been little in this account to indicate the function of RNA. It had been known for many years that some viruses contained RNA, so that it was clear that RNA as well as DNA might be able to carry coded information. But this provided no clue about the function of RNA in normal cells. It had also been recognized that RNA synthesis paralleled protein synthesis, but then so did the synthesis of many cellular components.

It is now known that there are at least three different classes of RNA in cells. One is present in a low concentration, as just a few percent of the total cellular RNA, and it has a high molecular weight of 1,000,000 to 2,000,000. The second class of RNA is of low molecular weight of about 30,000, and is about 10 percent of the total. The third class is found in the small subcellular bodies having sedimentation constants between 30 S and 80 S. These small bodies are called ribosomes because they are rich in RNA. They carry nearly 85 percent of the RNA of some cells, yet the function of this ribosomal RNA is not clear even today.

With this brief bit of hindsight, the experiments which suggested the function of the other two RNAs will be discussed.

Hershey (1953) noted that radioactive P^{32} phosphate, introduced briefly into *Escherichia coli* growth medium just as the cells were infected with phage, became incorporated in a small quantity of RNA. Volkin and Astrachan (1956) isolated this labeled RNA and found its base composition was quite

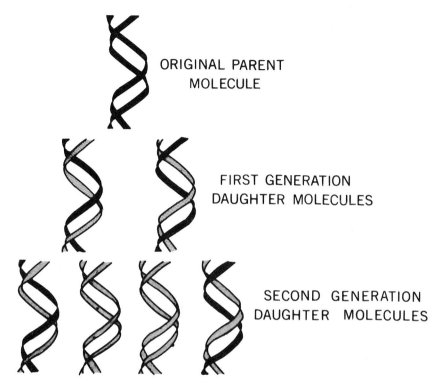

ORIGINAL PARENT
MOLECULE

FIRST GENERATION
DAUGHTER MOLECULES

SECOND GENERATION
DAUGHTER MOLECULES

Figure 11.5. Scheme of *in vivo* multiplication of DNA. Parent cells were grown in a medium containing the heavy isotope of nitrogen, N^{15}, and then in an N^{14} medium through two cell divisions. DNA strands marked with N^{15} are represented by heavy black lines.

different from that of the bulk of the RNA. It did not resemble the composition of the bacterial DNA but its composition showed some resemblance to that of the phage DNA. This suggested that it might have been copied from the phage DNA. Nomura *et al.* (1960) found that this newly formed RNA soon became associated with the ribosomes of the cytoplasm, and Jacob and Monod (1961) suggested calling this RNA "Messenger RNA." They visualized it as carrying the genetic message from the nucleus where this RNA is formed, into the cytoplasm, where this RNA is usually found, and where proteins are synthesized.

Studies with radioactively labeled amino acids showed that soon after they entered the cell they became associated with the second class of RNA, discussed earlier. This RNA has a low molecular weight. Each amino acid was bound to a slightly different (specific) RNA, but in each instance the binding was the same. It was an ester bond involving the carboxyl group of the amino acid and the terminal adenosine of the RNA molecule. Formation of each of the RNA-amino acid-complexes was found to be catalyzed by a separate enzyme, all of which have now been separated and purified. Each of these small RNAs is different. Each reacts specifically with but one amino acid, and also with specific sites on the messenger RNA. Soon after the amino acid-RNA complexes were added to *in vitro* protein synthesizing systems, the RNAs were found free in solution. Their function was considered to be one of transferring the amino acid from solution to a specific position on the messenger, and so they were designated "Transfer RNAs". The name "soluble" or "S" RNA is also used. The material discussed thus far is reviewed in Figure 11.6.

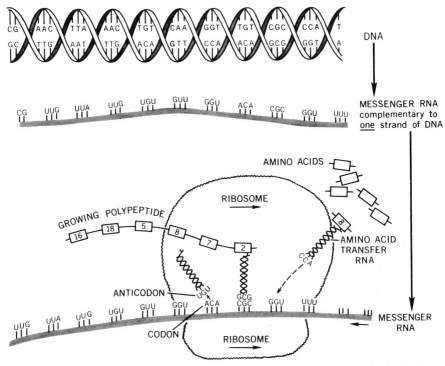

Figure 11.6. A scheme illustrating the role of DNA and RNA in protein synthesis. (Partly based on William D. McElroy, *Cell Physiology and Biochemistry*, 2nd. ed., ©1964. By permission of Prentice-Hall, Inc., Englewood Cliffs, N.J.)

In brief, DNA, the genetic material, is a coded message denoting the arrangement of amino acids in proteins. This information is first passed to the messenger RNA which is generated in the nucleus as a complementary copy of one of the DNA strands. Soon after its release from DNA, the messenger RNA is found in the cytoplasm. There the transfer RNAs with their amino acids and specific "anticodon" regions can interact (pair) with the corresponding "codons" of the messenger RNA. This places the acyl-amino acids in the correct sequence. Then by some mechanism, probably enzymatic, the amino acids are "zipped" together (linked) to form a long polypeptide or protein. Evidence is beginning to accumulate (Perutz, 1962) to show that the sequence of amino acids determines how the long polypeptide strand will fold up spontaneously into a uniformly specific shape. There are apparently so few alternate ways of folding, that the folding becomes relatively specific. This specific folding determines the distance between certain chemical structures of protein which in turn determine the nature of its enzymatic properties.

No substances other than proteins are coded by DNA, except the replicated RNA and DNA. That protein is the main product of genes may seem surprising. But all enzymes are proteins, and enzymes catalyze virtually all biological reactions. The rate of catalysis of the various cellular biochemical reactions will decide the quantity of substances formed or degraded in any cell. In this way virtually any biological variation can be visualized and thus the accumulated information is explained.

This has been a brief view of how DNA determines the properties of cells by dictating the proteins that are formed. Most of the proteins are enzymes that influence the various biochemical reactions of the cell. The basis of heredity rests on the *reliability* of the specific biochemical steps involved in copying the parental DNA and the transfer of the sequenced information to messenger RNA and thence to protein.

Not all genes are equally active all the time but are under some control. This cannot be dealt with in detail here. But it is worth noting that one or more mechanisms must exist to turn on and off the gene functions, these mechanisms being activated by environmental or temporal conditions. In higher organisms it is clear that since all the cells do not perform the same functions yet contain equal hereditary information, DNA, some of the information is suppressed. Liver cells always produce liver cells and not skin or kidney cells even though their DNA contains the information to make the proteins of all types of cells. During embryonic differentiation much if not most of the total potential of the hereditary material becomes suppressed. Each cell type is suppressed in a very specific manner. Some of the suppression is of a reversible type but most of it is probably irreversible. The reversible suppression permits regulation of cell activity by such signals as hormones, metabolic imbalances, or feedback mechanisms.

Interest in RNA viral infections of cells has recently been heightened by some beautiful experiments of Spiegelman and his associates. In a lucid account Spiegelman (1967) describes the steps and proof of the first *in vitro* synthesis of an infectious agent using the four ribonucleoside triphosphates, a little viral RNA as a primer, and some highly purified and specific RNA polymerase. This enzyme, isolated from infected cells, is very specific for the synthesis of each viral RNA. Spiegelman goes to some pains to demonstrate that it is the viral RNA, and not the enzyme, that carries the blueprint of the virus, and is the self-duplicating unit. That the infectivity in the synthesized product was not due to residual traces of initial primer RNA was established by showing that the initial primer would have been diluted out beyond detection and that the observed infectivity was therefore from newly synthesized material. Spiegelman also described an experiment with evolutionary implications. When he used a mixture of equal quantities of two distinguishable but closely related viral RNAs as primer, one viral RNA was formed faster than the other. The account of this work is recommended for those interested in the molecular biologists' method of analyzing subcellular processes.

THE BIOLOGICAL CODE

The following discussion will be concerned with the sequence of discoveries which led to the solution of the biological code, that is, the identification of the triplets of deoxynucleotides which code for each of the 20–25 amino acids. Before 1960 there was no indication of the code for even one amino acid. It was generally supposed that it would be necessary to correlate the sequence of bases in a gene with the sequence of amino acids in the protein produced from the gene, in order to solve the code. However, this formidable task was not required. From an unexpected experimental result, the solution was indicated, and in less than 4 months the code was largely understood.

A code composed from only two bases would be insufficient because it permits only 16 combinations, and there are over 20 amino acids for which there must be specific coding. A 3-unit code produces 64 possibilities. All evidence today supports the notion of a 3-unit code. The sequence of three bases responsible for an amino acid is a codon.

In August, 1960, Nirenberg reported to an international meeting that he used a synthetic polymer of uridylic acid in place of messenger RNA to study the synthesis of protein in a cell free-system. He found that only a polypeptide made up exclusively of phenylalanine was produced. To the "prepared minds" in the field, this was a very revealing fact. This meant to them that the code UUU (U for uridylic acid) coded for phenylalanine. It was only necessary to introduce small quantities, one at a time, of the other 3 nucleotides during polymerization of uridylic acid. This should then yield UUA, UUG, UUC, or some variation on them which would then code for other amino acids that would appear in the polypeptides formed. This scheme could then be extended to polymerized A, polymerized G, and polymerized C. Ochoa and those in his laboratory quickly took up the study and by the end of 1960, less than 4 months after the initial discovery, Nirenberg's and Ochoa's groups had solved the basic framework of the code. Much important extension, revision, and confirmation has gone on since then. It is worth noting as an aside that this is an instance of the unpredictable way in which solutions to important problems are sometimes obtained.

The code as understood today is shown in Table 11.2. It will be seen that more than one triplet sequence (codon) of ribonucleotides codes for a given amino acid. As many as four different codons exist for some amino acids. In the parlance of the specialists, this means the code is "degenerate". These multiple codons may have important biological value. It is not unreasonable to suppose that in assembling the various amino acid-transfer RNA units

TABLE 11.2

Nucleotide Sequences of RNA Codons Recognized by Amino Acid-Transfer RNAs from Bacteria and Amphibian and Mammalian Liver[a]

Second Base

First Base		U	C	A	G	Third Base
U	U	phenylalanine	serine	tyrosine	cysteine	U
U	U	phenylalanine	serine	tyrosine	cysteine	C
U	U	leucine	serine	terminator**	cysteine	A
U	U	leucine initiator*	serine	terminator**	tryptophan	G
C	C	leucine	proline	histidine	arginine	U
C	C	leucine	proline	histidine	arginine	C
C	C	leucine	proline	glutamine	arginine	A
C	C	leucine	proline	glutamine	arginine	G
A	A	isoleucine	threonine	asparagine	serine	U
A	A	isoleucine	threonine	asparagine	serine	C
A	A	isoleucine	threonine	lysine	arginine	A
A	A	methionine, initiator*	threonine	lysine	arginine	G
G	G	valine	alanine	aspartic acid	glycine	U
G	G	valine	alanine	aspartic acid	glycine	C
G	G	valine	alanine	glutamic acid	glycine	A
G	G	valine	alanine	glutamic acid	glycine	G

[a] After Nirenberg et al. (1966).
* Marks the beginning of a peptide (protein).
** Marks the termination of a peptide (protein).

together in the formation of protein, that certain combinations interact with difficulty. With several different units available carrying the same amino acid, such difficulty may be circumvented.

It has been possible to spot-check the validity of the code by observing amino acid differences in proteins of natural occurring and induced mutants. In mutant human hemoglobin (Ingram, 1957), in tobacco mosaic virus protein, and in mutants of *Escherichia coli* having different tryptophan synthetase, it has been seen that the change of but one nucleotide in the triplet codon in each instance would lead to the codon responsible for the amino acid in the mutant. This suggested that the code is universal. The DNA bases and the amino acids in all cells are, in general, the same, so it is perhaps not surprising that the code might be the same.

Three advances of unusual interest have appeared since this chapter was written. They can be discussed only briefly. The first is evidence that some multi-unit biological structures assemble automatically when the units are mixed together. This has been reported for tobacco mosaic virus by Lauffer (1966), for T_4 bacteriophage by Wood and Edgar (1967), and a beginning has been made with mitochondrial membrane components by Racker (1968).

The second advance is the test tube (*in vitro*) synthesis of infectious Q bacteriophage RNA by Spiegelman (1967). This required the four ribonucleoside triphosphates, an RNA polymerase isolated only from virus infected cells, and a small quantity of viral RNA as a primer or template. Spiegelman showed conclusively that a net increase in infectious RNA was produced. The mechanism of synthesis appears to be different from that described in this chapter for cellular DNA. Viral RNA is single-stranded and is designated as $+$. After it invades a cell a complementary strand ($-$) is formed, making a double-stranded replicating form (RF). Fenwick, Erikson and Franklin (1964) suggested earlier that active viral $+$ strands are synthesized on the $-$ strand with each succeeding $+$ strand pushing off the one ahead of it. (See also *Cold Spring Harbor Symp. on Quant. Biol.* **33,** 1969.)

The third and most recent advance is the *in vitro* synthesis of infectious viral DNA reported by Kornberg (1968). The DNA of bacteriophage ϕ X174 is circular as well as single-stranded, and completion of the synthesis required the ring-closing enzyme, ligase, as well as the polymerase. The evidence that the newly synthesized DNA was infectious was unequivocal.

CONCLUSIONS

The basic biological questions posed earlier in this discussion have been answered in part. The 1950's saw an enormous expansion of the understanding of hereditary mechanisms. Seldom have so many of nature's well kept secrets been laid bare in so short a period. C. P. Snow noted that the impact of the recent developments in the field of heredity on man's understanding of himself may be greater than the impact of Darwinism and the discovery of evolution.

For those contemplating the intriguing unsolved mysteries of cellular differentiation, malignancy, mental processes, and the properties of membranes, there is much to be learned from closer examination of the manner and circumstances that led to the rapid understanding of genetics and protein synthesis.

REFERENCES

Chargaff, E. 1951. Structure and function of nucleic acids as cell constituents. *Federation Proc.* **10**: 654–659.

Crick, F. H. C., and Watson, J. D. 1953. A structure of deoxyribose nucleic acid. *Nature* **171**: 737–738.

Fenwick, M. L., Erikson, R. L., and Franklin, R. M. 1964. Replication of the RNA of bacteriophage R-17. *Science* **146**: 527–530.

Hayes, W. 1964. *The Genetics of Bacteria and Their Viruses.* Wiley, New York.

Hershey, A. D. 1953. Nucleic acid economy in bacteria infected with bacteriophage T₂. Phage precursor nucleic acid. *J. Gen. Physiol.* **37**: 1–23.

Hoyer, B. H., McCarthy, B. J., and Bolton, E. T. 1964. A molecular approach in the systematics of high organisms. *Science* **144**: 959–967.

Ingram, V. M. 1957. Gene mutations in human hemoglobin: The chemical difference between normal and sickle cell haemoglobin. *Nature* **180**: 326–328.

Jacob, F., and Monod, J. 1961. Genetic regulatory mechanisms in the synthesis of proteins. *J. Mol. Biol.* **3**: 318–356.

Kornberg, A. 1959. Enzymatic synthesis of deoxyribonucleic acid. Harvey Lectures 1957–1958, Ser. 53, pp. 83–112. Academic Press, New York.

Kornberg, A. 1968. The Synthesis of DNA. *Sci. Am.* **219**: 64–79.

Lauffer, M. A. 1966. Polymerization-depolymerization of tobacco mosaic virus protein. VII. A model. *Biochemistry* **5**: 2440–2446.

McElroy, W. D. 1964. *Cell Physiology and Biochemistry.* 2nd ed. Prentice-Hall, Englewood Cliffs, New Jersey.

Marmur, J., and Lane, D. 1960. Strand separation and specific recombination in deoxyribonucleic acids. Biological studies. *Proc. Natl. Acad. Sci. U.S.* **46**: 453–461.

Marmur, J., Falkow, S., and Mandel, M. 1963. New approaches to bacterial taxonomy. *Ann. Rev. Microbiol.* **17**: 329–372.

Meselson, M., and Stahl, F. 1958. The replication of DNA in *E. coli. Proc. Natl. Acad. Sci. U.S.* **44**: 671–682.

Nirenberg, M., Caskey, T., Marshall, R., Brimacombe, R., Kellogg, D., Doctor, B., Hatfield, D., Levin, J., Rottman, F., Pestka, S., Wilcox, M., and Anderson, F. 1966. The RNA code and protein synthesis. *Cold Spring Harbor Symp. Quant. Biol.* **31**: 11–24.

Nomura, M., Hall, B. D., and Spiegelman, S. 1960. Characterization of RNA synthesized in *Escherichia coli* after T₂ infection. *J. Mol. Biol.* **2**: 306–326.

Perutz, M. I. 1962. *Proteins and Nucleic Acids.* American Elsevier, New York.

Racker, E. 1968. The membrane of the mitochondrion. *Sci. Am.* **218**: 32–39.

Spiegelman, S. 1967. An *in vitro* analysis of a replicating molecule. *Am. Scientist* **55**: 221–264.

Volkin, E., and Astrachan, L. 1956. Phosphorus incorporation in *Escherichia coli* ribonucleic acid after infection with bacteriophage T₂. *Virology* **2**: 149–161.

Watson, J. D. 1965. *The Molecular Biology of the Gene.* Benjamin, New York.

Wood, W. B. Jr., and Edgar, R. S. 1967. Building a bacterial virus. *Sci. Am.* **217**: 60–75.

Wilkins, M. F. H., Stokes, A. R., and Wilson, H. R. 1953. Molecular structure of deoxypentose nucleic acid. *Nature* **171**: 738–740.

CHAPTER 12

Microbial Resistance to Drugs

BYRON S. TEPPER

Microbial resistance is the capacity of a microbial cell and its progeny to be viable and multiply under environmental conditions that would destroy or inhibit others. Microbial resistance was recognized before the discovery of antibiotic drugs in microorganisms resistant to bacteriophage, phenol, toxic metal ions, dyes, and various radiations.

Resistance poses serious problems in the treatment of infectious disease. For example, only 6 years after the sulfonamides were first used for the treatment of gonorrhea, most new cases were caused by sulfonamide-resistant gonococci and so failed to respond to sulfonamide therapy. Only 1 or 2 years after extensive clinical use of penicillin there were reports from hospitals that over half of the strains of staphylococci causing infections were resistant to penicillin. In Southeast Asia *Plasmodium falciparum*, the cause of malignant tertian malaria, is frequently resistant to all drugs including quinine. So the emergence of drug-resistant strains of microorganisms continually thwarts the designers of antimocrobial agents.

THE ORIGIN OF RESISTANCE BY MUTATION

What role, if any, do the drugs themselves have in the genesis of resistance? It is possible to distinguish between changes affecting single cells and changes affecting all the cells in a bacterial population, and also between changes that are induced by the environment and those that are selected by it. In order to answer the question given above, the following possibilities need to be investigated.

Do resistant strains arise from genetic variants that are already present and that the toxic agents select?

Do resistant strains arise by direct interaction of the toxic agent and the microorganism?

Do resistant strains arise by some more general type of adaptation?

Three methods for determining the origin of resistant strains will be described. Phage resistance, the resistance of bacteria to lysis by bacteriophage, will be used as a model. However, each method has proved equally efficient in demonstrating the origin of drug-resistant strains.

The Statistical Clone Test of Newcombe

This was one of the more direct tests used to differentiate among the three possibilities listed above. Phage-sensitive bacteria were plated on agar and the plates were incubated so that the populations grew through several genera-

154

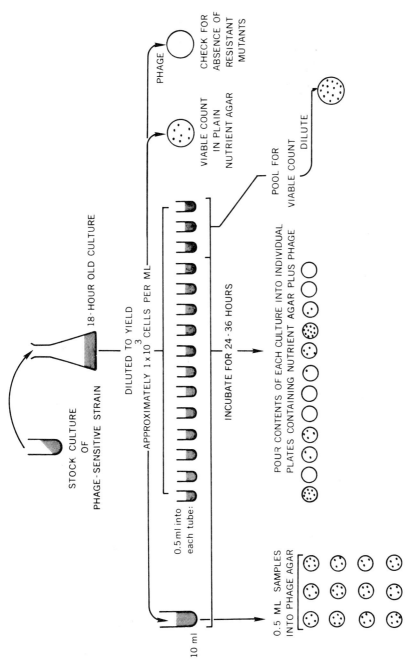

Figure 12.1. The fluctuation test. Numbers of mutants to phage resistance are much more variable in the series of plates from small tubes than in the series inoculated from one large tube. (After Braun, 1965.)

tions. Then bacteria in half the plates were respread and the rest of the plates were not. All plates were then sprayed with phage and incubation was continued so that colonies resistant to the phage could grow.

If resistance to phage arose after contact with phage, then only nonresistant bacteria were present at respreading, and respreading would not affect the final count of resistant colonies in the two plates. However, if phage-resistant bacteria were present before application of phage, respreading would increase the final count of resistant colonies. For instance, if mutation to phage resistance arose two generations before respreading, respreading would produce four resistant colonies from the mutant; and if it occurred three generations before, eight colonies would be produced instead of one.

Many more phage-resistant colonies occurred on the respread plates, showing that phage-resistant bacteria had been present in the absence of phage. The phage merely acted as a selective agent for phage resistance. Similar results were obtained when this test was used in a study of streptomycin-resistant variants.

The Fluctuation Test of Luria and Delbrück

In this experiment a young culture of bacteria was diluted in fresh medium and was divided into two 10-milliliter aliquots, each aliquot containing approximately the same number of cells. One aliquot was not subdivided, whereas the other was distributed equally among 50 tubes. All the cultures were incubated and after they had multiplied to a suitable density, they were plated on agar that had been coated with bacteriophage. From the large culture tube 50 equal samples were taken and plated separately. The contents of the 50 tubes were likewise plated separately. The results of this experiment (Figure 12.1) showed that roughly the same number of phage-resistant colonies were found on each of the 50 plates made from the large culture. If the phage were inducing the mutation after the samples had been plated, a similar distribution of numbers of resistant colonies should be found on the plates made from the 50 small cultures, for all the bacterial cells would be alike at the time of plating. If mutations occurred as chance events before the bacteria were plated with phage, some cultures might contain no mutants, whereas those aliquots in which mutations occurred early in the incubation period would contain many mutants. The numbers of mutants from the 50 small cultures would show a high degree of fluctuation, which was the case. Variance among samples from the large culture was the amount expected from sampling error, whereas variance among samples from the very small cultures was several hundred times higher.

This experiment showed again that mutations to resistance were rare chance events and were not in these cases directed by the environment.

Replica Plating, or the Indirect Selection Test of Lederberg and Lederberg (Figure 12.2)

In the two methods just described, selective agents were used (phage) which

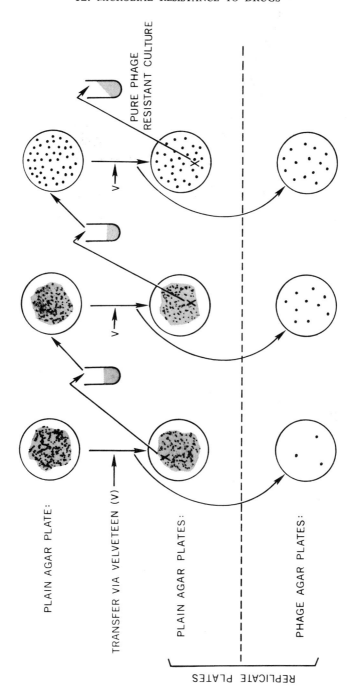

Figure 12.2. Indirect selection of mutants by replica plating. The pure phage-resistant culture (on the right) has not been in contact with phage during any of its development. (After Braun, 1965.)

prevented the multiplication of normal cells but allowed the growth of resistant mutants. More conclusive evidence for origin of resistance by spontaneous mutation would involve isolating mutants in complete absence of the selective agent. This ideal situation was achieved by means of the replica plating technique. A large number of cells from a bacterial population was plated on the surface of a nonselective medium. After incubation, replicate plates were made by transferring impressions of the superficial growth to another plate by means of a sterile velvet disc. The replicate plate contained the selective medium that permitted only mutants of the desired type to grow. By comparing the original with the replicate plate, it was possible to locate on the nonselective medium those colonies from which the mutant bacteria had been drawn. By plating from these colonies, again onto nonselective medium, the proportion of mutant cells was greatly increased. In a series of isolations, platings, and replica tests, a pure culture of mutant cells was obtained which had never been in contact with the selective environment.

To conclude, this evidence has shown that in the given conditions mutations to resistance are random events, unaffected by the toxic agent and that the drug or other toxic agent then serves as a selective agent of the resistant strain.

MUTATION RATES AND MUTAGENIC AGENTS

Proportions of mutants found after a given time are not reliable measures of mutation rates if the relative fitness of parent type and mutant is not known. Best measures of the probability of mutation per cell per generation can be derived if there is no differential selection. Mutation rates range from 10^{-10} to 10^{-2} per cell per generation, with most within the values of 10^{-8} to 10^{-6}. The spontaneous mutation rate can be raised 10- to 10^4-fold by mutagenic agents such as X-rays, gamma rays, ultraviolet light, alpha particles, chemicals such as nitrous acid, alkylating agents, acridine dyes, manganous chloride, purine and pyrimidine analogues as well as the metabolites formaldehyde, peroxides, adenine, and ferrous ion. These agents also kill bacteria and may affect both the parent populations and the mutants. Mutagens induce a great variety of changes in such characteristics of the bacteria as phage resistance, antibiotic and other drug resistance, nutritional requirements, and many others.

Understanding the cellular mechanisms involved in resistance to drugs requires knowledge of the modes of action of antimicrobial agents. Reviews of this topic can be found in the reference list. Among the possible mechanisms of resistance that have been suggested are: (a) decreased penetration of the drug; (b) increased destruction of the drug, or decreased conversion of an inactive precursor compound into an active compound; (c) increased concentration of a metabolite antagonizing the drug; (d) increased concentration of an enzyme utilizing a metabolite that the drug inhibits; (e) presence of an enzyme with decreased affinity for the drug compared with the metabolite. Two examples, resistance to sulfonamides and resistance to penicillin, illustrate the complexity of the biochemical mechanisms involved in the acquisition of resistance.

In some strains of bacteria it has been shown that a high degree of resistance to the sulfonamides is linked with an increased production of para-amino-benzoic acid (PAB). Early work suggested that sulfonamides inhibit an enzyme catalyzing an early step in the conversion of PAB to folic acid. It could be demonstrated that PAB overcame the bacteriostatic action of the sulfonamides, so resistance to sulfonamides was thought to be a result of increased synthesis of PAB. However, there were many instances in which it was impossible to show increased PAB synthesis in sulfonamide-resistant organisms. More recent work has focused on altered enzymes. In the pneumococci, resistance to sulfonamides can be explained by the production of enzymes with affinities for the drugs that are lower than the affinity exhibited by the enzyme system from sensitive strains. The enzyme of the mutant does not combine as readily with sulfonamides. A third type of sulfonamide-resistant mutant has been observed with neither increased PAB synthesis nor changes in either the amount or affinity of the sensitive enzyme. This mutation is believed to cause a difference in permeability so that sulfonamides cannot readily permeate the resistant cells.

The early observations on the resistance to penicillin were followed by the discovery of the enzyme penicillinase in a wide variety of bacteria. Penicillinase splits the beta-lactam ring of penicillin to form the antibiotically inactive penicilloic acid. All naturally occurring penicillin-resistant staphylococcal strains, as well as penicillin-resistant strains of many other species, owe their resistance to the formation of this enzyme. However, the action of penicillinase does not give the complete answer to the problem of penicillin resistance. No penicillinase can be detected in several species of gram-negative bacteria that are completely resistant to penicillin. Although it is possible to produce penicillin-resistant staphylococcal strains in the laboratory, it has not been possible to demonstrate mutations from complete absence of penicillinase to the presence of penicillinase. Penicillin resistance evoked in the laboratory is not related to penicillinase activity, but is based on the development of mechanisms for cell wall formation that are not susceptible to inhibition by penicillin.

The two examples illustrate that a single biochemical mechanism is not always enough to explain resistance to a single drug by all strains and species.

THE ESTABLISHMENT OF RESISTANT POPULATIONS IN THE LABORATORY

When individual resistant cells are present in a population they will better survive the specific toxic agent and the resistance of the population will increase.

Three principal patterns of increase of resistance are shown in the laboratory when bacteria are repeatedly subcultured through many generations in the presence of toxic agent (Figure 12.3): (A) the obligatory single-step pattern; (B) the multistep or multistep pattern with wide first step; (C) and the facultative single-step pattern. A pattern of increase of resistance in a

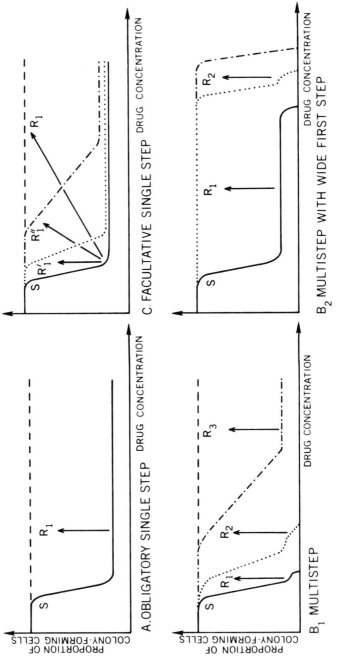

Figure 12.3. Patterns of change, in laboratory populations, from susceptibility, S, to resistance, R, in the presence of increasing concentrations of toxic agent. (After Bryson and Szybalski, 1955.)

population is characteristic for a particular organism in a particular environment, but the same organism exposed to other agents may show different patterns. The changes shown depend on the mode of action of the toxic agent and the number of mutations needed to produce the resistant phenotype.

The Obligatory Single-Step Pattern

In the obligatory single-step pattern the organisms are either sensitive or fully resistant. When the principally sensitive strain is exposed to the toxic agent a characteristic plateau of resistant survivors is obtained over a wide range of higher concentrations of the agent. Any clone grown from this plateau produces at once a fully resistant population. *Escherichia coli* resistant to bacteriophage follows this pattern, also *Bacillus megatherium* resistant to sodium para-aminosalicylate, isoniazid, erythromycin, and cinnamycin. Clinical information suggests that the resistance of *Mycobacterium tuberculosis* to thiosemicarbazone and sodium para-aminosalicylate may be of the same pattern.

The Multistep Pattern (Penicillin-Type)

This is characterized by a series of intermediate steps between sensitivity and a high degree of resistance. The average resistance of each new isolate in a series of consecutive experiments increases until a highly resistant strain is achieved. This multistep pattern appears very commonly and is shown by penicillin resistance in staphylococci and by neomycin resistance in mycobacteria. The last step is sometimes large, which suggests that accumulating mutations may have effects that are more than additive. *The multistep pattern with a wide first step* is not as common as a more regular multistep pattern. An example is a furodoxyl resistance in *E. coli*.

The Facultative Single-Step Pattern (Streptomycin-Type)

In this case, after exposure to low concentrations of the drug, resistant clones surviving from the initial population have a variety of grades of resistance to higher concentrations. Each of these grades of resistance may have arisen itself by a single step. This is a general pattern of resistance to streptomycin. *Mycobacterium ranae* shows it with isoniazid.

Induction

There is yet another way in which a resistant population can appear, and in this case all cells show a uniform change in the presence of the toxic agent. The drug itself causes specific increases in the rate of formation of an enzyme responsible for bypassing or eliminating the effects of the drug. This is a nongenetic adaptation.

Such nongenetic adaptation has been studied before, for instance in certain photosynthetic bacteria which use energy from chemicals when grown in darkness, but rapidly produce chlorophyll in light. Another carefully studied nongenetic adaptation is the induction in the bacterium *Escherichia coli*, of the enzyme β-galactosidase. This enzyme hydrolyses lactose into glucose and galactose. When *E. coli* is grown on glucose it lacks detectable β-galactosidase, but when the cells are placed in a medium containing lactose they begin to produce the enzyme. In about 30 minutes the *E. coli* cells are able to attack the lactose at a maximal rate. When the lactose in the medium is used up the synthesis of the enzyme stops abruptly. This induction occurs without genetic change. However, the ability to respond is gene-controlled. In *E. coli* the cell called lact$^+$ has the genotype conferring capability of inductive synthesis of the enzyme. By mutation a lact$^-$ cell may arise, and this cell together with all its progeny is incapable of synthesizing the enzyme even in the presence of the inducer.

Enzyme induction can be demonstrated in some cases of penicillin resistance. In *Bacillus cereus* cultures resistance to penicillin is due to the specific drug-destroying enzyme, penicillinase, and mutants vary in the concentration of penicillin which they can tolerate because of varying rates of penicillinase formation. There is, however, another type of mutation in *B. cereus* giving rise to highly resistant organisms in which penicillinase cannot be detected until penicillin is present. The penicillin substrate induces penicillinase production almost instantaneously and even small doses of penicillin will stimulate the cells to increase their rate of enzyme formation up to 300 times with a corresponding increase in resistance. There is good evidence that under most conditions all or most of the cells respond to the effects of the inducer. This phenomenon has also been demonstrated in many strains of staphylococci. Little is known about induced resistance to other drugs, but it is assumed that such changes do occur.

SELECTION IN NATURE

The patterns for establishment of resistant populations described above represent an extreme type of environmental pressure in which a selective agent inhibits the propagation of the original type. These highly selective environmental conditions of the laboratory and during drug therapy are probably rare in nature.

Several alternatives are possible for the resistant organisms which arise by spontaneous mutation in the absence of the specific toxic agent. The mutation to resistance may also confer some other adaptive advantage to the cell under the existent environmental conditions. The biochemical mechanisms resulting in resistance may increase the growth rate of the cells by increased synthetic or metabolic capabilities or prolong the growth by decreased sensitivities to normally inhibitory metabolites. The mutant may produce metabolites which inhibit the parent sensitive cells. By such mechanisms, resistant cells may be better suited to the environment or create an unfavorable environment for the parent cell, resulting in an increase in the resistant organisms in the population. Other changes in the functions of the resistant cells may offer a selective disadvantage resulting in the failure of resistance to be established as a population characteristic. When the resistant mutants have the same growth rate as the parent organisms and resistance confers neither a selective advantage nor disadvantage, mixed population of resistant and sensitive types can occur. The rate of establishment of the mutants is then proportional to the mutation rate of the organism. Back mutations to sensitivity can also occur. Such back mutation, occurring after the proportion of mutant cells is sufficiently large to yield mutants of their own, will reduce the increase of resistant mutants in the population. Since mutation rates to resistance and the back mutation rates to sensitivity may differ, the ratio of these rates will ultimately cause the attainment of an equilibrium level between the proportion of resistant and sensitive types.

Similar population changes can occur in a resistant population when the selective environmental condition is removed. Mutations to sensitivity can occur spontaneously and, in the absence of the toxic agent, these sensitive mutants can have an adaptive advantage, a selective disadvantage, or can be similar in all other aspects to the parent resistant strain. Thus, the number of resistant types in the population can either be maintained, diminished or excluded, or reach an equilibrium with the sensitive types.

The Hospital Environment

Drug-resistant staphylococci were clinical problems as early as 1944 when sulfonamide-resistant staphylococci were increasing and penicillin-resistant staphylococci appeared in war wounds. As penicillin was increasingly used, higher proportions of penicillin-resistant staphylococci were found in hospitals. In some hospitals the percent of resistant staphylococci rose from 14 in 1946 to over 40 in 1948. Similar increases were associated with the general use of other antibiotics. Hospital staphylococci have now been isolated which are resistant to penicillin, streptomycin, erythromycin, the tetracyclines, chloromycetin, bacitracin, neomycin, and other drugs.

Serious epidemics of virulent antibiotic-resistant staphylococcal infections occurred in hospital nurseries and surgical wards. High frequencies of virulent resistant staphylococci were found to be associated with hospital wards as compared with out-patient clinics and out-of-hospital communities; also with use of antibiotics in the ward and with the length of stay of patients and staff in hospitals. Resistant strains were found in infective lesions and in healthy carriers. Resistant staphylococci prevail among healthy carriers, healthy patients, and staff, irrespective of individual treatment.

Strains of staphylococci can be identified by bacteriophage typing. In most hospitals antibiotic-resistant staphylococci were usually lysed by group III phages, and the most common were type 80/81. This virulent penicillinase-producing strain of staphylococcus has also been identified in sublines of cultures that were isolated in 1928. It seems probable that there is some difference in virulence between the sensitive and resistant staphylococci, perhaps because over the years antibiotics have been directed principally at the more virulent ones in lesions, thus selecting the resistant variants of the more virulent strains.

In the Johns Hopkins Hospital in 1959 two-thirds of the staphylococcus infections were caused by phage-type 80/81. By 1960 type 54, also of group III, had become the most common and the cause of the change was investigated by Cohen *et al.* (1962). Type 54 was markedly more resistant than type 80/81 to erythromycin, chloromycetin, neomycin, and kanamycin, and these drugs were being used. The two strains differed most markedly in their resistance to neomycin and kanamycin, antibiotics which closely resemble each other. The type 80/81 was almost completely sensitive to these two drugs whereas type 54 was almost completely resistant to them. Kanamycin was rarely used, but there was a steady increase in the use of neomycin during the period when type 54 emerged as a frequent cause of infection. The vast majority of the neomycin used in the hospital was for preoperative sterilization of the bowel. Therefore it was not surprising that the new strain appeared to emerge primarily in the surgical services. So it appears that the use of neomycin in preoperative preparation of the bowel was a factor in the great increase of a strain of staphylococcus previously not known in the hospital. Only half of the patients infected by this strain had received neomycin. Type 54 was the commonest type of staphylococcus in the nose of surgical patients, but 1 of 100 of the personnel harbored type 54 and 5 of 100 still carried type 80/81. No infections in personnel were caused by the type 54 organism, and type 80/81 continued to be the most common cause. To explain why type 54 was frequent among patients and not personnel, one must assume that patients are more susceptible to colonization with staphylococci than normal persons working with patients.

These observations illustrate selection of antibiotic-resistant mutants and of different strains of bacteria. The introduction of new antibiotics and the varying use of others create selective environments that allow drug-resistant strains to establish themselves while drug-sensitive strains are inhibited.

INFECTIOUS HEREDITARY MULTIPLE-DRUG RESISTANCE

In Japan since 1957 cases of bacillary dysentery have increasingly yielded *Shigella* that are resistant to several or all of the following drugs: sulfonamides, streptomycin, chloramphenicol, and tetracycline—the *Shigella* show multiple-drug resistance. Resistance to these different drugs is usually determined by mutations at various loci. It was therefore surprising to find an extremely high prevalence of multiple-drug resistance, while the prevalence of resistance to one or two of these drugs was lower. By 1959, 86 percent of drug-resistant *Shigella* were resistant to at least three of these drugs. The usual mechanisms of mutation and selection could not explain the rapid increase in the incidence of multiple resistance or the epidemiology. Completely sensitive *Shigella* strains were isolated from some patients, but other patients in the same epidemic yielded strains of the same serotype with multiple-drug resistance. Moreover, some patients excreted both sensitive and multiple drug-resistant strains of the same serotype. After patients harboring sensitive organisms were treated with a single drug they often excreted *Shigella* resistant to multiple drugs but rarely resistant to one drug. Patients with *Shigella* resistant to multiple drugs often excreted *Escherichia coli* with the same patterns of resistance.

Subsequently it was demonstrated that multiple-drug resistance could be transferred between *E. coli* and *Shigella*, *in vitro*, in animals and also in human volunteers. The determinants for multiple resistance to unrelated drugs were found to be transmitted simultaneously during conjugation from resistant to sensitive cells. The drug resistance was determined by genes for host resistance that were attached to a transfer factor, or sex factor, analogous to the F factor of *E. coli* (Chapter 10). This transferable R factor, like the F, is an episome, a DNA element which can exist in an autonomous cytoplasmic state, or can enter into direct association with the bacterial chromosome. Episomes are not essential to the host bacteria and can be lost but cannot be acquired spontaneously. When autonomous the R factors can replicate on their own and can be transferred by conjugation independently of the bacterial chromosome. The resistance factors received by the recipient cells are rapidly expressed phenotypically. Most of the factors involved in multiple-drug resistance do not require cell division for their phnotypic expression.

There have now been isolations of enteric bacteria which carry resistance factors for as many as six drugs. Resistance factors can be transferred among all strains of enterobacteriaceae and even from enterobacteriaceae to other genera, including *Pasteurella*, *Vibrio*, and *Pseudomonas*. The recipient bacteria gain the capacity to become donors to other sensitive bacteria. Only gram-negative bacteria have been found to transfer resistance factors. Gram-positive bacteria do not conjugate.

Walton (1966) indicated that in Britain infectious multiple-drug resistance is widespread in *E. coli* in guts of healthy calves and pigs. He found association between the kinds of antibacterial drugs in their food and *E. coli* strains resistant to those drugs. Ninety-nine of the 134 multiple-resistant strains found were able to transfer their resistance to sensitive *Salmonella typhimurium* and *E. coli*, *in vitro*.

Serious public health problems could arise if normal human enteric bacteria became widespread carriers of R factors transferable to pathogens. Actually, the transfer of R factors normally occurs less readily in the intestinal tract than *in vitro*, because of the presence of fatty acids and other inhibitory agents. But it might be accelerated in the intestinal tract under selective pressures of chemotherapy.

MICROBIAL PERSISTENCE

When growing cultures of bacteria are exposed for several hours to many of the antibiotics, a small proportion of the organisms may survive, although on subculture the resistance of these organisms is no greater than in the original strain. The surviving organisms are thought to persist because they are not

multiplying. Nonmultiplying organisms have been shown to survive sulfonamides, penicillin, streptomycin, erythromycin, and isoniazid. The proportion
of these persisters is very small in actively growing cultures, but it may be
much larger when growth is prevented by unfavorable conditions. In man
microbial persistence is of great importance because it is probably responsible
for the fact that bacteriocidal activity of drugs on organisms in lesions is
slower than in actively growing cultures *in vitro*, resulting in the need for
lengthy periods of treatment.

DRUG RESISTANCE IN MALARIAL PARASITES

Human malarial parasites have shown single- or multiple-resistance to the
five main groups of antimalarial drugs: aminoacridines (Mepacrine), bi-
guanides (Proguanil), diaminopyrimidines (Pyrimethamine), 4-aminoquinol-
ines (Chloroquine, Amodiaquin, Sontoquine), and 8-aminoquinolines
(pamaquine, pentaquine, primaquine). This account will follow the chrono-
logical sequence of events in the recognition and increase of these phenomena.

Although decreased sensitivity of some malarial parasites to quinine,
pamaquine, and Mepacrine had been reported, the problem was insignificant
until 1948–1950 when Proguanil resistance in *Plasmodium falciparum* and
P. vivax was discovered in Malaya. This was followed by similar reports from
Indonesia, Assam, Vietnam, New Guinea, Indochina, Ghana, Kenya, and
Thailand. Resistance to Pyrimethamine first arose in East Africa and was
later reported from West Africa, New Guinea, Venezuela, Cambodia, and
West Irian. By 1953, *P. falciparum*, *P. malariae*, and *P. vivax* had demonstrated
resistance to Proguanil and Pyrimethamine, with most cases involving
P. falciparum.

Resistance to one malarial drug often confers resistance to another. For
example, resistance to Proguanil, Pyrimethamine, or the weak drug sulfa-
diazine may confer resistance to one or both of the others. Strains resistant to
Proguanil are likely to be resistant to chloroproguanil.

Chloroquine and other 4-aminoquinolines have proved to be the most
dependable for treatment of acute attacks and for suppression of parasitemias.
In 1960 a strain of *Plasmodium falciparum* from Colombia was found to be
resistant to Chloroquine. Cases responding poorly to treatment with the
4-aminoquinolines and, in particular, Chloroquine, have been reported
during the past few years also from Brazil, British Guiana, Cambodia, Colombia, Malaya, Thailand, and Vietnam. Cross-resistance of Chloroquine-
resistant strains to other drugs such as Mepacrine has also been reported.
General spread of resistance to 4-aminoquinolines could have serious consequences. In Southeast Asia resistance in *P. falciparum* has progressed to the
point where treatment with Chloroquine and other synthetic drugs frequently
fails to control the clinical disease. Even with quinine, which has been useful
in controlling fever, recrudescences have approached the level of appearing in
almost 50 percent of the cases.

The seriousness of the problem of drug resistance of malarial parasites has stimulated a great deal of work aimed at discovering new chemotherapeutic drugs. Many new drugs are promising in the laboratory, used either singly or with other antimalarial drugs; however, they await evaluation in the field.

CONCLUSIONS

Several approaches can be used to limit the increase of drug resistance. For instance, penicillin homologues methicillin and oxacillin have been synthesized and are not attacked by penicillinase. Although less efficient than penicillin they have proved effective against penicillin-resistant organisms. Homologues of other drugs may prove equally useful. Development of resistance can be reduced by avoiding indiscriminate use and low, frequently administered doses and favoring high doses of drugs to which populations develop multistep patterns of resistance. Also, simultaneous use of two unrelated drugs should help, since chances of mutations to resistance at two loci in one cell are likely to be in the order of 10^{-12}. Also one drug might inhibit the metabolic pathway within the cell which is responsible for resistance against the other.

The problem of resistance to antimicrobial drugs is a major one, providing a warning against complacency in the indiscriminate and uncritical use of antimicrobial agents.

REFERENCES

Braun, W. 1965. *Bacterial Genetics.* Saunders, Philadelphia, Penn.

Bryson, V., and Szybalski, W. 1955. Microbial drug resistance. *Advan. Genet.* **7**: 1–46.

Cohen, L. S., Fekety, F. R., and Cluff, L. E. 1962. Studies of the epidemiology of staphylococcal infection. IV. The changing ecology of hospital staphylococci. *New Engl. J. Med.* **266**: 367–372.

Moyed, H. S. 1964. Biochemical mechanisms of drug resistance. *Ann. Rev. Microbiol.* **18**: 347–366.

Newton, B. A. 1965. Mechanisms of antibiotic action. *Ann. Rev. Microbiol.* **19**: 209–240.

Walton, J. R. 1966. Infectious drug resistance in *Escherichia coli* isolated from healthy farm animals. *Lancet* **1**: 1300–1302.

Watanabe, T. 1963. Infective heredity of multiple drug resistance in bacteria. *Bacteriol. Rev.* **27**: 87–115.

Wolstenholm, C. E. W., and O'Connor, C. M. (eds.) 1957. *Drug Resistance in Microorganisms.* Ciba Foundation Symposium. Little, Brown, Boston, Mass.

World Health Organization. 1965. Resistance of malaria parasites to drugs. *World Health Organ. Tech. Rept. Ser.* **296**, WHO, Geneva.

CHAPTER 13

The Cellular Basis of Heredity

FREDERIK B. BANG

Populations of animals, their growth in various environments, the limits to their growth, and their interactions with other populations have been discussed. How heredity and the environment interact, how populations produce similar populations, and how this sequence is modified will now be considered. The task here is to bring together three major themes that are essential to population genetics. These themes are (a) chromosome behavior in mitosis and meiosis; (b) the localization of hereditary units, the genes (or whatever name they may be given), on chromosomes so that there is an inseparable relationship between the morphological behavior of genes, and the patterns of inheritance; (c) and finally the Darwinian ideas of evolution.

CHROMOSOME BEHAVIOR

(See White, 1961, for detailed description and illustrations)
Most animal and plant cells are diploid—there are two sets of like or homologous chromosomes in the nucleus. Each of these two homologous sets has a number of chromosomes, called n, that is typical for the species or race. For instance, for man n is 23, and the full complement of the nucleus, $2n$, is 46. Gametes, and cells in the gametophyte stage of plants, have n chromosomes (are haploid). Bacteria and viruses have only single sets of genetic information.

Chromosomes are not visible microscopically during interphase, the time between cell divisions.

Mitosis

Mitosis involves one division of a diploid cell to produce two diploid cells. Between the fertilized egg and a fully grown man there are some fifty cell generations, derived by mitotic divisions; moreover, cell replacement continues throughout life in most tissues.

Before division of the whole nucleus each chromosome becomes a pair of chromatids that lie together along their whole length. In mitosis the chromatids separate. They pass to daughter nuclei, and each becomes a chromosome of the daughter cells.

In a little more detail, the events are as follows. There is an increase in nucleic acid—there is concentration of DNA within the nucleus. The chromosomes become visible and as they appear they begin to uncoil. They turn and twist and finally line up on the axis of the spindle. The spindle has fibers that pull apart the two chromatids of each chromosome. Each $2n$ group of these chromatids becomes a $2n$ group of chromosomes in a daughter nucleus. The

168

walls of the nucleus and of the cell are recreated and the chromosomes are again lost to view in the two daughter cells. Thus a series of mitotic divisions involves repeated separation of two chromatids from each of the chromosomes, and the passage of these separated parts into daughter cells. There is no reduction of the $2n$ number of the chromosomes.

Meiosis

In fertilization there is a fusion of gametes, yet the chromosome number for a species or race is almost always constant. A reduction in chromosome number from $2n$ to n occurs before fertilization. This takes place in meiosis, which immediately precedes gamete formation in animals, and precedes the gametophyte stage of plants.

Meiosis involves two divisions of the nucleus. In the process of the first division there is a preliminary pairing of the two sets of homologous chromosomes. Since each chromosome consists at that time of two chromatids, the pairing of homologous chromosomes results in four strands lying together along their whole linear arrangement. In this first division of meiosis the homologous chromosomes separate on the spindle and pass into the two daughter cells. Thus by the first division, the chromosome number in each new nucleus is reduced to n.

The second division of meiosis involves a separation of the two chromatids of which each chromosome still consists. By this division the chromosome number n is maintained.

In gametogenesis in the female, only one of the four daughter cells of meiosis is retained as the gamete, and the other products of the division are small polar bodies that are lost from the cell line.

A biologically important reshuffling of chromosome material, called crossing-over, occurs among the four closely associated chromatids in the early stages of the first division of meiosis (Figure 13.1). Pieces of chromatids

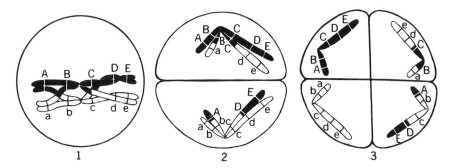

Figure 13.1. (1) A pair of chromosomes consisting of a total of 4 chromatids is lying together early in meiosis, and two chiasmata are shown. (2) At the end of the first division, the two chromatids in each segregating chromosome are no longer alike. (3) End of the second division of meiosis. (After Swanson, 1964.)

are exchanged between pairs of homologous chromosomes. Thus the eventual chromosomes in the new four daughter cells are reconstituted so as to be partly maternal and partly paternal in origin. The mechanism is as follows. As the two homologous chromosomes come together early in meiosis they touch and synapse in places. This involves a breaking apart of two adjacent chromatids and a rejoining with the other. So a cross-formation, or chiasma, occurs. The chiasmata can be seen as little knobs in stained preparations at this stage, and there may be several in any one pair of homologous chromosomes. As the chromosomes separate in the first division of meiosis, the actual points of crossing-over can be seen to slip along to the ends of the separating pair.

Autoradiography allows incorporation of radioactive components such as thymidine into chromosomes, and the subsequent distribution of the radioactive material can be followed. Its final distribution among chromosomes of the new nuclei support other observations that crossing-over occurs at high frequency in most chromosome pairs.

The reconstitution of chromosomes by crossing-over enormously increases genetic variability in populations. The basic biological significance of sexual reproduction is that it allows exchange and recombination of the genetic material that is the basis of heredity. In no group of organisms is sexual reproduction known to be absent. For instance, bacteria obtain a rearrangement of genetic material by conjugation. Variability from sexual reproduction is discussed further in the next chapter.

Effects of Radiations, Other Mutagens, and Some Viruses on the Chromosomes

Today the chromosomes of all sorts of animals can be fairly easily examined by a simplified tissue culture technique. Cells are grown by tissue culture and are exposed to an inhibitor in the final stages of mitosis. Then the cells are blown open by a change in osmotic pressure. Under these conditions the chromosomes can be spread out and seen and counted. The carefully arranged picture of the separated chromosomes in mitosis is presented in Plate 13.1.

Puck has shown that radiation doses of about 50 r will produce visible morphological changes in the chromosomes. Other substances, such as bromuridine, which interfere with the nucleic acid metabolism of the cell, may produce other morphological changes in the chromosome. Recent work has shown that a virus, such as herpes, which under certain conditions can grow within cells without destruction of the cells, may also change the morphology of the chromosomes.

Polytene Chromosomes

Nuclei of the secretory cells of dipterous larvae, such as *Drosophila*, have very thickened chromosomes, each with many strands of the individual chromosomal material. When stained, these polytene chromosomes show banding, which has a pattern typical for each chromosome pair of the species. Polytene

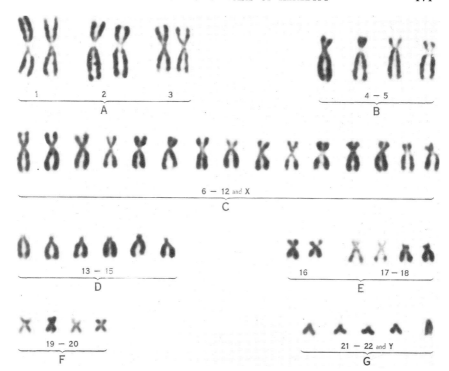

Plate 13.1. Karyotype of a normal human male. The notation given to the chromosome pairs, including the X and Y, is shown. (Provided by D. S. Borgaonkar, the Johns Hopkins Hospital.)

chromosomes of irradiated flies show damage and consequential reorganizations of the banding patterns, such as deletions and inversions. The morphological aberrations of the chromosomes can be correlated with phenotypical variance in the flies. From such studies of series of chromosomes and mutants functions of specific chromosomal regions and specific arrangements can be recognized. This aids in the preparation of chromosome maps, that are discussed under "linkage groups."

To summarize this section on chromosome behavior—there are two basic mechanisms for the division of cells of multicellular organisms. In one of these, meiosis, there is a process for reduction of the chromosome number from $2n$ to n, and for exchange of material among homologous chromosomes. Some factors that can influence the heredity of the cell, and of the individual, have their effects directly upon the chromosomes.

THE PARTICULATE NATURE OF INHERITANCE

The individual has his own characteristics, resulting from his genetic constitution, the expression of that constitution, and the environment. To clarify these ideas, the terms genotype and phenotype are used. The genotype is the hereditary constitution of an individual; the phenotype is the individual manifesting a set of characteristics.

Hybridization and Heterozygotes

When Mendel did his famous experiments with peas, he was working with self-fertilizing plants that had developed certain standard characters. He picked some seven of these characters. Two of them, color and roundness of the peas, will be considered here. In one experiment he used two strains as parents, one producing yellow peas and one producing green peas, crossing them with each other. The first filial generation (F_1) consisted of 258 plants with phenotypically yellow peas and none with phenotypically green peas (Table 13.1). By crossing F_1 plants with other F_1 plants, he obtained a second

TABLE 13.1

Results of Two Sets of Mendel's Experiments, Starting with Inbred Parents with one Character Difference in Each: Yellow or Green Peas, and Round or Wrinkled Peas

	Phenotype	
	Yellow peas	Green peas
Parents	√	√
F_1	258	0
F_2	6,022	2,001
Repeated self-fertilization of the F_2	1/3 of the F_2 yellow peas produced all yellow peas	All the F_2 green peas produced green peas
	Phenotype	
	Round peas	Wrinkled peas
Parents	√	√
F_1	253	0
F_2	5,474	1,874
Repeated self-fertilization of the F_2	1/3 of the F_2 round peas produced all round peas	All the F_2 wrinkled peas produced wrinkled peas

filial generation, F_2, of 8,023 plants, of which 6,022 were phenotypically yellow and 2,001 phenotypically green. This is the now classical phenotypic ratio of 3:1 in the F_2 generation. When the phenotypically green group was examined by further breeding experiments, it was shown to be a pure green strain, producing only green peas when crossed with other plants with genotypically pure green peas. Analysis of the phenotypically yellow in the F_2 generation showed that one-third of them continued to breed as pure yellow peas, but two-thirds of them were hybrids, genotypically a combination of green and yellow. So while the phenotypic ratio of the F_2 is 3 to 1, the genotypic ratio is 1:2:1.

A similar set of figures was obtained for the characters of round peas and wrinkled peas.

These experiments illustrated heterozygosis and dominance. The F_1 generation consisted of plants that were heterozygous for the genetic factor for green and yellow pea color. Yellow color was dominant over green color, so all the peas were phenotypically yellow. Half the plants of the F_2 generation were also heterozygous. They were phenotypically yellow because of the dominance of the yellow character.

A whole series of these experiments was carried out for a long time. Mendel chose seven characters, and it happens that there are seven chromosomes in the pea, and he had chosen factors that were all on different chromosomes.

The Segregation of Independent Characters

Another phenomenon that Mendel demonstrated was the segregation of independent characters in the F_2 generation. Inbred pea plants that had the two independently determined characters of yellow peas and round peas were crossed with inbred plants that had green and wrinkled peas. The F_1 plants had all yellow round peas. Figure 13.2 shows the ratios in the F_2 generation. The characters segregated independently. The 1:2:1 ratios occurred in the F_2 generation just as if these characters of color and shape had no relation to each other.

Mendel said:

My experiments with single characters all lead to the same result: that from the seeds of the hybrids, plants are obtained half of which in turn carry the hybrid character (Aa); the other half, however, receive the parental characters A and a in equal amounts. Thus, on the average, among four plants two have the hybrid character Aa, one the parental character A, and the other the parental character a. [These would now be written AA and aa.] Therefore 2Aa+A+a or A+2Aa+a is the empirical simple, developmental series for two differentiating characters. Likewise it was shown in an empirical manner that, if two or three differentiating characters are combined in the hybrid, the developmental series is a combination of two or three simple series. Up to this point I don't believe that I can be accused of having left the realm of experimentation. If then I extend this combination of simple series to any number of differences between the two parental plants, I have indeed entered the rational [i.e., theoretical] domain. This seems permissible, however, because I have proved by previous experiments that the development of any two differentiating characters proceeds independently of any other differences.

Mendel argued with a Swiss botanist, Nägeli, and wrote in a letter that he regretted "not being able to send your Honor the varieties you desired."

As I mentioned above, the experiments were conducted up to and including 1863; at that time they were terminated in order to obtain space and time for the growing of other experimental plants... I had to abandon the experimentation the following year because of devastation of the pea beetle, *Brucus pisi*. In the early years of experimentation this insect was only rarely found on the plants; in 1864 it caused considerable damage, and appeared in such numbers in the following summer that hardly a fourth or fifth of the seed was spared. In the last few years it has been necessary to discontinue the cultivation of peas... (Mendel, *in* Gabriel and Vogel, 1955.)

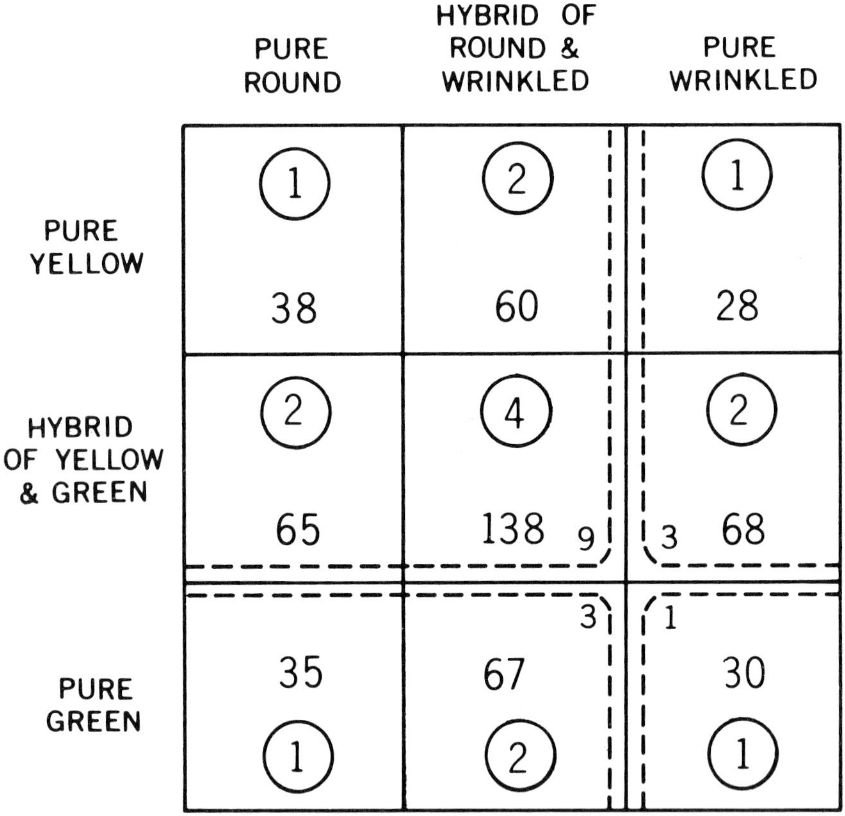

Figure 13.2. Genotypic characters of plants in the F₂ generation in an experiment of Mendel's. The parent generation was inbred plants with yellow round peas (both dominant characters and determined by factors on separate chromosomes) or with green wrinkled peas (both recessive). Plain figures in the squares show number of F₂ plants with the particular combinations of genotypic factors—e.g., there were 38 plants with genotypes for pure yellow pure round peas. Ratios of the genotypes are shown in the circles. Ratios of the phenotypes were 9:3:3:1.

In 1868 Mendel was appointed head of his monastery and shortly he became embroiled in administrative battles over the problems of taxation. He died some 16 years later, not having carried his experiments any further. Thus, major reasons for the work not having been recognized at that time were that an epidemic of a beetle killed all of his subjects and that he changed into the area of administration.

Linkage Groups

Years later when Mendel's findings had been rediscovered, and a variety of

animals and plants had been studied, Morgan found that two pairs of characters did not always follow the 9:3:3:1 ratio of the F_2. His considerations were
brought about to a large extent by technical advances that he had introduced
by starting to work on *Drosophila*, a fly which allowed for larger numbers to
be bred.

> Mendel's law of inheritance rests on the assumption of random segregation of the factors
> for unit characters. The typical proportions for two or more characters, such as 9:3:3:1
> that characterize Mendelian inheritance depend on assumptions of this kind. In recent
> years a number of cases have come to light in which when two or more characters are
> involved the proportions do not agree with Mendel's assumption of random segrega-
> tion . . . I venture to suggest a comparatively simple explanation based on results of
> inheritance of eye color, etc. If the materials that represent these factors are contained in
> the chromosomes, and if those factors that couple be near together in a linear series, then
> when the parental pairs (in the heterozygote) conjugate like regions will stand opposed.
> . . . In consequence, the original materials will, for short distances, be more likely to fall
> on the same side of the split. . . . In consequence, we find coupling in certain characters,
> and little or no evidence at all of the coupling in other characters. . . . Such an explana-
> tion will account for all of the many phenomena that I have observed and will explain
> equally, I think, the other cases so far described. The results are a simple mechanical
> result of the location of the materials in the chromosomes and of the method of union of
> homologous chromosomes, and the proportions that result are not so much a result of
> an expression of a numerical system as of the relative location of the factors in the
> chromosomes. Instead of random segregation in Mendel's sense, we find association of
> factors that are located near together in the chromosomes. Cytology furnishes the mech-
> anism that the experimental evidence demands (Morgan, *in* Gabriel and Vogel, 1955).

Individual genetic factors are morphologically linked in sequence on the
chromosome. It can be shown that they are tied together closely if they are
frequently associated in individuals, or are farther apart on the chromosome,
depending upon the degree to which they approach the segregating 9:3:3:1
ratio, without actually reaching it. Characters separated by a long distance
may be separated if the chromosome crosses over. If they are close together,
crossing-over is not so likely to separate them. They may indeed be so close
as to be thought to be a single factor.

This linear array is detailed to an even greater extent in bacterial genetics,
as described in Chapter 10.

Two further aspects of genetic control will be mentioned, because of the
difficulty they introduce into the interpretation of human defects and disease.
This will be followed by brief comments on a few more genetical terms that
will be used in the rest of the text.

Variable Penetrance

In man occasionally there are individuals with a stiff fifth finger that they
cannot bend. Sometimes this is on the right hand, sometimes the left, sometimes on both, and sometimes on neither although the individual has the
requisite genotype. It is a dominant character that can skip a generation or a
hand. One is forced to conclude that this dominant character has variable

penetrance—it may or may not appear phenotypically. Other such examples are known.

Phenocopies May Be Difficult to Distinguish from Mutants

A phenocopy is a morphological manifestation that has the same form as a particular genetically determined character, and one of the nicest of these is rumplessness in chickens, the lack of a tail. It has been studied for several centuries and observed to be an inherited recessive character. On the other hand, perfectly normal chickens may be induced to have such changes by the administration of insulin or some other substance to the egg, while the embryo is developing. In this case it is not a genetic character, for these rumpless chickens yield normal offspring.

Reflections on how genes act and also how a metabolic inhibitor may act make the relationship easier to understand. The genetic factor influencing the cytoplasm, and thereby the development of the embryo, may act at the same point as an introduced metabolic inhibitor. If they interfere at the same stage of development of a particular character, then the morphological manifestation will be the same.

Genes

Genetic information is studied at many levels, such as DNA arrangement and its direct chemical consequences, differentiation during ontogeny, and grosser variations among phenotypes in breeding experiments and populations. The investigator may find himself working with units of genetic information that are not of the same degree as those of another investigator. Until there is detailed knowledge of the constitution and eventual manifestation of many genetic factors in many organisms, it is convenient to use "gene" as a loose term.

Alleles

Alleles are alternate forms of genes that occupy the same locus (place) in homologous chromosomes, such as the pairs studied by Mendel in pea plants, controlling round or wrinkled, or yellow or green peas. Alleles are specifically discussed further in Chapter 14.

Dominance and Recessiveness

When the presence of an allele causes development of a character that is dominant, the gene itself is not dominant, only the character. This can be illustrated by the case of the gene influencing development of sickle-cell hemoglobin. When a heterozygous individual is tested in the three ways listed in Table 13.2, it can be concluded that the character is recessive, is dominant, or is codominant with the normal character. That dominance can evolve and is dependent on the rest of the genome is discussed in Chapter 14.

TABLE 13.2

A Genotype May Have Phenotypic Effects That Vary in Their Degrees of Dominance[a]

	Genotype			
Test	$Hb^S Hb^S$	$Hb^S Hb^+$	$Hb^+ Hb^+$	Dominance of character
Disease	√	—	—	Recessive
Sickling of cells	√	√	—	Dominant
Hemoglobin	S only	$S+$	$+$ only	Codominant

[a] Three genotypes are shown, the individual homozygous for sickle-cell anemia, the heterozygote, and a homozygous normal individual. They are scored for the three phenotypic tests of the left-hand column, and conclusions are made about the dominance of the character tested.

Pleiotropism

Although genes have single immediate biochemical products, all seem to have multiple ultimate effects in the phenotype. Several examples and consequences of this are discussed in the next chapter.

Polygenes

These are genes with individually small and similar effects that influence quantitatively varying characters, such as height in man. For example, resistance to DDT in some *Drosophila* populations was shown to vary in degree according to presence or absence of factors located on the six different chromosomes. Spotting on the underside of the hindwing of the meadow-brown butterfly is a quantitatively varying character known to be under polygenic control.

Supergenes (see Ford, 1964)

Series of very closely linked genes are involved in determining color and banding patterns of the shells of the snails, *Cepea nemoralis*. The genes within these series are so close together that crossing-over within each series is extremely rare. Thus the alternate series are inherited as blocks, called supergenes. As mentioned previously, it may be difficult to distinguish between genes with such very close linkage, and alleles—alternative forms of genes at the same locus. This difficulty occurs concerning many of the loci where alternate types of human blood-groups are determined. In the process of evolutionary adjustment, polygenes can also be accumulated into a supergene. This has probably been the case with factors controlling the coagulation time of human blood.

There are several ways in which a group of genes can be kept together in a chromosome. One is by being situated so very close together that crossing-over within the group is rare. Another is by inversions and translocations of pieces of chromosomes, which can develop from breakages that occur while chromatids are separating in meiosis. These inverted or differently located pieces do not pair properly with their homologues in meiosis, and opportunity for crossing-over within them is lost. Also, chiasmata do not occur with equal frequency in all chromosomes and species, in fact, crossing over is suppressed in most Y chromosomes. Haldane (1930) recognized that crossing-over is under genetic control, and that such control can have great adaptive value by keeping groups of characters together in the individual.

EVOLUTION

Darwin's theory was that nature selects variants occurring in populations, and thereby causes the adaptive change that is evolution. He had no information about one large part of the theory—how variance was maintained.

Variation Is Dependent on the Fact That Genetically Controlled Characters Have a Basis of Separateness, Not Blending

If, during the process of mating, there occurred a blending that produced off-spring of intermediate types, then variation would decrease. Darwin did not know of Mendel's demonstration of the segregation of characters. Genetic factors are phenotypically manifested as combinations, but they are geno-typically distinct, as shown by segregation in the F_2 generation. Fisher (1930) studied the particulate nature of inheritance and introduced the statistical concepts necessary for its understanding.

The Genome

The genome is all the genetic material of a nucleus. Recombination produces new variants, and variants are selected. But selection does not act on the phenotypic effects of single genes. A gene is manifested against a background of other genetic material. This can be clearly demonstrated by some experiments on inbred mice. A mutant that produces a spot on the coat in a certain strain of mice can be introduced into a second strain. If the mutant character is dominant, the F_1 mice showing it are retained and backcrossed with the second strain. Offpsring with the character are again selected and again back-crossed with the second strain. This is repeated and eventually there is an almost homozygous strain of mice of the second strain, but it contains the mutant from the first. This type of experiment is done repeatedly by geneti-cists, and it is found that the mutant's manifestations themselves may have changed, because the mutant genes are working against a new genetic back-ground.

Dobzhansky (1951), also, has introduced genes into new backgrounds, using those favoring growth at a particular temperature, in certain inverte-brates. He found that the factor which before favored growth at the particular temperature might no longer do so.

The selective pressures on a population do not act on the effects of single genes by themselves. They act on the phenotypic effects of total genomes. Genetic variants are being repeatedly reassorted and recombined, through sexual reproduction. Selection is in terms of the total phenotype, in relation to the total population, rather than in terms of individual traits. There are many who believe that the most powerful effects of selection in evolution are for the heterozygous conditions. Some of the reasoning for these statements will be developed in the next chapter.

CONCLUSIONS

There are unique mechanisms for separation of genetic material within cells. Genetic information is in a linear sequence in chromosomes, and the fine structure of chromosomes is being defined chemically. Variation is maintained by the separateness of genetic factors, and the variation is continually resorted and modified within the nucleus.

However, there are still enormous gaps in understanding, from the rapidly increasing information about coding in the DNA molecule, to the knowledge of the classical geneticist, and to the description of evolution.

Evolution has been mentioned here in terms of microchanges, and such changes are occurring continually. Every doctor, particularly every public health worker, who causes a change in a population of animals or plants or microorganisms is in some way dealing with evolution. Toxins, such as DDT when used on mosquitoes, modify the genetic constitution of populations. The proportions of mosquitoes that are susceptible or resistant to DDT are changed. When penicillin is administered in a hospital and penicillin-resistant staphylococci appear, it is because the populations have been modified by selection. Indeed, whenever the environment is changed in some way for public health, the selective values of that particular environment have been changed, and proportions of genes in populations are changed. Inevitably one is involved with population genetics.

The next chapter will analyze more closely the factors that maintain proportions of genes in populations and the forces that modify these proportions from one generation to another.

REFERENCES

Dobzhansky, T. 1951. *Genetics and the Origin of Species*. 3rd ed. Columbia Univ. Press, New York.
Fisher, R. A. 1930. *The Genetical Theory of Natural Selection*. Oxford Univ. Press. Reprinted in 1958 by Dover Publications, New York.
Ford, E. B. 1964. *Ecological Genetics*. Wiley, New York.
Gabriel, M. L., and Vogel, S. (eds.). 1955. *Great Experiments in Biology*. Prentice-Hall, Englewood Cliffs, New Jersey.
Haldane, J. B. S. 1930. A note on Fisher's theory of the origin of dominance, and on a correlation between dominance and linkage. *Am. Naturalist* **64**: 87–90.
Swanson, C. P., 1964. *The Cell*. 2nd ed. Foundations of Modern Biology Series. Prentice-Hall, Englewood Cliffs, New Jersey.
White, M. J. D. 1961. *The Chromosomes*. 5th ed. Methuen, London.

CHAPTER 14

Animal Population Genetics

BRENDA K. SLADEN AND BERNICE H. COHEN

The preceding chapter has discussed three of the levels of biological organizations that need to be considered in genetics: chromosome behavior in the nucleus; genetic segregation affecting the phenotypic characters of individuals from generation to generation; and processes of genetic change in populations. This chapter is concerned again with the population level—with the origin of new genetic types, with genetic variation in populations, and with further causes of change. The sequence of the discussion is:

1. The origin of new genetic types, including new combinations
   a. By mutation
   b. By new arrangements of genetic material that is already present. This is by:
      (1) Random assortment of chromosomes in meiosis
      (2) Crossing-over between parts of homologous chromosomes in meiosis
      (3) New combinations in fertilization.
2. Changes in relative frequencies of genetic types in populations. In a large randomly mating population if there were no mutation, no selection, and no migration, relative frequencies of alleles and genotypes would be constant from generation to generation.
   a. Mutation rates change relative frequencies of genetic types very slowly.
   b. Migrations bring new material into populations.
   c. In very small populations
      (1) atypical relative frequencies occur when new isolates are established;
      (2) there can be a drift into atypical relative frequencies, and loss of variants, by chance.
   d. Selection. Survival of individuals from conception through to the end of their own reproductive period and degrees of reproductive success play the largest role in determining relative frequencies of genetic types from one generation to the next. Selection of any genetic unit or character is measured by this relative representation from one generation to the next.
      (1) Genetic variation in the population increases when selection pressure is reduced.
      (2) The rate of change in relative frequency of an allele made by selection depends particularly on the dominance of the character and prevalence of the allele and the strength of the selective force.
      (3) Genes have single immediate biochemical products but these usually

180

have multiple phenotypic effects—they are pleiotropic. The various effects can have different degrees of dominance and different selective values. This helps to maintain variation in the population. There is discussion of evolution of dominance, of cases of selective advantage of the heterozygote over homozygotes, and of transient and balanced polymorphisms.

(4) Characters controlled by polygenes are influenced by selection.
(5) Relative frequencies of chromosome variants in wild *Drosophila* populations are highly sensitive to the ecological environment.

A *Mendelian population* is a sexually reproducing population with a common gene pool, and it is only this kind of population that is discussed in this chapter. Many animals at some time reproduce asexually, for instance aphids (which include greenflies) and the sporocyst stage of schistosome parasites; these populations are excluded from the discussion. When bacteria and viruses exchange genetic material, their populations do not follow patterns of inheritance of Mendelian populations. For one thing, bacteria and viruses are not diploid and thus do not follow the Mendelian patterns of assortment of characters during reproduction. They do not carry the loads of alleles for recessive characters in heterozygous state that give populations of diploid organisms great potential for phenotypic variability.

Much of the research described here, that by Ford and by Kettlewell and their associates, and by Dobzhansky, is described in Ford's (1964) book *Ecological Genetics*.

THE ORIGIN OF NEW GENETIC TYPES
Mutation

Mutation is any alteration in the genetic material which is heritable. It may involve a gene, a chromosome section, or the number of chromosomes. The recognizable spontaneous mutation rate in most organisms is about 10^{-5} to 10^{-6} per locus per cell generation. Chapter 11 discusses the biochemical basis of some mutations and Chapter 13 briefly discusses chromosomal changes.

Mutation is a very important source of variation. Indeed, it is the basic material of evolution.

New Arrangements of Genetic Material That is Already Present in the Population

The gene pool is all the genetic material of a population. A population of the moth *Panaxia dominula* of Cothill, England, studied by Ford (see Ford, 1964) will be used to illustrate what is meant by the genetic pool in terms of the genetic material that can occur at one locus. The moth's habitat is 15 acres of marsh, and the population is separated from other populations of the species and immigrations are rare. In 1961 the alleles *medionigra* and *dominula* were present in the population in relative frequencies of 0.02 and 0.98, and no other alleles for that locus were known to be present. These particular alleles can be estimated by direct count because the phenotypes of the two alleles, the homozygous *dominula* form, the heterozygous *medionigra* form and the homozygous *bimacula* form are visibly different. The adult population size was estimated to be 1,200 to 1,600. Thus the gene pool of the adults contained approximately 60 *medionigra* genes and 2,740 *dominula* genes, all being potential sources for the next generation.

Many alleles for one locus may exist in a single population, and combined pairs of these alleles give rise to many phenotypic variants.

It is unknown how much genetic information is relatively fixed because variation would be incompatible with life. Some new tests for genetic similarity among species, genera, orders, and classes of animals were discussed in Chapter 11. The pairing compatibility of DNA of related forms suggests that large amounts of DNA of a nucleus might be concerned with common basic properties, and it might be further speculated that these properties are not subject to much variation. However such estimates are only beginning to be made. Undoubtedly a species nevertheless has a large number of loci at which series of alleles can occur. So, without considerations of changes of relative frequencies of genetic types in the gene pool, there is material for enormous amounts of genetic variation. How genetic variables get recombined so that different alleles at the various loci can come to lie in the same nucleus is discussed next.

RANDOM ASSORTMENT OF CHROMOSOMES INTO GAMETES IN MEIOSIS
(Figure 14.1)
There are 23 pairs of chromosomes in the nucleus in man. Let it be assumed that each chromosome pair is heterozygous by being different at at least one locus. Since in gametogenesis one member of each pair of chromosomes assorts randomly with one representative from each of the other 22 pairs, there are 2^{23} possible variants in the gametes from combinations of these 46 chromosomes.

CROSSING-OVER BETWEEN PARTS OF HOMOLOGOUS CHROMOSOMES IN
MEIOSIS
This number of 2^{23} variations for man is the minimal number. The chromosome pairs in a large heterogeneous population differ from each other in their alleles at many loci. This would not affect the number of possible combinations, were it not for crossing-over within chromosome pairs which separates genes within a chromosome. In the absence of crossing-over a gamete receives by meiosis a particular chromosome of the type of either the individual's father or his mother. With crossing-over a gamete has the further possibilities of receiving chromosomes that are in part paternal and in part maternal.

A new allele arises by mutation in a chromosome that is made up of a sequence of coded genetic information. Crossing-over allows alleles to switch over to become part of homologous chromosomes bearing information that

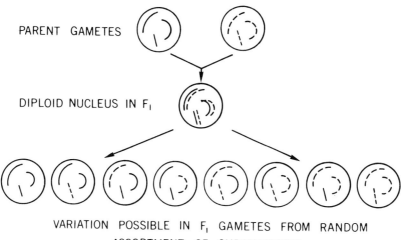

PARENT GAMETES

DIPLOID NUCLEUS IN F_I

VARIATION POSSIBLE IN F_I GAMETES FROM RANDOM
ASSORTMENT OF CHROMOSOMES

Figure 14.1. The F_1 individual represented has 3 chromosomes that were paternally derived, and 3 that were maternally derived. The 3 pairs of homologous chromosomes can assort in meiosis into 2^3 combinations of 3. In man, the 23 pairs can assort into 2^{23} combinations of 23. Crossing-over is not considered here.

may be different in many respects. Thus an allele is able to perform against various genetic backgrounds.

For precise experiments that measure the extent of variation from crossing-over in laboratory *Drosophila* populations, Dobzhansky's report (1955) and the papers to which he refers are recommended.

NEW COMBINATIONS IN FERTILIZATION

In most populations, most or all of the individuals are unique in their particular combinations of genetic variables. When their gametes combine in fertilization, opportunity for new combination occurs again. The possibility for association of differing alleles is greater when parents do not have recent common ancestors. This is the same as saying that outbreeding results in more heterozygotes than does inbreeding.

Thus one of the major functions of the sexual mechanism in a population is provision of the recurring reassortment and rearrangement of the genetic material during meiosis and fertilization.

CHANGES OF RELATIVE FREQUENCIES OF TYPES
IN THE POPULATION

In a large, randomly mating population, if there were no mutation, no selection, and no migration, relative frequencies of alleles and genotypes would be constant from generation to generation. This concept is known as the Hardy-Weinberg principle.

The mechanism of gamete production by meiosis and recombination cannot of itself alter the relative frequencies of alleles in the gene pool from generation to generation. For example, consider two alleles, A and a, and let them be the only ones present for that locus in the population. Let allele A have a relative frequency of q, and let $q=0.4$, and so the relative frequency of a is $1-q=0.6$. Let the population be large and let there be no forces giving one allele any survival or reproductive advantage over the other. In a new generation as shown in Table 14.1 the relative frequencies of the genotypes are 0.16 AA: 0.48 Aa: 0.36 aa. So allele A in the new generation has a relative frequency of $0.16+0.24=0.4$; and a has a relative frequency of $0.24+0.36=0.6$. This is the same ratio of the alleles as in the parent population.

TABLE 14.1

A Hypothetical Case of Relative Frequencies of Alleles A and a in Parental Gametes and the Relative Frequencies in the Offspring of Genotypes AA, Aa, and aa, if conditions stated in the Hardy-Weinberg principle obtain. (It has also to be assumed that no other alleles at that locus are present in the population.)

			Relative frequencies of alleles in male gametes	
			A $=0.4$ $=q$	a $=0.6$ $=1-q$
Relative frequencies of alleles in female gametes		A $=0.4$ $=q$	AA $=0.16$ $=q^2$	Aa $=0.24$ $=q(1-q)$
		a $=0.6$ $=1-q$	Aa $=0.24$ $=q(1-q)$	aa $=0.36$ $=(1-q)^2$

The relationships are summarized as follows: Let the ratio of two alleles, no other alleles for that locus being present, be q and $1-q$. When no forces for change exist, the genotypes in the next generation are in the ratio of:

$q^2:2q(1-q):(1-q)^2$,

and the ratio of alleles continues to be $q:1-q$.

The Hardy-Weinberg principle establishes what is static in Mendelian populations. The ratios of relative frequencies of alleles and genotypes that are derived from the principle are based on populations that do not change the relative frequencies of their alleles from generation to generation. So when the Hardy-Weinberg ratios are used for estimations of numbers of genotypes and alleles there must be assurance that the forces of mutation, migration, selection, and chance are unimportant at that time, or that adjustments can be made to account for these forces. The types of estimations are as follows. When recessive characters are being considered, heterozygous individuals usually cannot be distinguished on sight from homozygous dominant individuals. The Hardy-Weinberg ratios indicate the expected relationship between the homozygous recessives, which can be counted in samples, and the other genotypes. An example follows, and another is given later for the peppered moth.

Question: A dose of larvicide of 0.1 part per million kills all *Anopheles* mosquito larvae that are homozygous for susceptibility to the larvicide (genotype *rr*), but none that is homozygous for resistance (*RR*) and none that is heterozygous (*Rr*). This dose is applied to a laboratory population of *Anopheles* larvae, and 24 hours later the mortality is 9 percent. What were the probable relative frequencies in that population of *rr*, *Rr*, and *RR* forms? It is assumed that the population is large and that mutation and migration do not occur, and that in the time between reproduction of the parents and testing of the larvae there was no selection affecting the relative frequencies of these alleles. Then the relative frequencies of the genotypes of the larvae can be taken to be $q^2 : 2q(1-q) : (1-q)^2$. The larvae that died, *rr*, represented one of the homozygous genotypes—let it be the q^2 group. Then

$$q = \sqrt{q^2} = \sqrt{0.09} = 0.3.$$

The *RR* group has à relative frequency of $(1-q)^2 = 0.7^2 = 0.49$. The *Rr* genotype is the remainder of the population, 0.42. If the conditions for the Hardy-Weinberg principle held in this case, the genotypes *RR*, *Rr*, *rr* would have the relative frequencies 0.49 : 0.42 : 0.09.

MUTATION RATES

Spontaneous mutation alone can change alleles at a rate of about one in a million cells per locus per cell generation. At very low relative frequencies the mutants are likely to be lost from the population. If a mutation occurs only once in the germ line of a single individual, it has very little chance of surviving through several generations. However, with repeated mutation, mutants can accumulate if there is no selection against them. If mutation went in one direction only, say from allele A to allele A_1 and there were no selection, then over a *very* long time A_1 could replace A. But it appears that mutations are usually able to occur in the opposite direction also, from A_1 to A. In that case, if mutation alone were acting the relative frequencies of A and A_1, after a long time, would depend on the mutation rates. This can best be seen in bacterial populations when they are kept in chemostats with carefully regulated environments.

Chapter 16 discusses the accumulation of mutants in populations and loss of individuals because of deleterious mutants, when mutation rates are increased by mutagenic agents.

MIGRATION AND GENE FLOW

A population is changed by invasion of new genetic material from other populations. Glass and Li (1953) have studied changes in the American Negro

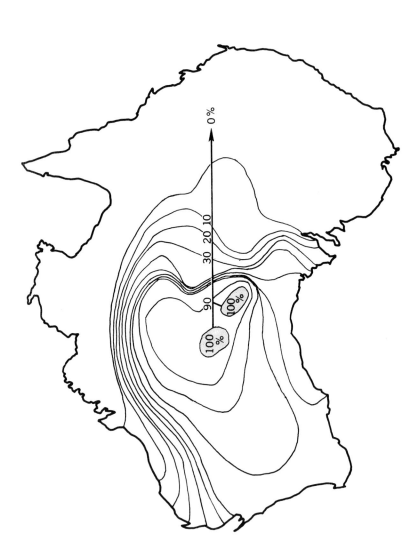

Figure 14.2. The distribution of tawny hair in Australian Aboriginals, from 100 percent tawny haired, to its absence. Lines are drawn through areas where the percentages of tawny haired people are 90, 80, 70, 60, etc., in order to show the gradient of change. (After Birdsell, 1950.)

from mixture with other Americans. The relative frequency of the Rh_0 "allele" (for simplicity the term "allele" is used here for a gene complex) is about 0.02 in west European white populations and also 0.02–0.03 in New York City white populations. In the African Negro the relative frequency of this allele is 0.63–0.64. In the North American Negro the frequency is 0.41–0.45. From such figures Glass and Li estimated that about 30 percent of the genetic material of the American Negro population is of white origin. Investigations of other loci in American Negroes have led to similar conclusions.

The Negro-White mixture began with migrations of various populations to the same place; then it depended on gene flow between two interdispersed stationary populations. Gene flow may also result in a gradient of relative frequencies of a character across a geographical area. For example, tawny hair color in Australian aborigines probably originated in the tribe of the Pitjandjara in Central Australia and is thought to be a semidominant character controlled at a single locus. From this center there is a gradient of decreasing relative frequencies of tawny haired individuals (Figure 14.2). The gradient is no doubt due to distance from the place where the allele originated, and the distance has determined the amount of passage of the allele from that place by interbreeding.

CHANCE EFFECTS IN SMALL POPULATIONS

One of the conditions of the Hardy-Weinberg principle is that the population be large. Atypical relative frequencies can occur when small segments of a population become isolated.

Also, from generation to generation a very small population can lose some of the genes in the gene pool just by chance. Consider one male and one female, producing two offspring, and one locus for which both parents are heterozygous for the same pair of alleles (A and a). Let there be no selective forces acting differentially on these alleles.

| | | Male gametes | |
		A	a
Female gametes	A	AA	Aa
	a	Aa	aa

The probability of both offspring being AA is $1/4 \times 1/4 = 1/16$. This is the chance of losing the a allele from the population of two individuals. The probability of both offspring being aa is also 1/16, in which case A would be lost. Therefore the chance of losing one or the other of these alleles from the population is 1/8. If four offspring live instead of two, the probability that one of the alleles is lost in the first filial generation becomes 1/128. Perhaps the best examples of small inbred populations are found in laboratory stocks where inbreeding may be practised to deliberately lose genetic variability (see Chapter 23).

Some small human populations such as religious isolates in the United

TABLE 14.2
Relative Frequencies of ABO Blood Groups[a]

	Number of people tested	A (%)	B (%)	AB (%)	O (%)
West Germany	5,036	44.6	10.0	4.7	40.7
Dunker isolate in the United States	228	59.3	3.1	2.2	35.5
New York City and North Carolina	30,000	39.5	11.2	4.2	45.2

[a] Glass et al. (1952).

States show deviations from the rest of the population. The Dunkers of Franklin County, Pennsylvania, are an inbred group of about 300 persons, whose ancestors came from Western Germany. Table 14.2 shows that the distribution of ABO blood groups in this population is different from West Germans and from other Americans. Proof that drift due to chance has caused such deviation has to depend on showing that no selective forces have caused the unusual distribution. Also the composition of the founding colony should be known. One can perhaps generalize that in small populations any character is always under the influence of chance *and* selection. For some alleles in some places, chance may be the greater force.

Ford worked with unusual distributions of characters in very small populations of *Maniola jurtina* butterflies on the Scilly Islands. For that particular case he was able to show that selection was so great that drift caused by chance was unimportant.

SELECTION

Mutation probably seldom has a neutral effect, but confers upon the cell or organism some disadvantage or advantage relative to other members of the population. This is probably true also for all new variations caused by re-arrangement of genetic material in meiosis and fertilization.

Survival of individuals from conception through to the end of the reproductive period, and degrees of reproductive success, play the largest role in determining relative frequencies of characters with genetic bases from one generation to the next. Selection is measured by this relative representation from generation to generation, where the change is not due to mutation, migration, chance, and assortative mating.

If a character increases in relative frequency from one generation to another, except by mutation, chance, migration, and assortative mating, it is said to have positive *fitness* and a favorable selection coefficient. Given successful cell division, it is the phenotypic properties of gametes, cells, individuals, groups, and populations in relation to any aspect of the environment, that are the qualities upon which selection can act. Selection does not act directly upon the genotype, only upon its manifestations.

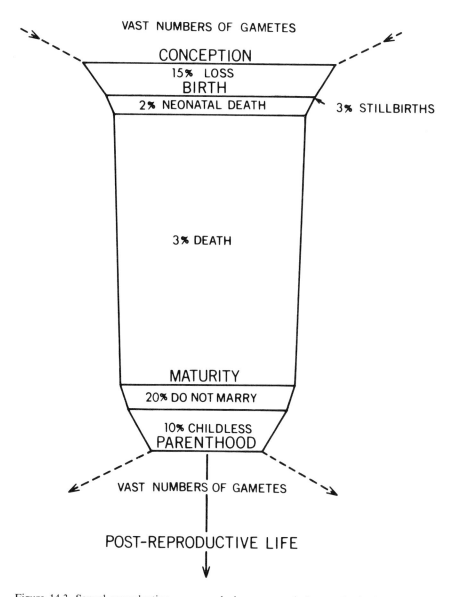

Figure 14.3. Sexual reproduction enormously increases variation, and selection tends to narrow it down again. For the white population of the United States it is estimated that only half the individuals conceived become parents. The genotypes of those individuals that do not contribute to the next generation are being selected against. Selection acts also through relative fertility. It cannot act directly on characters that appear only in post-reproductive life.

Even in modern man there is much opportunity for genetic variants to contribute differentially to the next generation. A clear example is selection by falciparum malaria for the sickle-cell trait in Negroes (Chapter 1). Probably the famines that periodically killed one-third of the population, particularly children, in areas like Bengal, have some considerable selective action, as cholera probably has in India and tuberculosis in Alaska. Within a nation, genetically distinct races frequently have differing rates of increase, and so do genetically differing nations.

In countries where death rates are lowest, there is still considerable loss of individuals between the zygote and the parent (Figure 14.3). The rates used here to demonstrate this point are from Penrose (1959), confirmed by current United States Government demographic data. (They would require detailed qualification if used for anything beyond the following gross presentation.) In the white population in the United States 50 percent of all human zygotes are unfit in the Darwinian sense that they do not contribute their gene patterns to the next generation. Prenatal loss is not all measurable, but here it is taken to be about 15 percent. About 3 percent of the remaining individuals are stillborn and 2 percent of the remainder are lost in neonatal death. Three percent more die before sexual maturity, and of the survivors 20 percent do not marry before the end of childbearing age. Of those who do marry, 10 percent are childless. To some extent the loss of contributors to the next generation is based on genetics, that is, some of the individuals do not reproduce or do not survive to reproduce because of their genotypes. They are being selected against.

There are two other parts to Figure 14.3. First *postreproductive life* will be considered. How much of the survivorship in the postreproductive life span is genetically determined (Chapter 29)? How much of the malfunctioning leading to death is caused by genetic determinants which have no phenotypic effect until after reproduction? These are questions of some significance because any purely postreproductive effect does not directly come under the action of selection, because the contribution to the next generation has already been made. There are, however, some indirect ways in which postreproductive effects could come under selective pressures. The numbers of postreproductive individuals, their food needs, their social roles, affect the economy and efficiency of the whole population and, in this way, presumably contribute to survival of whole populations, though one cannot give documented examples. Care of the offspring might be one of the social roles involved. A clearer way for postreproductive effects to come under selective pressure is by pleiotropism. Genes have multiple phenotypic effects and many of the less immediately obvious effects are on fertility and general viability. So a gene concerned with manifestation of a disease in later life might be under selective pressures because of other phenotypic effects in earlier life.

The remaining part of Figure 14.3 concerns the *gametes*. They are produced in large numbers and suffer a very large loss. Probably almost all of this loss is due to chance. But there must be some operation of relative ability to survive based on genetic constitution, if only elimination of the gametes with genomes* that are lethal at that stage.

Two conclusions will be made. All the way along the life span there is loss of individuals by lack of viability, and from the group of potential parents there is a loss by failure to mate and failure to produce zygotes. When characters that lead to this loss from the reproducing segment of population have a genetic basis, the genes and genomes helping to determine the loss are being selected against. They will be represented in the succeeding generation to a lesser degree. They are less fit. Similarly, genes and genomes contributing to an ability to reproduce, and to reproduce to a more successful degree, are being selected for, are more fit.

Genetic Variation in the Population Increases When Selection Pressure Is Reduced

The second conclusion from the discussion of Figure 14.3 is that while meiosis and fertilization increase variation, selection tends to decrease it again. The decrease is in part a matter of the reduction in numbers. So variation could increase when a population expands its numbers. As an example of this, a population of the marsh fritillary butterfly, *Melitaea aurinia*, in Cumberland, England, will be discussed. The population inhabits eight fields surrounded by woods, which are barriers to migrations. The populations had been observed by amateur naturalists for 36 years. Samples were in butterfly collections and notes on relative abundance were available. Then Ford and his father studied the population for 19 years and thus 55 years were documented in all. The butterfly has an annual life cycle. In 1881 the butterflies were abundant; in 1894 they were reported to be "in clouds". After 1897 there was a decline. In 1906 to 1917 there were small numbers and from 1912 to 1920 they were rare. One or two specimens were found in a day's collection where previously thousands could have been taken. From 1920 to 1924 they increased to vast quantities, so that one sweep of a butterfly net would catch several. From 1925 to 1935 numbers continued to be high. Up to 1920 specimens in collections were very uniform in pattern and no deformities were seen, and this was confirmed by the notes. During the increase of 1920–1924 there were marked departures from the normal in color pattern, size, and shape, and these differing forms were common. Some were deformed, some even unable to fly. After 1924 these "undesirable types" disappeared. The population settled down to an approximately constant form, which was, however, recognizably different from the form of the earlier population. So here has been described an increase in variability when forces opposing population increase were relaxed and a decrease in variability as the forces increased again. The form of the butterfly changed perhaps by chance, perhaps as an adaptation to some aspect of the internal or external environment. Such change in time is, in a small way, part of the process of evolution.

* The genome refers to all the genetic material of an individual nucleus.

As the human population expands at this present time, it frequently includes variants who could not have survived in previously more rigorous environments. Environmental changes include changes in medicine, food supply, and sanitation. That many people with disorders of various types can now live successful lives, is, indeed, a response of a population to a changed environment. Some observers fear, however, that the human race is weakening because some human disorders that surely have genetic bases are losing their selective disadvantages. The argument supposes that these disabilities will noticeably increase in relative frequencies in the following generations. Before such conclusions are attempted, many factors that influence rates of change of genotypes must be considered. ("Weakening the human race" is discussed also in Chapter 1).

Rates of Change Caused by Selection Are Dependent on, Among Other Things, Dominance of Characters and Prevalence of Alleles

Several features of the rates of change of relative frequencies of genotypes in populations will be discussed.

Consider a randomly mating population maintaining a size of about 10^6 individuals. For one particular autosomal locus let there be only two alleles, and let them determine one dominant character and one alternative recessive character. Let one of the phenotypes be more fit, having a selective advantage of 5 in 100. Then for every 105 individuals produced from the more fit phenotypes there will be 100 produced from the same number of the less fit phenotypes.

Let there be initially present ten of the alleles for the more fit character. Then the following approximate numbers of generations will pass to bring the relative frequencies of the more fit alleles first to 0.01, then to 0.05:

If the *dominant character* has the selective advantage, the allele for that character changes from

0.00001 —in 141 generations————→0.01
0.01 ——in 35 generations————→0.05

If the *recessive character* has the selective advantage, the allele for that character changes from

0.00001 —in 10^5 generations————→0.01
0.01 ——in 1,600 generations————→0.05

Figure 14.4 takes the comparison of rates of change further, with the allele for the more fit character starting at a relative frequency of 0.05. The figure shows that if the *dominant* character is more fit considerable change of relative frequencies occurs at once. By the 100th generation the allele has increased from 0.05 to 0.5 and the dominant phenotype from 0.0095 to 0.75; at the 200th generation the allele has a relative frequency of 0.84 and the phenotype 0.95. By the 300th generation the allele for the dominant character is almost at 0.89 and the phenotype is 0.99. After this, for a very long period the allele for the alternate, less fit recessive character declines very slowly; indeed, it is virtually impossible for it to disappear. Figure 14.4 does not go beyond the

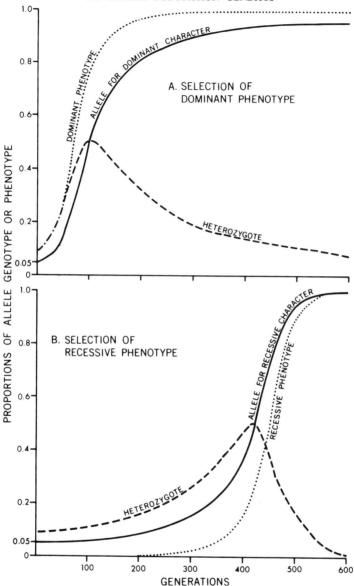

Figure 14.4. Two alternative characters with a complete dominant-recessive relationship and controlled by a single pair of alleles, are given selective advantages of 5 in 100. The population is large and mating is random; the selective advantage is not allowed to vary as the population changes, is not influenced by other genetic factors, and the heterozygote has no other selective values. Changes of relative frequencies are shown for (1) the allele controlling the character with the selective advantage; (2) the character itself; (3) the heterozygote. The calculations were made by G. S. Watson and G. Chase.

600th generation, but it has been calculated that at the 2,000th generation from the beginning of the figure, the relative frequency of the allele for the recessive less fit character is still as high as 0.009.

On the other hand, Figure 14.4 shows that when the allele for the *recessive* character has the selective advantage and has an initial relative frequency of 0.05, there is little immediate change in relative frequencies. After 300 generations the allele is only at 0.15. However, during the next 200 generations there is a change from 0.15 to 0.95. By generation 600 the population has virtually no remaining alleles for the dominant character.

The Slow Change of the Recessive Character at Low Relative Frequencies

When the allele is rare, the rise of a recessive character is extremely slow because recessive phenotypes occur only when the responsible alleles are homozygous. If the allele has a relative frequency of 0.05 (and the population is large and mating at random) the recessive phenotype has a frequency of 0.05^2, or 0.0025. Because the phenotype is so very rare, even very high selective advantage results in only very slow increase. Selective mating, including inbreeding, can increase the rate of appearance of homozygotes and thus increase the rate of change brought about by selection.

The decline of a rare allele for a recessive character is also very slow for the same reason, and the allele can be lost only by chance; and chance has greater effect in smaller populations.

This very slow decline in relative frequency of rather rare alleles for recessive characters, even in the presence of high selective disadvantage, is an important principle in public health. It is illustrated again in Figure 14.5. The figure shows that complete selection, such as sterilization, against a rare recessive character is not very effective in removing the responsible genotype and that the method is least effective against the rarest characters. However, it should not be forgotten that if the character is dominant, and the fitness is 0, removal is achieved in one generation.

It is interesting to note from Figure 14.4 that the *heterozygotes* do not ever reach a proportion greater than 0.5. Should a population sample show them to be present in a proportion greater than half, then homozygotes of the same generation have already been removed.

Modifications of These Patterns

These simple patterns can be modified in many ways. For instance, it is known that the expression of characters can be altered by the nature of the rest of the genome. This has been shown for the character of melanism in the peppered moth, discussed later in the chapter. The significance of such a change can be seen from consideration of the extremely slow rise in relative frequency of a recessive character. If against a new genetic background the character became wholly or partially dominant its increase would be enormously enhanced.

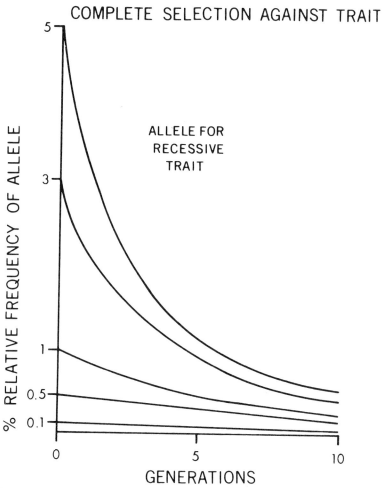

Figure 14.5. As an allele for a recessive character becomes rarer in a population, selection has less and less effect upon its relative frequency.

Probably most alleles have multiple effects, and the different effects have different selective values and different degrees of dominance or recessiveness. Such differences change the patterns of change and can cause balanced polymorphism (see later discussion in this chapter).

Another modification is caused by the fact that sex chromosomes have different exposures to selection. The Y chromosome in many animals carries little genetic information. When the accompanying X chromosome carries alleles for recessive characters, the recessive effects are shown in the phenotype in spite of being present only in a single dose. Therefore rare recessive

X-linked characters are under the influence of selection more frequently than are rare recessive autosomal characters. This is probably the basis of much of the differential mortality rates between the sexes. Males, when they are the heterogametic sex, usually have higher mortality rates because deleterious recessive mutants on the X chromosomes show their phenotypic effects.

PLEIOTROPISM AND SELECTION

As stated above, genes are pleiotropic and the various phenotypic effects can have differing dominance and differing selective values. Among other consequences, this can lead to heterozygotes having selective values differing from either homozygote. These cases are illustrated by Kettlewell's study of melanism in moths in industrial areas, particularly the peppered moth, *Biston betularia* (Kettlewell, 1958, 1965; and Ford, 1964).

The Evolution of Industrial Melanism in the Peppered Moth

What causes the irregular distribution of two or more forms of this moth? In industrial areas the populations are 98 percent melanic (the black, or *carbonaria* form) and in some other areas 100 percent peppered (the pale, or typical form). The geographical distribution is associated closely with the distribution of soot in Britain. Near industrial areas and down-wind from them, soot lies heavily on the leaves and trunks of the trees that are the microhabitat of this moth. Melanic lepidoptera occur in industrial areas in many countries, and their history corresponds with the history of industry.

In the peppered moth one locus has alleles responsible for much of the melanism: C controls the *carbonaria* form, which is dominant; c_i controls another dark form, *insularia*, which is dominant over the typical form but recessive to the *carbonaria* form. But there are other *insularia* forms that are indistinguishable visually which are determined by alleles on other loci; c_i is known only from the Oxford district. The typical form is recessive, and is controlled by the allele c_t. Results of breeding experiments by Clarke and Sheppard (1964) in Table 14.3 show the relationships of these three alleles.

TABLE 14.3
Biston betularia—Three Alleles at the Locus for Melanism[a]

| | | Parent heterozygous for *carbonaria* and typical | |
		C	c_t
Parent heterozygous for *insularia* and typical	c_i	Cc_i *carbonaria*	$c_i c_t$ *insularia*
	c_t	Cc_t *carbonaria*	$c_t c_t$ typical

NOTE: The ratios of phenotypes in the offspring were 2 *carbonaria* : 1 *insularia* : 1 typical.
[a] Clarke and Sheppard (1964) bred heterozygous forms together, demonstrating the dominance relationships of the characters produced.

The *carbonaria* allele has been shown to have pleiotropic effects. *Carbonaria* larvae grow more slowly than do larvae of the typical form. Kettlewell suggested that slow feeding could allow more opportunity for elimination of toxic materials in soot. The mating behavior of the two forms is slightly different and this varies with temperature. On cold nights pale females receive more visits from males than do dark females; on warm nights dark females receive more visits than do pale females.

The *carbonaria* allele seems to confer some physiological advantage on the heterozygote that makes it more viable than the homozygous typical form, but this effect has probably not been true for all its history, as indicated by the proportions of offspring produced from experimental crosses that were made in 1900.

If an allele is new in the population (e.g. a mutation) and it happens to confer some advantage in its main effects, it is not likely that all its pleiotropic effects will also give advantage to the bearer. It is believed that the beneficial effects evolve into dominant traits, and the nonbeneficial effects are recessive traits. One consequence of this difference in dominance is the *vigor of heterozygous individuals*. Highly inbred laboratory animals have reduced fertility and reduced longevity. In these animals the recessive pleiotropic effects of major genes are in homozygous doses and they demonstrate their deleterious consequences. When the inbred individuals are crossed with animals of different strains, much of the fertility and longevity can be restored in one generation by having deleterious manifestations masked in the heterozygous state again (Chapter 30).

Kettlewell's studies of the peppered moth illustrate the fact that *dominance evolves*. Dominance depends on the rest of the genome. It can be changed by breeding experiments and changes with time in wild populations. In 1880 the *carbonaria* allele was rare, and so the melanic forms in the collections that were wild-caught then were almost certainly heterozygotes. They do not look the same as modern heterozygotes, having white patches and being paler. It is concluded that the *carbonaria* character was not completely dominant then. Breeding the modern form with typical form moths from Canada where there is no *carbonaria* allele present has shown that dominance can be broken down again when the allele occurs in another genetic background. Dark heterozygotes from Birmingham were crossed with the pale form from Canada. By the fourth generation the heterozygotes carried the *carbonaria* allele in a genome that was mostly Canadian. There were all gradations of lightness and darkness and the heterozygotes could not be clearly defined as a class. From these heterozygotes dominance of the *carbonaria* character could be completely reestablished by breeding back with the peppered form from Britain. Kettlewell has shown that the modifying effects are themselves dominant, but are not linked to the *carbonaria* locus. The modifiers are probably universal in Britain regardless of the presence of *carbonaria* alleles. This is thought to reflect the fact that melanism would have adaptive value in other circumstances also, and alleles conferring melanism may have spread at other times.

One of Kettlewell's greatest contributions to population genetics has been that he showed selection acting upon this moth in the field. He showed that predatory birds were very selective in their choice of the forms of the moth. A greater proportion of pale forms was taken by birds from the soot-covered trunks where no lichens grew, and more black forms were taken from pale trunks where there was no soot and there was good growth of lichens. This work was very carefully documented in series of experiments, and selective advantage was assessed for each form. In sooty areas the dark adult moth had twice the chance of survival of the pale form.

When the melanic form is collected in the field today it is not possible to tell without breeding whether it is a heterozygote. However, the Hardy-Weinberg formula can be used in an attempt to determine the relative frequencies of the alleles in a typical industrial habitat. The peppered moth is censused and late in the adult season there may be 1 percent pale forms there. This is the part of the population that is homozygous recessive at the locus for melanism ($c_t c_t$). The possibility is examined that the two alleles are present in the ratio of $q^2 CC : 2q(1-q)Cc_t : (1-q)^2 c_t c_t$. (This assumes that no other alleles of that locus are present.) This ratio is true if the population is large and is breeding at random in respect to the alleles in question, and if there has been no migration and no selection. Some of these possibilities can perhaps be ignored, but it is known that the adults have different selective values because of predation. Dark adults have twice the probability of surviving predators that the pale adults have and, to adjust for this, the pale proportion can be called 2 percent instead of 1 percent. No allowances can be made for relative survival rates at other stages of development because they are not known. At the locus for melanism, the proportion of c_t alleles in the gene pool of the parent population is the square root of $(1-q)^2$, which is $\sqrt{c_t c_t}$, or $\sqrt{0.02}$, or 0.14. The proportion of the C allele is therefore 0.86. The assumption is, then, that the population at the beginning of each generation contains 14 percent c_t alleles and 86 percent C alleles. The probable proportions of the genotypes can thus also be calculated. The proportion of

$CC = q^2 = 0.86^2 = 0.74.$

The proportion of

$Cc_t = 1.00 - (c_t c_t + CC) = 1 - (0.02 + 0.74) = 0.24.$

The relative distribution of genotypes, then, is 74 percent CC, 24 percent Cc_t, and 2 percent $c_t c_t$. This balance is constant from generation to generation at the present time in habitats contaminated with soot from industry.

In the Manchester area from about 1810 to 1848, the allele C probably changed from a very low frequency to 0.01, and in the next 50 generations to 0.86 (Figure 14.6). Such a rate of change is consistent with an overall selective advantage of 30 percent for the melanic form in the sooty environment. From 1898 to the present time the C allele has constantly been in the proportion of 0.86.

Why do the alleles stay in these proportions? In fact, why does the c_t allele

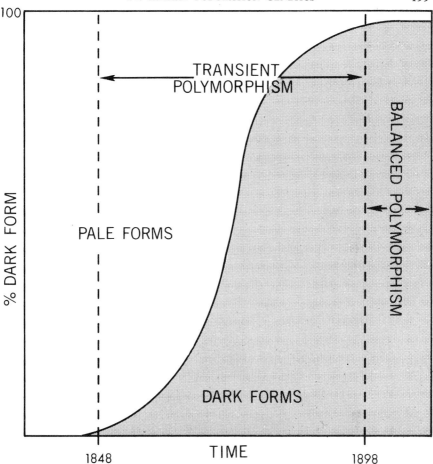

Figure 14.6. The increase of melanism in the peppered moth in an industrial area, such as Manchester, England. The allele C was assumed to be initially present in the proportion of one in a million. By about 1848 it would have had a relative frequency of 0.01. By 1898 in Manchester the dark forms were 0.98 and the allele, C, 0.86. Balanced polymorphism has persisted therefore at least 68 years. The rate of change is based on an assumed constant selective advantage of the heterozygote, which in practice would not occur, of 30 percent over the typical form. (Based on Kettlewell, 1958; and Haldane, 1924.)

remain at all? The balance of proportions in industrial areas is made by a balance of forces, which can be summarized as follows:
1) Mutation rates are very low and are insignificant in this case.

$$C \underset{\xrightarrow{\hspace{4cm}} c_t}{\overset{\xleftarrow{\hspace{4cm}}}{}}$$

2) Predators select against $c_t c_t$ in the dark ⎤ This is the main cause of
environment   ⎬ change of relative frequencies
$C \longleftarrow\hspace{-0.5cm}\underline{\hspace{6cm}}c_t$ ⎦ in recently industrialized areas.

3) Cc_t is more viable than $c_t c_t$ in breeding ⎤
experiments |
$C \longleftarrow\hspace{-0.5cm}\underline{\hspace{6cm}}$ ⎬ The selective advantage of the
Physiological advantages of Cc_t over CC | heterozygote is largely respon-
$\underline{\hspace{6cm}}\longrightarrow c_t$ ⎦ sible for the eventual balance.

4) Advantages in attracting mates, varying
with temperature

$$C \overset{\longleftarrow}{\underset{\longrightarrow}{\rule{6cm}{0pt}}} c_t$$

5) Quicker larval development of $c_t c_t$ may be
an advantage in short summers
$C \underline{\hspace{6cm}}\longrightarrow c_t$

6) Slower larval development of Cc_t and CC
may cope better with toxic materials in
soot
$C \longleftarrow\hspace{-0.5cm}\underline{\hspace{6cm}}c_t$

This summary of the balance of forces acting upon two alleles, though doubtless incomplete, shows how complex and how responsive can be the adaptation of characters to the ecological environment.

The existence of two or more forms of a species in a population is called a polymorphism. When the relative frequencies of the forms are changing, it is a *transient polymorphism*. When they are stable it is a *balanced polymorphism*. "Genetic polymorphism is the occurrence together in the same locality of two or more discontinuous forms of a species in such proportions that the rarest of them cannot be maintained merely by recurrent mutation ... The most general basis of genetic polymorphism is the balance of opposed advantage and disadvantage such that the heterozygote is favored compared with either homozygote, as originally suggested by Fisher in 1927" (Ford, 1964). In the peppered moth the period of change from high relative frequency of the c_t allele to low relative frequency was the time of transient polymorphism. In long-established industrial areas there is now balanced polymorphism. The greatest cause of the balance is the selective advantage that the heterozygote has over the homozygotes. The advantage is due to pleiotropic effects that are recessive and so exert their deleterious effects only in homozygotes. These effects are concerned with viability and other physiological properties that are not understood. The deleterious effects have changed in time; they have evolved so that they no longer are manifested in the heterozygote.

Studies of balanced polymorphisms in human populations are making important contributions to human genetics. They make research easier for the following reason. When the relative frequencies of two or more alleles are high in the population, then selection is most likely to be a significant force. The existence of a polymorphism, then, advertises selection. The investigator is able to ask specific questions about the selective properties of characters. Particularly he can ask: "Is presence or absence of a particular allele correlated with disease?" Allison and Raper were able to ask, "Is the high relative frequency in some human populations of the obviously deleterious allele of sickle-cell anemia associated with resistance to malaria?" (Chapter 1). This allele was indeed proved to give selective advantage to the heterozygote in the presence of falciparum malaria. The polymorphism may be transient in American Negroes; apparently the sickle gene is declining. The selective advantage of the heterozygote has gone because falciparum malaria is not significantly present in the North American environment. Many other such human polymorphisms are being investigated.

Balanced polymorphisms help to retain alleles* in populations, and so contribute to the maintenance of genetic variability. They are homeostatic mechanisms. Genetic variability of any sort is "insurance" against such environmental changes that are deleterious to the population. For instance, the genetic information for melanism was almost certainly already present in peppered moth populations although in low proportions. Similarly mosquito species had the genetic constitution, in low proportions, that enabled them to avoid complete destruction by modern insecticides (Chapter 15). Haldane's (1949) essay on the role of disease in evolution is recommended for, among other things, its discussion of the survival value of rare forms in disease agents and in hosts.

The peppered moth research illustrated (1) dependence of dominance on the genome, (2) pleiotropism, (3) many selective forces acting on forms that differed basically in one pair of alleles, (4) polymorphisms, and (5) adjustment to a changed environment by means of selection. These are basic in human population genetics. Concerning the fifth point, it is interesting to use the model of the melanic moths to speculate how human populations might respond in their genetic makeup to certain environmental changes, and how long such changes would take. More important, perhaps, is that such changes would be brought about partly by changes in death rates, some of which may be apparent now. Environmental changes to be considered might be industrialization and increased crowding.

The fifth point also concerns important problems in the control of disease vectors.

CHARACTERS CONTROLLED BY POLYGENES ARE INFLUENCED BY SELECTION

Ford's studies of the meadow-brown butterfly, *Maniola jurtina*, showed the effects of selection on polygenic characters in a wild animal population. Also, a general statement can be made from this example—the irregular distribution of a phenotypic character in a population may have hidden causes of great

* Polymorphisms based on bigger units of genetic material, such as chromosomes, occur also, as discussed later in the chapter.

interest. A hidden cause in this particular case was, very probably, differential susceptibility to disease.

The underwings of these butterflies have small, inconspicuous spots that vary in number from 0 to 5 and also vary in arrangement. Breeding experiments show that the patterns and numbers are determined by polygenes. Field surveys of many populations for very many years showed that populations have remarkably constant frequency distributions of the various spot types. Most of the males (about 68 percent) have 2 spots, but a few populations have been found with the mode at 1 or 3 spots. However, the distributions of frequencies of spot types in the females are varied throughout the wide range of the species in the British Isles. But whatever the local pattern is, it remains constant from year to year. This great local stability indicates that powerful selective forces are involved. Dowdeswell (1961) investigated in more detail some of the populations in isolated areas of chalk downs. Females laid approximately 250 eggs each. Dowdeswell brought into the laboratory young larvae taken from the field. Indoors there was 10 percent loss in the larvae and a 6 percent loss of pupae. The laboratory environment had allowed a much greater proportion of females with 2 and more spots to emerge. From comparison with the laboratory females it was estimated that, in the wild, young-stage females destined to become adults with 2 and more spots experienced a 69 percent mortality and the 0-spot females a 26 percent mortality. When Dowdeswell brought in late-stage larvae from the field he had two interesting results. (1) There was 77 percent parasitism by a wasp that caused death at the end of the larval period, and (2) the adult female pattern of spot frequency was for the first time the same as the wild pattern (Figure 14.7). A third piece of interesting evidence came in a summer with very good weather when the populations were all denser. The frequency distribution of spotting had changed so that more of the females with more than one spot were present. This all suggested that number of spots in the adult is correlated with mortality from parasitism in the young; that 2-spot females may be more susceptible to the disease than 0-spot females. Differential susceptibility of the larvae presumably is not directly associated with spotting, but is indirectly correlated through the pleiotropism of the controlling polygenes. Thus from the constancy of the spot-frequency distribution in habitats from year to year, its change from habitat to habitat, and its change in the laboratory, it may be concluded that relative frequencies of these polygenes are under the influence of selective forces. It is probable that a large part of the selective force is susceptibility or availability to the parasite.

POLYMORPHISMS OF CHROMOSOME VARIANTS IN WILD DROSOPHILA POPULATIONS

Dobzhansky (1961) studied chromosomes in wild populations of *Drosophila pseudoobscura*. The salivary glands of the larvae have giant (polytene) chromosomes, and inversions of sequences of parts of the chromosomes can be seen from their patterns of cross-striations. Inversions can also be identi-

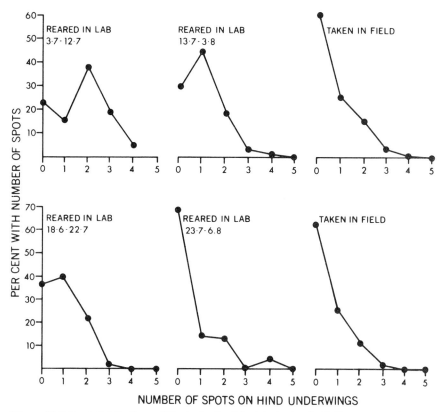

Figure 14.7. Spot-distributions of the butterfly, *Maniola jurtina*, in 1958 upper graphs, and 1959 lower graphs. Specimens reared in the laboratory from young wild larvae, and emerging early or late in the season, are compared with wild adults from the same locality. Spot-frequencies of the specimens reared throughout 1958 and early 1959 are similar, but differ from those of the wild adults; however the spotting of those reared late in 1959 is very like that of the wild butterflies. (After Ford, 1964.)

fied in meiosis, because pairs of chromosomes that are heterozygous for inversions have loop formations when they lie together. The third chromosome of *D. pseudoobscura* has 16 known morphological forms* occurring in high relative frequencies in wild populations of western United States and Mexico. The phenotypes are indistinguishable. Some of these forms occur in high relative frequencies over great areas. This suggests adaptive significances of unknown nature. *Drosophila* has from 6 to 8 generations a year there. At one location in Southern California in March the relative frequencies of the *ST* and *CH* forms of the third chromosome are 53 percent and 24 percent. In

* Three of the most common inversions of the third chromosome are known as Arrowhead (AR), Standard (ST) and Chiricahua (CH).

June they have changed to 28 percent and 40 percent. From June to August these changes are reversed to the proportions of early spring, and these remain through the winter. This cycle is accurately repeated each year at that place. Proportions of the *AR* form also fluctuate regularly in the same area. Other cycles occur in other localities, sometimes with quite different patterns. In the Sierra Nevada, at 850 feet the *ST* form is 46 percent and the *AR* form is 25 percent. At 10,000 feet *ST* has declined to 10 percent and the other risen to 50 percent. There are many other examples of these chromosome polymorphisms in various species and places, studied by many workers.

In order to understand these variations Dobzhansky did a series of laboratory experiments that started with populations of *D. pseudoobscura* with known relative frequencies of chromosome types. In the laboratory there are 14.5 generations in a year. In uniform environments 11 percent *ST* and 89 percent *CH* changed in 6 months to 52 percent *ST* and 48 percent *CH*, and in 4 months more to 70 percent *ST* and 30 percent *CH*. They stabilized at this relative frequency. This was a balanced polymorphism of chromosome types. *ST/CH* was superior to *ST/ST* and *CH/CH*, so all three genotypes were permanently maintained. This proportion was influenced by temperature, by changes in food, and by competition with added flies of different inversion patterns. Dobzhansky pointed out that in laboratory populations, by changing the temperature he could parallel certain aspects of the seasonal variation in chromosome types that occurred in the wild. On the Piñon Flats, *ST* increases in the hot summer while *CH* declines, and little change occurs in winter. In the laboratory, *ST* had selective advantage over *CH* at 25°C, while the two had equal advantage at 15°C. In spring in nature there is a decline of *ST* and a rise of *CH*. This part of the changes was never detected in the population cages, whatever environmental changes were introduced. Then it was pointed out that *Drosophila* in the cages was limited in numbers by larval competition for food, while in nature in spring food was unlimited, and the population expanded in numbers. This was imitated in the cages and *CH* increased at the expense of *ST*. This is another example of relative frequencies in the gene pool being influenced by change of numbers.

Malariologists might find such cyclical changes of chromosome types in mosquito populations throughout the year; and such changes could be involved in vertebrate populations that fluctuate widely and regularly. They could even be part of the regulatory mechanisms. The relevance of this kind of observation to human genetics is unknown.

CONCLUSIONS

To understand human genetics, models from the entire living world are used; and Chapters 10 and 11 indicated that the genetic code and its biochemical interpretations have a remarkable amount of uniformity throughout most of the forms that have been studied. Population genetics is a fairly young subject whose exploration has hardly begun. Man is difficult to study genetically. For instance, chromosome inversions could be studied in meiosis, but living cells in meiosis for tissue culture are hard to obtain. They come from testes and from ovaries of early fetuses. However, the latter should be available where induced abortion is legal. Animal, plant, and microbial genetics have preceded most aspects of human genetics and will continue to supply experimental material and models.

Examples have been chosen here from animal populations in the wild, as far as possible, and with the aim of avoiding concepts of single causes and simple reactions. The reader is referred to the summary of contents at the beginning of the chapter for a synopsis of the ideas presented.

REFERENCES

Birdsell, J. B. 1950. Some implications of the genetical concept of race in terms of spatial analysis. In *Cold Spring Harbor Symp. Quant. Biol.* **15**: 259–314.

Clarke, C. A., and Sheppard, P. M. 1964. Genetic control of the melanic form *insularia* of the moth *Biston betularia* (L.). *Nature* **202**: 215–216.

Dobzhansky, T. 1955. In *Population Genetics: The Nature and Causes of Variability in Populations. Cold Spring Harbor Symp. Quant. Biol.* **20**: 1–15.

Dobzhansky, T. 1961. On the dynamics of chromosomal polymorphism in *Drosophila. Symp. R. Ent. Soc. Lond.* **1**: 30–42.

Dowdeswell, W. H. 1961. Experimental studies on natural selection in the butterfly *Maniola jurtina. Heredity* **16**: 39–52.

Ford, E. B. 1964. *Ecological Genetics.* Wiley, New York.

Glass, B., and Li, C. C. 1953. The dynamics of racial intermixture—an analysis based on the American Negro. *Amer. J. Human Genet.* **5**: 1–20.

Glass, B., Sachs, M. S., Jahn, E. F., and Hess, C. 1952. Genetic drift in a religious isolate: an analysis of the causes of variation in blood group and other gene frequencies in a small population. *Am. Naturalist,* **86**: 145–159.

Haldane, J. B. S. 1924. A mathematical theory of natural and artificial selection. *Trans. Cambridge Phil. Soc.* **23**: 19–40.

Haldane, J. B. S. 1949. Disease and evolution. *Ric. Sci. Suppl.* A **19**: 68–75.

Haldane, J. B. S. 1959. Natural Selection. In *Darwin's Biological Work. Some Aspects Reconsidered.* P. R. Bell (ed.). Cambridge University Press, Cambridge.

Kettlewell, H. B. D. 1958. A survey of the frequencies of *Biston betularia* (L.) (Lep.) and its melanic forms in Great Britain. *Heredity* **12**: 51–72.

Kettlewell, H. B. D. 1965. Insect survival and selection for pattern. *Science* **148**: 1290–1296.

Penrose, L. S. 1959. Natural selection in man: some basic problems. In *Natural Selection in Human Populations.* D. F. Roberts and G. A. Harrison (eds.). *Symp. Soc. Study Human Biol.* **2**. Symposium Publications Division, Macmillan, New York.

CHAPTER 15

The Genetics of Insecticide Resistance

LLOYD E. ROZEBOOM AND FRANCIS S. L. WILLIAMSON

Since the early 1940's new toxicants for the control of arthropods of agricultural and medical importance have been intensively developed. There have been many beneficial results but also certain problems have arisen. One problem is the appearance and spread of arthropod resistance to the toxicants and the purpose of this chapter is to present the genetic aspects of this. Thorough reviews of the physiology and genetics of insecticide resistance have been published by Brown (1958), Hoskins (1967), O'Brien (1966), Oppenoorth (1965), Perry (1964), Roan and Hopkins (1961), and Wright and Pal (1967).

Chemicals used to kill insects enter through the gut (stomach poisons), the respiratory system (fumigants), and through the external surfaces (contact poisons).

Stomach poisons, especially arsenicals and cryolites, were used in former years against agricultural pests. One of these, Paris green, was used extensively in malaria control as an *Anopheles* larvicide.

Fumigants are used against insects in grain storage bins, holds of ships, and similar structures where access is difficult or the pests are protected by unusual opportunities for concealment. Fumigants are not very useful in public health procedures.

Contact poisons—*The space sprays and aerosols* are dispensed as fine mists into the atmosphere to kill flying insects or those resting on walls or ceilings. These sprays are most familiar as household pest killers. They contain pyrethrum or a thiocyanate as a knockdown (paralysis) agent, combined with a killing agent, which may be one of the chlorinated hydrocarbons or an organophosphorus compound. *Residual insecticides* are applied to surfaces on which insects come to rest. Several kinds of such chemicals are employed for control of disease vectors and pests, and the first was DDT, one of the chlorinated hydrocarbons. In World War II, DDT was shown to be highly effective against lice for several weeks if dusted into clothing. When sprayed on surfaces it was toxic for weeks to houseflies, mosquitoes, and other insects. This was one of the most significant discoveries for public health of recent years. It permitted a new approach to the control of diseases transmitted by insects in the home. DDT sprayed in houses caused such a dramatic reduction in malaria that in 1948 the World Health Organization adopted for their international programs a policy of malaria eradication. It was predicted with great confidence that malaria would disappear within a few years. Although malaria is no longer the serious problem it was, eradication has not been achieved except in a few areas. There are several reasons for this, the greatest of which is the development of resistance by the vectors.

206

RESISTANCE AND ITS SPREAD IN WILD POPULATIONS

In 1946 in Italy and Sweden and in 1947 in the United States houseflies were found to be no longer affected by the usual doses of DDT. Other chlorinated hydrocarbons were effective for a time but then the flies became resistant to these substitutes. Other species of insects such as mosquitoes, fleas, and lice followed in developing populations resistant not only to the chlorinated hydrocarbons but to the carbamates and even to some extent to the organophosphorus compounds. In 1965 Georghiou listed 186 arthropod species with resistant populations, including 34 species of *Anopheles*, and the number still increases.

INSECTICIDES USED IN PUBLIC HEALTH PROCEDURES

The Chlorinated Hydrocarbons

DDT is a slow-acting poison probably acting most on the peripheral sensory system. DDT is highly soluble in lipids and is absorbed in the lipoprotein surface of nerves, altering their permeability. A single stimulus of a DDT-treated sensory nerve produces a patternless series of impulses, resulting in twitches and spasms. The insect becomes hyperactive, then shows lack of coordination followed by prostration, convulsions, paralysis, and death. The actual causes of death are not known but may include exhaustion, accumulation of toxic metabolites, and disruption of hormonal and enzyme systems. Other chlorinated hydrocarbons in use are benzene hexachloride (BHC) and two cyclodiene compounds, chlordane and dieldrin.

The Organophosphorus Insecticides

Many synthesized derivatives of phosphoric acid are inhibitors of cholinesterase. Those used as insecticides are Malathion, diazinon, and DDVP as well as certain others more dangerous to man and animals such as parathion.

The Carbamates

The carbamates are esters of carbamic acid. Certain of these have been used as substitutes for the chlorinated hydrocarbons as residual household sprays, the most common being known as Sevin.

THE MECHANISMS OF RESISTANCE

An animal with behavioral resistance avoids lethal contact with the insecticide. This subject is discussed in Chapter 28.

Physiological resistance may protect an insect against a specific insecticide by one or more of the following mechanisms.

Reduced Permeability Caused by Increased Thickness of the Cuticle

It has been suggested that DDT resistance in the housefly is correlated with darker pigmentation of the cuticle, stiffer tarsal bristles, and thicker and more pigmented tarsi, pulvilli, and membranes at the tarsal joints. There are great differences in cuticle permeability having nothing to do with insecticide resistance. Perry and Sacktor (1955) could find little difference in amounts absorbed from topical applications of DDT to DDT-resistant and susceptible strains of houseflies. Furthermore, several investigators have shown that resistant insects are resistant to injected insecticides as well as to those applied topically.

Storage of Toxicants in Nonactive Tissues

The chlorinated hydrocarbons have an affinity for lipids, and in some studies a positive correlation has been observed between resistance and higher fat content in houseflies and mosquitoes. Higher lipid content may make it possible for the insect to dissolve more DDT and to store it in various tissues and thus reduce its concentration at the site of action. It also has been suggested that the DDT molecule may be destroyed enzymatically more readily when it is dissolved in lipoprotein. There have also been many negative correlations. For example, Perry et al. (1963) could find no difference in the distribution of stored aldrin and dieldrin in susceptible and resistant flies.

Excretion

Resistant strains may be able to eliminate toxicants more rapidly. In some resistant larvae of *Aedes aegypti* DDT is taken up by the peritrophic membrane which lines the gut. This membrane is then excreted in long streamers (Abedi and Brown, 1961).

Decreased Sensitivity

Pratt and Babers (1953) observed the duration of leg tremors induced by applying DDT to the thoracic ganglia. Ganglia and legs from DDT-resistant houseflies recovered more rapidly than preparations from normal flies. It is doubtful that the recovery was the result of enzymatic breakdown of DDT in resistant flies because the recovery occurred during the initial 5 minutes.

Bypassing a Blocked Physiological Pathway

DDT can cause partial inhibition of the cytochrome oxidase system in flies, and DDT-resistant flies may have higher cytochrome oxidase activity. However, there is no proof that the chlorinated hydrocarbons in general inhibit enzyme systems, or that bypass systems are of significance in insecticide resistance.

Detoxication of the Insecticide

This appears to be the most important mechanism whereby an insect overcomes the lethal action of a poison. The resistant insect produces a specific enzyme which breaks down the molecule. Chamberlain (1950) demonstrated that although flies could recover from low doses of pyrethrum, prior or simultaneous treatment with piperonyl butoxide destroyed this ability. Sternberg, Vinson, and Kearns (1953) showed that in resistant flies DDT was converted to DDE (dichlorodiphenyldichloroethylene) by DDT-dehydrochlorinase in the presence of glutathione.

Enzymatic breakdown of the cyclodiene compounds also occurs. Thus BHC is converted to pentachlorocyclohexane, dichlorothiophenol, and other metabolites. Chlordane, heptachlor, and aldrin are changed but the relationships to resistance are not clear. Aldrin is changed to dieldrin which is not further broken down despite the fact that individuals in which this has taken place may be strongly resistant to dieldrin.

Resistance to DDT protects to limited extents against analogues of DDT, but usually not to cyclodiene compounds. Conversely, resistance to BHC, dieldrin, and related insecticides gives little or no protection against DDT and its analogues.

The highly effective organophosphorus insecticides may undergo enzymatic detoxication. Parathion is converted to the more deadly paraoxon, but supposedly this compound is hydrolyzed further. Malathion is converted to malaoxon.

The carbamate insecticides also are attacked by insect enzymes. Georghiou et al. (1966) treated groups of larvae of *Culex fatigans* that were susceptible or resistant, with C^{14}-labeled ortho-isopropoxydiphenyl methyl carbamate. Scanning for radioactivity showed that susceptible larvae had metabolized 12.4 percent of the insecticide, and resistant larvae 30.4 percent. This is a 2.5-fold difference in carbamate detoxication.

The correlation of knowledge of enzymatic detoxication and resistance is far from perfect. It was noted above that dieldrin is not broken down although insects develop strong resistance to it. One analogue of DDT is prolan-bulan which cannot be dehydrochlorinated; yet flies and mosquitoes become resistant to these compounds. Experiments with C^{14}-labeled TDE (tetrachlorodiphenylethane) indicated that *Culex tarsalis* larvae metabolized the compound by oxidation as well as by dehydrochlorination. Plapp et al. (1965) therefore suggested that DDT and related compounds may be destroyed by mechanisms other than dehydrochlorination.

Vigor Tolerance

Well-nourished large insects, containing ample nutritional reserves and in full possession of normal enzymatic and other physiological functions may be able to survive contact with an insecticide that would kill weaker individuals. Selection for these characters, not necessarily through contact with an insecticide, may further increase general vigor and with it greater insecticide tolerance. The effects of such selection on insecticide resistance are usually small.

To conclude, several systems appear to be involved in insecticide resistance. Despite inconsistencies such as those referred to above, the most important appear to involve specific enzymes. This conclusion is supported by studies of the inheritance of resistance.

THE ORIGIN OF RESISTANCE

There is no evidence that insecticide resistance arises by mutation caused by insecticides. Rather the evidence indicates that there is selection for genetic factors that are already present in populations.

Beard (1952) subjected wax moth larvae, *Galleria mellonella*, and the milkweed bug, *Oncopeltus fasciatus*, to doses of various toxicants which caused 30 to 70 percent mortality (called the pretreatment). The survivors were given the same or another insecticide. Pretreatment with nicotine, DDT, pyrethrum, and arsenic did not make the surviving individuals more resistant to these insecticides. In fact those pretreated with DDT appeared to be somewhat more susceptible to subsequent treatment with DDT and nicotine. The pretreated *Galleria* larvae were also more susceptible to the second treatment than were control larvae. Similar results have been obtained with houseflies. Individuals treated with sublethal concentrations become more rather than less sensitive to subsequent treatment.

Individual insects thus appeared to lack the ability to acquire a tolerance for an insecticide during a lifetime. Nor was there evidence that heritable changes were induced among populations of adults or larvae so treated. For instance, *Drosophila* larvae have been reared in media containing sublethal concentrations of DDT and BHC (benzene hexachloride) without enhancement of resistance in the adults. *Drosophila melanogaster* was reared for 50 generations on media containing sublethal concentrations of DDT and there was no increase in resistance.

TABLE 15.1

Theoretical Kills by Discriminating Doses of Insecticide[a]

| Dominance of the character | Dose of insecticide | Percent kill of genotypes | | | Percent kill of populations derived from crosses of homozygous resistant and susceptible parents | | Backcross of F1 with: | |
		Homozygous for: Resistance	Homozygous for: Susceptibility	Heterozygous	F_1	F_2	Resistant parents	Susceptible parents
Resistance dominant	High	0	100	0	0	25	0	50
Resistance semidominant	High	0	100	100	100	75	50	100
	Low	0	100	0	0	25	0	50
Resistance recessive	High	0	100	100	100	75	50	100

[a] Susceptible insects have been bred with resistant insects, and resistance is determined by two alleles at one locus.

A field population that is strongly selected by an insecticide will eventually attain a certain level of resistance. This plateau remains rather constant if the relative frequencies of genes influencing the resistance do not change. If it could be shown that continued treatment caused another rise in the level of resistance, mutation would be indicated. No such evidence is available. Efforts to bring about resistance in populations by application of insecticides to successive generations have met with failure when the genes for this resistance were not already present.

The conclusion from these studies is that insecticides are not mutagenic agents. Field and laboratory experience supports the theory that populations become resistant to an insecticide through selection for genetic factors already present. The rapidity with which resistance appears depends among other things upon the initial frequency of these factors, the dominance of the characters, and the intensity of selection. The level of resistance attained is determined by the degree of protection conferred by the genotype, and in the case of polyfactorial inheritance, upon the number of factors present.

RESULTS OF BREEDING EXPERIMENTS
Musca domestica

Lichtwardt (1956) compared parent and hybrid populations of houseflies resistant and susceptible to DDT. Theoretical kills for such experiments are given in Table 15.1. Several concentrations of DDT were applied topically to adults. Discriminating doses were 0.608 and 7.6 mg of DDT per fly. Results presented in part in Table 15.2 showed that mortality patterns in the F_1, F_2,

TABLE 15.2

Percentage Mortalities of Adult Houseflies, *Musca domestica*, Twenty-four Hours after Topical Application of DDT[a]

Generation	Dose of DDT in μg per fly				
	0.456	0.608	7.6	15.2	38.0
	Percent mortality				
Susceptible parent SS[b]	98.9	100	—		—
Resistant parent RR[c]	—	—	1.9	1.2	3.2
F_1 SR	—	—	0	—	24.2
F_1 RS	—	—	0.9	1.1	11.1
F_2 from $SR \times SR$	—	22.2	22.0	—	—
F_2 from $RS \times SS$	—	—	28.1	—	—
Backcross $SS \times SR$	—	51.4	47.7	—	—
Backcross $RS \times SS$	—	50.4	48.9	—	—
Backcross $RS \times RR$	—	—	0	—	—
Backcross $SR \times RR$	—	—	0	—	—

[a] Lichtwardt (1956).
[b] S represents the allele conferring susceptibility.
[c] R represents the allele conferring resistance.

and backcross populations were consistent with the hypothesis that there was a single factor for resistance to DDT and that resistance was dominant.

Harrison (1953) showed that an allele at another locus in *Musca domestica* controlled resistance to knockdown by DDT. In this case resistance to knockdown was recessive. Georghiou *et al.* (1963) showed that resistance to dieldrin in *M. domestica* was semidominant and controlled at a single locus. Resistance to Malathion in the housefly and in *Culex tarsalis* and to parathion in the housefly appears to be inherited as single factors. Continuing studies with marker genes are showing that there are several factors for resistance of the housefly to the chlorinated hydrocarbons, the carbamates, and organophosphorus compounds. Some are major genes, others have only small effects and the character is dominant, recessive, or intermediate.

Mosquitoes

Qutubuddin (1958) tested crosses of a DDT-resistant strain of *Aedes aegypti* from Trinidad and a susceptible laboratory strain. Larval mortalities were read after 24 hours of exposure to from 0.1 to 10.0 parts per million (ppm) of DDT (Table 15.3). The results suggested that there is control by a single allele

TABLE 15.3

Percentage Mortalities of Larval Mosquitoes, *Aedes aegypti*, Twenty-four Hours after Exposure to DDT[a]

Generation	Parts per million of DDT in water				
	0.1	0.5	2.5	5.0	10.0
	Percent mortality				
Resistant parent RR[b]	—	—	—	—	5
Susceptible parent SS[c]	71	92	—	—	—
F_1 RS	—	9	21	58	92
F_1 SR	—	—	36	67	93
F_2 from $RS \times RS$	—	27	—	—	64
F_2 from $SR \times SR$	—	20	—	—	65
Backcross to SS female	—	46	—	—	—
Backcross to RR female	—	—	—	—	45

[a] Qutubuddin (1958).
[b] R represents the allele conferring resistance.
[c] S represents the allele conferring susceptibility.

with resistance dominant at concentrations of 0.5 ppm. At higher concentrations a heterozygous population is only partially resistant. In a more detailed reinvestigation of DDT-resistance in the Trinidad strain Wood (1965) concluded that in addition to a major gene conferring a low degree of resistance there is a modifier close to the locus for yellow on chromosome 2, affecting dominance of the resistance, and that there is also a sex-dependent factor.

Klassen and Brown (1964) used mutant *Aedes aegypti* to show linkage among genes conferring resistance to DDT and dieldrin, and other phenotypic characters. The following sequence was observed in linkage group 2: spot—yellow—black pedicel—gold—dieldrin resistance—DDT resistance. The distance from yellow to dieldrin was 18.6 cross-over units and from yellow to DDT 22.1. This is additional evidence that although closely linked there are separate loci for control of DDT and dieldrin resistance.

Davidson (1956, 1958) crossed a strain of *Anopheles gambiae* from Northern Nigeria that was resistant to cyclodiene compounds with susceptible strains from Lagos and Diggi. Adults were exposed for 1 hour to papers impregnated with dieldrin or benzene hexachloride. Exposure to 0.33 percent dieldrin for 1 hour killed the homozygous susceptible *SS* adults, but not the heterozygous *SR* or the homozygous resistant *RR* adults. A 2-hour exposure to 4.0 percent dieldrin killed both *SS* and *SR*, but not the *RR* adults. Similar discriminating dosages for BHC were 1 hour of exposure to 0.025 percent and 0.1 percent concentrations. Results of tests of backcross populations (Table 15.4) were

TABLE 15.4

Results of Tests of *Anopheles gambiae* Backcross Populations to Low and High Dosages of Dieldrin[a]

Backcross	Percentages of genotypes[b]		
	SS	SR	RR
$F_1 \times SS\male$	57	43	—
$F_1 \times SS\female$	49	51	—
$F_1 \times RR\male$	—	52	48
$F_1 \times RR\female$	—	49	51
$F_1 \times SS\male$	49	51	—
$F_1 \times SS\female$	56	44	—
$F_1 \times RR\male$	—	55	45
$F_1 \times RR\female$	—	51	49

[a] Davidson (1956).

[b] Percentages of genotypes were derived from deaths of adults following exposure to low or high doses.

S=allele conferring susceptibility.

R=allele conferring resistance.

close to those expected for single factor control. Resistance was considered to be semidominant, as a high dose killed the heterozygous *SR* but not the homozygous *RR* adults, while a low dosage killed the *SS* adults but not the heterozygotes.

There have been many other studies of the genetics of resistance in mosquitoes. It appears that, in general, resistance to DDT is recessive and monofactorial and to the cyclodiene compounds it is dominant or semidominant and monofactorial. There is also evidence for ancillary genes.

Polyfactorial Inheritance

It is difficult to distinguish polyfactorial from semidominant monofactorial resistance. However, with advances in the genetics of houseflies, mosquitoes, and other insects, marker genes are recognized and experiments with these confirm that most resistance is monofactorial, but that some is polyfactorial.

Crow (1957) demonstrated polygenic control of specific insecticide resistance in *Drosophila*. The resistant strain was developed in a large genetically variable population kept in cages with inside surfaces painted with irregularly increasing amounts of DDT in an attempt to simulate conditions in nature. Selection for resistance was increased by exposing adults for 18 or 24 hours to DDT on filter paper. This strain was crossed with a susceptible strain with marker alleles on all the chromosomes, and populations were derived that had 1 chromosome from the resistant parent, 2 chromosomes, 3 chromosomes, and so on. Tests showed that resistance correspondingly increased from less than 1 percent survival in the parent susceptible strain to almost 70 percent

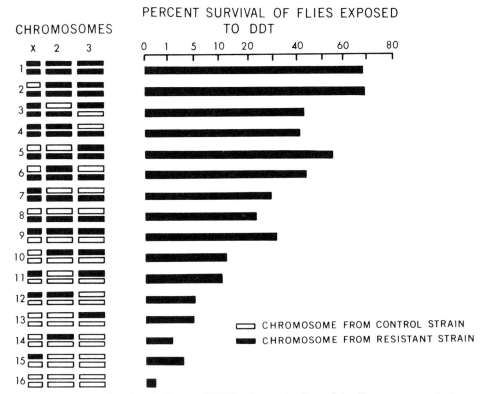

Figure 15.1. Polygenic inheritance of DDT resistance in *Drosophila*. Chromosome content of flies is shown against the percent survival of such adults when exposed to a standard concentration of DDT on filter paper. (After Crow, 1957.)

survival in animals with 6 chromosomes from the resistant strain (Figure 15.1) Most of the factors that conferred dominant resistance were in chromosome 2, in two regions, with probably more than one locus in each region. For chromosome 3 and the X-chromosome there was less information but these did contribute to resistance.

Harris, Weardon, and Roan (1961) with Malathion-resistant houseflies and Bragassa and Brazzel (1961) with endrin-resistant boll weevils showed F_1 and F_2 generations to be resistant to an extent intermediate between the resistant and susceptible parent populations. This was interpreted as being due to poly-factorial control of resistance. Similar results have been obtained by Geor-ghiou (1965) with *Culex fatigans* resistant to ortho-isopropoxyphenyl methyl carbamate.

SELECTION BY INSECTICIDES

Selection changes the relative frequencies of genotypes so that a largely susceptible population by increase of its survivors is replaced by a resistant population.

In the Field

In 1945 DDT began to be used extensively for housefly control in Illinois. Decker and Bruce (1952) determined the LD_{50}'s (dose lethal to 50 percent of the population) of samples of wild populations through 1951 and observed a very gradual increase in resistance for the first 4 years. In 1949, 86 percent of the samples were showing a significant degree of resistance; in 1950 and 1951 all populations sampled were moderately or highly resistant. Meanwhile as DDT began to fail, farmers used other insecticides and quickly the flies showed resistance to these as well. In 1951, 95 percent of the field populations were resistant also to methoxychlor and 50 percent to benzene hexachloride and dieldrin. The reason for the more rapid rise of resistance to newer insecticides is discussed under "lag phase."

In wild populations an insecticidal program may cause a very heavy mor-tality of the target species and even eliminate it locally. But wild populations are vastly larger than laboratory colonies and small numbers of individuals with high degrees of protection may survive. Such resistance is likely to be controlled at a single locus. Genes with moderate or low effects, such as poly-genes, are likely to exist alone and so confer small degrees of resistance and be eliminated. Thus in the field there tends to be selection for monofactorial control of resistance.

In the Laboratory

Selection for resistance in laboratory populations ordinarily is not as severe as in wild populations with the result that genes with small effects tend to survive. Several or many such genes may accumulate through continued selection to give rise to polyfactorial inheritance of resistance.

Laboratory tests showing changes of populations through many genera-

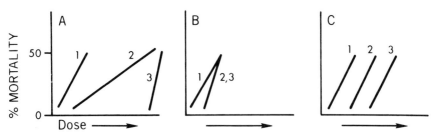

Figure 15.2. Dosage–mortality curves shown here are percentage mortality plotted against increasing dose of insecticide to the level that kills 50 percent of the population (LD$_{50}$). The tests were made as the population passed through several generations, each exposed to insecticide.

Population A started by being fairly homogeneous for susceptibility (curve 1) but contained some individuals with genetically based resistance to the specific insecticide. So the selective agent at some point began to rapidly change the relative frequencies of the genotypes. During the period of change, or transient polymorphism, the mortality–dosage curve was flatter (2) because of the presence of all kinds of the genotypes. When most of the population was resistant, the curve was again steep (3).

Population B was all susceptible. There were no individuals with genetically based resistance that could be increased in relative frequency by selection. The later generations indicated loss of their most susceptible types (curves 2, 3).

In *Population C* specific genetic resistance to insecticide was not involved, but the genetics of the population changed so that the surviving population became more vigorous in a general fashion. Such changes might be induced by many kinds of environmental extremes, and to be due, among other things, to increase in weight, fat content, and cuticle thickness. The genetic factors involved are likely to be polygenes. (After Hoskins and Gordon, 1956.)

tions, each exposed to insecticide, have been summarized by Hoskins and Gordon (1956). They recognize three kinds of changes of dosage–mortality curves (Figure 15.2), representing changes due to selection for specific resistance with genetic basis, changes where there is no selection, and changes due to selection for vigor tolerance.

Rates of Change of Relative Frequencies of Genotypes

The rates and the patterns of change under selective pressure follow the same principles as in other animals and with other selective agents (Chapter 14 and Figure 15.3). In laboratory populations of houseflies subjected to DDT, Decker and Bruce (1952) found that the change in relative frequencies of phenotypes was slow at first (the lag phase of 6 to 20 generations), rapid later (6 to 20 generations), and then slow again (the plateau). The fewer the factors conferring resistance, the more rapid will be the rise in resistance in the entire population (see the mathematical model of Grayson and Cochran, 1955). At the plateau the population may be resistant to doses that are 100 to more than 1,000 times greater than doses killing the original strain.

The Lag Phase

In untreated populations most factors conferring resistance are present in low

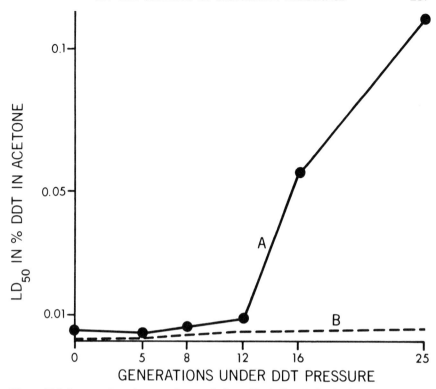

Figure 15.3. Increase in resistance of a population of body lice, *Pediculus humanus corporis*. Population A was reared through 25 generations on cloths impregnated with 0.001 percent DDT-acetone solution; B was a regular laboratory population. (After Eddy *et al.*, 1955.)

relative frequencies, and the rarer the resistant animals the longer is the lag phase. Moreover, if resistance is recessive and the genetic determinants are rare, resistant phenotypes (homozygous recessives) are extremely rare and thus the lag phase is extended further.

There is no doubt that there is also selection for fittest genetic combinations, for modifiers, and for vigor tolerance. Hoskins and Gordon (1956) believe that in populations previously heavily selected such advantages tend to be already assembled in those rare individuals that have resistance to other insecticides determined at other loci. This would explain the greatly decreased lag phase in the spread of resistance to second insecticides.

Balanced Polymorphisms

A population can reach a level of resistance at which a high percentage survives a large dose of insecticide and then shows no further change. This occurs despite attempts to continue selection by breeding from the survivors of high doses. This may mean that all factors for resistance are already homozygous, or there may be a balanced polymorphism because of selective advantage of the heterozygote. In insects homozygotes for the most resistant factors are frequently less viable, less fertile, and have longer developmental periods than heterozygotes. Indeed, the most resistant individual may be the heterozygote (Crow, 1957).

Reversion to Susceptibility

In the laboratory where vigorous selection has produced resistant populations, when the selection pressure is relaxed, reversion of the population to susceptibility can occur. In treated extensive wild populations complete homozygosity for resistance cannot be attained. Susceptible individuals tend to compete better by being more viable and more fertile and perhaps by developing more rapidly than resistant individuals. This may well be because their particular genetic combinations have already existed for long times. Because most natural populations have a very low frequency of factors for specific resistance, it may perhaps be assumed that these factors are deleterious in normal environments, by means of pleiotropic effects or by linkage with other, less favorable factors. Wild populations have been reported to change from DDT resistance to susceptibility in *Anopheles sundaicus* in Java, *A. stephensi* in Iran, *A. sacharovi* in Greece, and the housefly in Denmark, and from dieldrin resistance to susceptibility in *Anopheles culicifacies* in India and *A. gambiae* in Nigeria (Brown, 1963; Keiding, 1963).

CONCLUSIONS

Resistance to an insecticide is not absolute. A strongly resistant insect can be killed by continued contact with a toxicant. But even a moderate degree of resistance may be the cause of failure of residual DDT to stop transmission of malaria. Partial resistance combined with increased irritability can prevent the mosquito from contacting a lethal dose. Thus it is possible for some mosquitoes to feed inside houses treated with DDT and then survive. DDT is the preferred insecticide for control of mosquitoes that carry malaria, but when it becomes ineffective it can be replaced by dieldrin. Then liberal use of dieldrin causes rapid rise of very resistant mosquito populations. So dieldrin should be used only for short times in small problem localities. After this it may be possible to revert to DDT or to replace dieldrin by one of the organophosphorus compounds.

It should be apparent from this example that for efficient control programs knowledge of the status of resistance in vector populations is indispensable. With understanding of the probable modes of inheritance of the resistance, and with field surveys using the Hardy-Weinberg principle where appropriate, it may be possible to get crude but rapid estimates in the field of the extent of resistance and the probabilities of spread.

Knowledge of the physiology and genetics of resistance may suggest better ways of control. For instance, an enzyme may be produced to protect an insect against a certain poison. If its production is genetically determined in a way that precludes the production of another enzyme affording protection against another insecticide, selection for resistance to the first toxicant might give rise to a population unable to adjust to the second poison. That such a phenomenon of negative correlation may be of practical significance is indicated by some work of Georghiou and Metcalf (1963), in which selection of *Anopheles albimanus* by a carbamate was accompanied by a loss of resistance to dieldrin.

The search for better insecticides continues, as does exploration for biological and other methods of control. Liberation of sterile males and application of chemosterilants have given good or promising results in certain instances. It has been suggested that insect growth hormones might be exploited as insecticides. However, it is evident that for some time chemical insecticides will have to be used for control of pest and disease-bearing arthropods.

REFERENCES

Abedi, Z. H., and Brown, A. W. A. 1961. Peritrophic membrane as a vehicle for DDT and DDE excretion in *Aedes aegypti* larvae. *Ann. Entomol. Soc. Am.* **54:** 539–542.

Beard, R. L. 1952. Effect of sub-lethal doses of toxicants on susceptibility of insects to insecticides. *J. Econ. Entomol.* **4:** 561–567.

Bragassa, C. B., and Brazzel, J. R. 1961. Inheritance of resistance to Endrin in the boll weevil. *J. Econ. Entomol.* **54:** 311–314.

Brown, A. W. A. 1958. *Insecticide Resistance in Arthropods.* World Health Organ., Geneva. 240 pp.

Brown, A. W. A. 1963. Meeting the resistance problem. *Bull. World Health Organ.* Suppl. **29:** 41–50.

Chamberlain, R. W. 1950. An investigation on the action of piperonyl butoxide with pyrethrum. *Am. J. Hyg.* **52:** 153–183.

Crow, J. F. 1957. Genetics of insect resistance to chemicals. *Ann. Rev. Entomol.* **2:** 227–246.

Davidson, G. 1956. Insecticide resistance in *Anopheles gambiae* Giles: a case of simple Mendelian inheritance. *Nature* **178:** 863–864.

Davidson, G. 1958. Studies on insecticide resistance in anopheline mosquitoes. *Bull. World Health Organ.* **18:** 579–621.

Decker, G. C., and Bruce, W. N. 1952. House-fly resistance to chemicals. *Am. J. Trop. Med. Hyg.* **1:** 395–403.

Eddy, G. W., Cole, M. M., Couch, M. D., and Selhime, A. 1955. Resistance of human body lice to insecticides. *Publ. Hlth. Rep.* (Washington, D.C.) **70:** 1035–1038.

Georghiou, G. P. 1965. Insecticide resistance with special reference to mosquitoes. *Proc. 33rd Ann. Conf. Calif. Mosquito Control Assoc.:* 34–40.

Georghiou, G. P., and Garber, M. J. 1965. Studies on the inheritance of carbamate-resistance in the housefly (*Musca domestica*). *Bull. World Health Organ.* **32:** 181–196.

Georghiou, G. P., March, R. B., and Printy, G. E. 1963. A study of the genetics of dieldrin-resistance in the housefly (*Musca domestica* L.). *Bull. World Health Organ.* **29:** 155–165.

Georghiou, G. P., and Metcalf, R. L. 1963. Dieldrin susceptibility: partial restoration in *Anopheles* selected with a carbamate. *Science* **140:** 301–302.

Georghiou, G. P., Metcalf, R. L., and Gidden, F. E. 1966. Carbamate resistance in mosquitoes. Selection of *Culex pipiens fatigans* Wiedemann (= *C. quinquefasciatus* Say) for resistance to Baygon. *Bull. World Health Organ.* **35:** 691–708.

Grayson, J. M., and Cochran, D. G. 1955. On the nature of insect resistance to insecticides. *Virginia J. Sci.*, July: 134–145.

Harris, R. L., Weardon, S., and Roan, C. C. 1961. Preliminary study of the genetics of house fly resistance to Malathion. *J. Econ. Entomol.* **54**: 40–48.

Harrison, C. M. 1953. DDT-resistance and its inheritance in the housefly. *J. Econ. Entomol.*, **46**: 528–530.

Hoskins, W. M. 1967. *Advances in the Understanding of Insecticide Resistance in Insects Affecting Human Health*, 1960–1966. *World Health Organ.*, WHO, 168 pp.

Hoskins, W. M., and Gordon, H. T. 1956. Arthropod resistance to chemicals. *Ann. Rev. Entomol.* **1**: 89–122.

Keiding, J. 1963. Possible reversal of resistance. *Bull. World Health Organ.* Suppl. **29**: 51–62.

Klassen, W., and Brown, A. W. A. 1964. Genetics of insecticide-resistance and several visible mutants in *Aedes aegypti. Can. J. Genet., Cytol.* **6**: 61–73.

Lichwardt, E. T. 1956. Genetics of DDT resistance of an inbred line of house fly. *J. Hered.* **47**: 11–16.

O'Brien, R. D. 1966. Mode of action of insecticides. *Ann. Rev. Entomol.* **11**: 369–402.

Oppenoorth, F. J. 1965. Biochemical genetics of insecticide resistance. *Ann. Rev. Entomol.* **10**: 185–206.

Perry, A. S. 1964. The physiology of insecticide resistance by insects. In *The Physiology of Insecta.* Vol. 3. M. Rockstein (ed.). Academic Press, New York.

Perry, A. S., and Sacktor, B. 1955. Detoxification of DDT in relation to cytochrome oxidase activity in resistant and susceptible house flies. *Ann. Entomol. Soc. Am.* **48**: 329–333.

Perry, A. S., Miller, S., and Buckner, A. J. 1963. The absorption, distribution and fate of C^{14}-aldrin and C^{14}-dieldrin by susceptible and resistant houseflies. *J. Agr. Food Chem.* **11**: 457–462.

Plapp, F. W., Jr., Chapman, G. A., and Morgan, J. W. 1965. DDT resistance in *Culex tarsalis* Coquillett: cross resistance to related compounds and metabolic fate of a C^{14}-labelled DDT analog. *J. Econ. Entomol.* **58**: 1064–1069.

Pratt, J. J., and Babers, F. H. 1953. The resistance of insects to insecticides. Some differences between strains of house-flies. *J. Econ. Entomol.* **46**: 864–869.

Qutubuddin, M. 1958. The inheritance of DDT-resistance in a highly resistant strain of *Aedes aegypti* (L.). *Bull. World Health Organ.* **19**: 1109–1112.

Roan, C. C., and Hopkins, T. L. 1961. Mode of action of insecticides. *Ann. Rev. Entomol.* **6**: 333–346.

Sternberg, J., Vinson, E. B., and Kearns, C. W. 1953. Enzymatic dehydrochlorination of DDT by resistant flies. *J. Econ. Entomol.* **46**: 513–515.

Wood, R. J. 1965. A genetical study on DDT-resistance in the Trinidad strain of *Aedes aegypti* L. *Bull. World Health Organ.* **32**: 563–574.

Wright, J. W., and Pal, R. 1967. *Genetics of Insect Vectors of Disease.* American Elsevier, New York.

CHAPTER 16

Genetic Effects of Radiations in Human Populations

WILLIAM D. HILLIS

It seems highly probable that man and his fellow species will be exposed to increasing amounts of artificial ionizing radiations for a minimum period of the next several generations. In all likelihood man will exhaust all his known and even now unknown fuel deposits on the earth within a few hundred years, with one exception, the incredibly large source available in nuclear energy. Unfortunately, use of nuclear energy is accompanied by the release of ionizing radiations, which have mostly been found to be detrimental to man's health and his progeny. Until man learns to use energy more directly from the sun, he will become ever increasingly dependent on nuclear energy and will expose himself to larger and larger dosages of artificial ionizing radiations. Moreover, X-rays and other radiations in medical and industrial diagnostic and therapeutic procedures are being increasingly used.

Increased exposure is thus probable even if nuclear weapons are not used. The probable effects of such radiation on human populations must therefore be carefully considered.

The physiological effects of radiations on somatic cells of individuals will not be discussed here, although the genetic effects and somatic effects of radiation are difficult to separate. The somatic effects are indeed important and interesting. The great fields of radiation therapy, of so-called "radiation sickness," of the development of leukemia as a result of radiation exposures, and a great many other aspects of these considerations are subjects of major importance to public health. But this chapter will be restricted to a consideration of radiation effects upon the genetics of Mendelian populations.

THE KINDS OF IONIZING RADIATIONS

The "high-energy" or "ionizing" radiations, of both spontaneous and artificial origins will be briefly discussed.

Alpha particles consisting of aggregates of two protons, a configuration identical with the nucleus of a helium atom, vary in velocity according to their source. Usually they are too ponderous to be good penetrators, but where they do penetrate, their ionizing effect is quite large. Alpha particles emanating from external sources have little effect on soft tissues of the body because of their poor penetration, but those emitted from radioactive sources within the body are quite potent in ionizing the immediately surrounding tissue.

Beta particles are high-speed electrons, which vary in their penetrating power according to their energy level. They are more penetrating than alpha particles and can ionize soft tissues to a depth of about 1 cm.

221

Gamma rays are electromagnetic radiations that constitute the shorter wavelength portion of the same spectrum of which ultraviolet and visible light, infrared and radio waves are parts. They are of very high energy and are emitted by atomic nuclei. They have great penetrating ability equivalent to very high-energy X-rays. They can penetrate the whole body of human beings with relatively little absorption.

X-rays are similar to gamma rays, except that they are usually produced by electrical machines. Their energy levels are somewhat lower and they vary in their penetration according to the electrical energy used to produce them. X-rays, in passing through tissues, cause the liberation of high-energy electrons, so that biologically, X-rays behave similarly to beta particles, except that they are much more penetrating.

Neutrons are uncharged particles ejected during the disintegration of radioactive substances which find their way into the nuclei of other atoms, where unstable configurations then occur and further disintegration ensues. This fission may result in the liberation of many other neutrons and the production of new elements.

MEASUREMENTS OF RADIATIONS

Radiations may be measured in several ways. The *roentgen, r,* which measures exposure to X-radiation or gamma radiation, is the amount of radiation which produces about 2×10^9 ion pairs per cubic centimeter of air, or 1.8×10^{12} ion pairs per gram of living tissue.

The *rad* is a measurement of the amount of radiation absorbed. When 1 rad of radiation is absorbed, it delivers 100 ergs of energy per gram of matter.

The biological effects of radiation may be measured by equating them with the effects produced by a given dose of X-rays in roentgens. Thus the unit called the *rem* is the roentgen equivalent for man. One-thousandth portions of the rad and the rem are called millirads and millirems, respectively.

SOURCES OF RADIATIONS: NATURAL SOURCES

Where do these radiations come from? There are two broad categories of sources: naturally occurring radiations, called "spontaneous," and artificially induced ones. Naturally occurring radiations, which occur everywhere in man's environment, are generally very low in intensity. They vary from area to area, depending on a number of factors and have probably existed since the beginning of life. No environment has yet been created which is completely free of this background radiation. Cosmic radiations are the most omnipresent. They result in a total dosage ranging from 500 to more than 1,000 millirads per generation, depending on geographic location and altitude. On the average in the United States each person receives some 870 mrads per 30 years.

Another source is the radiation emitted from radioactive substances in the earth and in the materials which compose the buildings in which people live and work. The intensity of these radiations also varies greatly from area to area. In Maryland the background radiation level, including that of both cosmic and radioactive element origins, is something less than 3,000 mrads per 30 years. In some areas of the world, however, where the soil is laden with great amounts of radioactive thorium-containing sands, background levels are above 84,000 mrads per 30 years.

There are also atomic disintegrations that occur within the body because of minute quantities of radioactive elements which are normal body constituents. Radioactive potassium (K^{40}) and carbon (C^{14}) are the primary ones. The total average dose of natural radiations received by human gonads all round the world, from both internal and external sources, probably does not exceed 3,000 mrems per generation or per 30 years, according to a report by the United Nations Scientific Committee on the Effects of Atomic Radiation. These "natural" dosages are unavoidable and relatively constant.

SOURCES OF ARTIFICIAL RADIATIONS

Man-made, or artificial, radiations, on the other hand, are those which will continue to increase. One source is "fallout", resulting from the testing of nuclear weapons. Despite its popularity in contemporary literature, its contribution to the total dose of man-made radiation is rather small, compared with radiations produced for medical and industrial uses. Radioactive fallout is, nevertheless, a source of added radiation to human populations, primarily through the inhalation and ingestion of radioactive strontium (Sr^{90}) and cesium (Ce^{137}). Since these substances become incorporated in bone, they form a source of continuous internal radiation. Until 1958, all the weapons testing had resulted in an increase of man's radiation dose equal to only one-third of 1 percent of the amount received from natural sources. However, this relatively minor increase will eventually result in increased mutation which man may not be able to afford. Should nuclear war occur, the radiation dose would be increased to several times the level of background radiations.

X-ray machines are the greatest single source of ionizing radiations in man's present environment. The 1956 report of the Radiation Committee in Great Britain estimated that *diagnostic* X-ray procedures in practice at that time amounted to as much as 22 percent of the background radiation which the reproductive organs received. A recent survey by the United States Public Health Service revealed that 58 percent of all the civilian, noninstitutional population of the United States had one or more X-ray visits in a single year of study (1964). These visits amounted to 173,000,000 X-ray examinations or procedures during that year, involving 506,000,000 X-ray films or exposures. Although calculations of total genetic and organ doses on data obtained from this study have not yet been published, it is probable that previous estimations of radiation exposures from these sources fall far short of actual dosages received by the population. In addition, a number of medical procedures call for increasing diagnostic and therapeutic uses of radioactive substances. All of these sources of radiation, though for the most part not intentionally directed at the reproductive organs, nonetheless do impinge to varying extents upon the gonads. Superimposed upon the increases resulting from more frequent use of these procedures is the continuously increasing exposure resulting from industrial uses of X-rays and gamma rays for inspection of products and for research.

A number of estimations of total doses of man-made radiations have been made by various agencies. The figures range from 720 to 5,500 mrems per generation of 30 years. People in technologically advanced countries experience, on the average, a total artificial radiation dose about twice as great as the average natural background. Such increases above natural radiations result inevitably in increased genetic change.

MUTATION PRODUCTION BY IRRADIATIONS

Mutations are occurring spontaneously in populations at rather low rates. In every species thus far experimentally studied, ionizing radiations have been shown to induce further mutations. The mutation-inducing effect of radiation was first demonstrated by Muller in 1927, using *Drosophila* (Table 16.1 and Figure 16.1). The curve in Figure 16.1 shows intensity of radiation and the

TABLE 16.1

The Effect of X-ray Irradiation on Production of Lethal Chromosomes in *Drosophila*[a]

X-ray dose	Number of chromosomes	Number of lethal mutant chromosomes[b]	Mutant chromosomes / Total chromosomes
Massive	1,777	143	0.12
1/4 Massive	741	59	0.08
None	6,016	5	0.0008

[a] Muller, from Wallace and Dobzhansky (1959).

[b] Lethal chromosomes are those carrying any sort of change resulting in death of offspring. Mutation-rate to lethal conditions is the number of lethal chromosomes produced, divided by the number of chromosomes irradiated.

frequency of mutations induced. When the curve is corrected for the probability of two or more lethal mutations being induced in the same chromosome and being erroneously scored as a single lethal, it becomes straight (the dotted line in the figure).

Every species studied, including mammals, has been shown to demonstrate essentially the same curve. There are no data of this type for man, but surely the curve for man would be identical. Glass and Stephenson have concentrated on the shape of this curve at much lower doses, those in fact which man is more likely to experience. In Stephenson's studies the doses have reached as low as 0.01 r. In every case, the relationship of mutation rate to intensity of radiation is linear. These curves have tremendous implications. First, the linear function indicates a direct relationship of increase in dose to increase

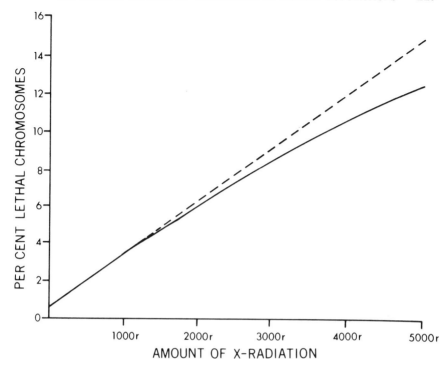

Figure 16.1. Mutation in *Drosophila* induced by X-ray irradiation, scored as percentage of chromosomes carrying any sort of change that resulted in death of offspring. The dotted line is corrected for the probability of two lethal mutations occurring in one chromosome. (After Wallace and Dobzhansky, 1959.)

in mutation rate. Every increase of irradiation, no matter how small, results in a corresponding increase in mutation rate. Second, the curve starts slightly above the horizontal axis. Some mutations, spontaneous ones, are occurring even in the absence of radiation. Third, there is no known threshold below which irradiation is ineffective in producing mutations. It is important to re-emphasize that no matter how small the dosage of radiation to which an organism is exposed, a certain amount of increase in mutation rate will be induced by it. There is no "safe" or "tolerable" level of radiation as far as genetic effects are concerned. There is a definite threshold in individuals for producing physiological or somatic effects of radiation below which no detectable damage occurs. But this in no way holds true for genetic effects.

RELATIONSHIP OF RADIATIONS TO SPONTANEOUS MUTATIONS

Can the spontaneous mutation rate be accounted for by the background radiation? The answer is "no." It has been experimentally demonstrated, for instance, that considerably less than 1 percent of spontaneous mutations in *Drosophila* can be accounted for by background radiation. The radiation-induced fraction of human spontaneous mutations is unknown but probably it is a small proportion. Radiation-induced mutations, then, are added upon spontaneous mutations.

The Doubling Dose

This dosage is the amount of radiation required to double the spontaneous rate of mutation. The doubling dose for *Drosophila* has been demonstrated to be from 30 to 100 r. Bacteria have a range from 16 to 100 r for different loci. Because of the absence of a number of important data for human genes, the doubling dose is not accurately known, but estimates range from a most pessimistic estimate of 3 r, up to 150 r. The lowest estimate of 3 r assumes, probably erroneously, that the background, natural radiation induces 100 percent of the spontaneous mutations. The likely value for a human doubling dose is between 30 and 80 r.

MECHANISM OF ACTION

The exact mechanisms of action of radiations upon genes are not understood. The former notion that radiation quanta merely acted as "bullets" that shot "holes" into genes was a gross oversimplification, and probably wrong. It appears that radiations, because they induce ionization of the molecules through which they pass, probably produce highly reactive chemical substances in the chromosomes. The reactive chemical configurations produced by ionization persist for only very brief times, but introduce errors in the nucleic acid codes which are recognized as genetic mutations. The idea of chemical actions of radiation offers greater promises than the old bullet-hole hypothesis for the discovery of "anti-mutagens."

It was formerly believed that the effects of radiation upon the genes was independent of the rate at which the radiation was administered. However, Russell and his colleagues have demonstrated in mice that large, single doses of radiation produce quantitatively more genetic damage than the same total dose given in small, divided amounts. They have shown that the differences observed are probably based on intracellular recovery and not on differential cell killing. They suggest that small divided doses allow for repairs of some sort of premutational changes that do not occur with massive, acute radiation doses. Recent evidence has suggested that specific enzyme systems in the cell are available for keeping the genetic mechanism in repair, to a limited degree. The repair mechanism appears to be related to the ability of cells to excise pyrimidine dimers from DNA molecules, but just how the integrity of the DNA is restored after the excision has occurred is not presently understood.

One action of large doses of radiation can be morphologically followed. X-ray and other radiations may produce chromosome fractures. If the breakages allow reconstruction to take place, chromosome aberrations may occur.

Most breakages act as dominant lethals so that the next generation of the cell dies. The action of radiation is direct in that its effects occur in the cells irradiated and not at some distant point. Its action does not apparently take place through some intermediary which is created at one point and circulates to another point where it manifests itself. To obtain mutations in sex cells, one must irradiate sex cells and not the head or thorax.

ACCUMULATIONS OF MUTANT GENES AND THE ROLE OF NATURAL SELECTION

Increased radiation produces increased mutation and natural selection removes deleterious mutants from populations. Damage to biological populations depends on many factors. Two of them are the kinds of mutations produced and the total dose of radiation. The effects of rates of administration will be disregarded here although they may play a significant role in human populations.

The following calculations about genetic damage are simplified. They are based upon theoretical computations presented by Wallace and Dobzhansky (1959). Damage will be discussed in terms of loss by natural selection, some of this loss being by death and some by failure in reproduction.

Consider a gene pool with two alleles at one locus—the normal allele and a deleterious mutant that arises at a rate, U. When the mutation rates and selection rates affecting each of the two alleles have been constant through a number of generations, the alleles reach an equilibrium of relative frequencies. At equilibrium let the relative frequency of the normal allele be q and that of the mutant will be $1-q$ (Chapter 14). Then the relative frequencies of the genotypes are:

q^2 (homozygous normals): $2(1-q)$ (heterozygotes): $(1-q)^2$ (homozygous mutants).

It can be shown that at equilibrium

$$(1-q)^2 = \frac{\text{the mutation rate } (U)}{\text{the selection rate } (S)}$$

Let the mutant be entirely recessive, so that selection operates against only homozygous mutants. If the homozygous mutant is lethal, the negative selection coefficient is 1. These homozygotes do not contribute to the next generation because of death or failure to reproduce. In this case:

$$(1-q)^2 = \frac{U}{1} = U.$$

The relative frequency of mutant homozygotes at equilibrium in this case is equal to the mutation rate.

If a mutant is only slightly deleterious, it might have a selection coefficient of, say, 0.01. This means that the homozygous mutants leave 99 offspring where an equal number of normals leaves 100. In this case,

$$(1-q)^2=\frac{U}{0.01}$$

which is a hundred times larger number than the relative frequency of the lethal mutant discussed above.

Total genetic loss from recessive mutants is the product of (1) the selection rate against the homozygous mutant, (2) the relative frequency of homozygous mutants, and (3) the total population size (N).

Total genetic loss$=S\times(1-q)^2\times N$

But $\qquad (1-q)^2=\dfrac{U}{S}$

So total loss$=S\times\dfrac{U}{S}\times N=UN$

when the mutant allele is totally recessive.

With similar simplified calculations the total genetic loss from incompletely dominant mutants, or completely dominant mutants is taken to be 2 *UN*. Therefore, when the dominance of the mutants is not known, it can be said that:

Total genetic loss$=N\times$a value ranging somewhere from U to $2U$.

One conclusion that can be drawn from this is that the final genetic loss from the population is independent of the degree of damage caused to individuals. The mutation rates and the population size determine the amount of genetic loss from deleterious mutants, not the selection rates.

What might occur if irradiation resulted in increased production of genes which would achieve greater fitness in heterozygotes than in either normal homozygotes or mutant homozygotes? Fitness of the whole population might conceivably be augmented, even if the mutant in homozygous condition resulted in incapacitation or death. There is some experimental evidence by Dobzhansky and others with *Drosophila* that the viability of some flies carrying irradiated chromosomes was greater than that of nonirradiated controls. In other words, these workers demonstrated the production of vigor in heterozygotes by the use of ionizing radiation.

Another factor of importance is this: the total increase in the amount of radiation administered to the population equals the product of the increased level of radiation dose and the amount of time during which the increase occurs. If man's need for energy production will necessitate a permanent increase in radiation received by the population, then the damage resulting from infinite increases will in time amount to infinite damage. If man discovers in 5 or 10 generations how to harness solar energy, the lower total dose will result in far less total genetic damage.

Suppose the human population to be continuously exposed to 10 r of radiation, which is the mean maximum permissible dose before age 30 that is recommended by the Genetics Committee of the National Academy of Sciences. What may be expected to happen, by the time genetic equilibrium occurs, if the mutations induced are all recessive and deleterious? Data are lacking for man for the most part, but data for *Drosophila* will be considered. The figures will not necessarily be the same for man, but his genetic damage would be likely to be greater.

In *Drosophila* the average spontaneous mutation rate per gene for all types of deleterious recessives may be considered to be 5×10^{-5}. The radiation-induced mutation rate for recessive deleterious genes has been shown experimentally to approach 5×10^{-7} per gene per 1 r of radiation. If this is multiplied by 10 r, 5×10^{-6} is obtained. This value is 10 percent of the spontaneous mutation rate, 5×10^{-5}. Then at equilibrium, the increase in radiation to 10 r per generation will have resulted in a 10 percent increase in loss from genetic causes, in terms of death and reduced reproduction.

Calculations for recessive lethal mutants show that only after 100 generations, about 3,000 years, of such continuous radiation, more than half of the final equilibrium frequency would have been attained. It may be noted that a larger number of generations would be required to attain half the equilibrium frequency for a nonlethal recessive gene. Unfortunately, these calculations cannot be made because the selection coefficients and mutation rates for human genes are not known. Some additional genetic loss will occur at the very beginning of the increase in radiation.

If the mutations occurring as a result of continuous increased radiation produce effects in heterozygotes, then equilibrium frequencies of such genes would be much lower than if the genes were completely recessive, but the genetic loss, by virtue of its effects on heterozygotes would be twice as great. The greater the degree of injury to the heterozygote, the more rapidly would the equilibrium frequencies be obtained. Consider, for example, a dominant lethal mutant which kills every individual who carries it, homozygotes and heterozygotes alike. Such lethals are eliminated from the population in the same generation in which they appear, and equilibrium occurs in one generation.

Let us now consider the human population receiving an average dose of 10 r each generation for 10 consecutive generations, after which radiation returns to its normal "background" level. Such a situation might occur if man discovered, say in 10 generations, a safe substitute for atomic energy that satisfied his energy requirements. Again, let us substitute data for *Drosophila* in the absence of data for man. Ten roentgens for 10 generations may be expected to induce 1×10^{-5} recessive lethal mutants in every gene capable of such mutation. Every individual homozygous for the mutant will be eliminated. If the population is 3×10^{9} individuals, 30,000 deaths caused by irradiation-induced mutants would occur for each gene locus, or 30,000,000 deaths for 1,000 loci. These deaths would occur over a huge span of time. Within the first 20 generations, or about 600 years for man, 180,000 deaths from this cause would occur each generation, or 6,000 deaths per year for 600 years. Some 90 percent of the induced mutants would still have to be eliminated. Now, these figures deal only with recessive lethal mutants. Dominant mutants would result in twice the number of deaths, although elimination would take place much more rapidly.

The discussion may seem somewhat unreal as it has been based on *Drosophila* data. There are no data for man, since no great numbers have been given large doses of radiation. Nor can genetic markers be deliberately bred, as with flies and mice. Nonetheless, some limited data are available for genetic effects of radiation in mankind.

EVIDENCE FOR GENETIC EFFECTS IN HUMAN POPULATIONS

Neel and Schull (1956, 1962; Neel, 1958) studied survivors of the atomic explosions at Hiroshima and Nagasaki. These are the most extensive data available. Among 20,000 women and 14,000 men who received from 8 to 200 r, there were 33,181 parturitions in which either or both parents had received radiation. Among these, 546 stillbirths occurred, a frequency of 1.64 percent. Stillbirths occurred among matched controls at a frequency of 1.29 percent. This is not statistically significant at the 0.05 probability level. Among 33,527 live births, of which one or both parents were irradiated, 300 congenital malformations occurred, opposed to 294 congenital malformations among 31,904 live births from non-irradiated (control) parents. There was no indication in this study that irradiation caused an increase of congenital malformations.

Studies conducted among 6,000 American radiologists by Macht and Lawrence (1955) and by Crow (1955) showed the results given in Table 16.2.

TABLE 16.2

Data on Births Where Fathers Were Radiologists[a]

Fathers	Number of births	Number of stillbirths and abortions	Number of congenital malformations
Radiologists	5,461	766	328
Controls	4,484	528	216

[a] Macht and Lawrence (1955) and Crow (1955).

There is a statistically insignificant increase in the incidence of abortions and stillbirths and an apparently significant increase in congenital malformations in the offspring of radiologists if the rates are adjusted for parity. The total dosages of irradiation are unknown.

INDICATIONS TO LOOK FOR

An interesting observation has come from radiological experiments with rodents. In these studies irradiated males have been demonstrated to father litters which are smaller than usual and even the apparently normal young in these litters have reduced life spans. These effects, however, are not well-marked hereditary changes which can be ascribed to identifiable mutations.

One likely indicator of possible change in mutation rates in man is a change that might occur in the sex ratio at birth. Mothers with new mutation of the X chromosome could have too few sons. Mutants on the single X chromosome of the sons have no opportunity to be masked by an allele, because the Y chromosome carries little genetic information. Evidence collected by the Atomic Bomb Commission in the studies with survivors at Hiroshima and Nagasaki who became parents shows that there did occur a statistically significant difference in sex ratios between offspring of irradiated and control parents.

Since the gonads of female fetuses are in meiosis during early months of gestation, if mothers receive irradiations during that time, the fertility rate of the daughters may be affected, and also the sex ratios of the grandchildren.

CONCLUSIONS

There is a linear relationship of mutation frequency to radiation dose with no demonstrable threshold. Because of this, it appears that man should restrict his exposure to ionizing radiations to the very minimum. Some radiological practices that are not necessary should clearly be eliminated wherever possible and for the most part are being eliminated. One such practice is sterilization by irradiation of gonads. Other, preferable methods are available. One might also consider the consequences of radiation administered to the fetus during gestation. Here, direct radiation effects may be induced not only in the mother, but also in the sex cells of the fetus, which are probably especially susceptible to radiation effects because of their stage of differentiation. X-ray pelvimetry, then, should not be regarded as an entirely innocuous procedure.

REFERENCES

Crow, J. F. 1955. A comparison of fetal and infant death rates in the progeny of radiologists and pathologists. *Am. J. Roentgenol.* **73**: 467–471.

Macht, S. H., and Lawrence, P. S. 1955. National survey of congenital malformations resulting from exposure to roentgen radiation. *Am. J. Roentgenol.* **73**: 442–466.

Medical Research Council. 1956. *The Hazards to Man of Nuclear and Allied Radiations.* Her Majesty's Stationery Office, London.

Medical Research Council. 1960. *The Hazards to Man of Nuclear and Allied Radiations, a Second Report.* Her Majesty's Stationery Office, London.

Muller, H. J. 1950a. Our load of mutations. *Am. J. Human Genet.* **2**: 111–176.

Muller, H. J. 1950b. Radiation damage to the genetic material. *Am. Scientist* **38**: 33–59, 126.

Neel, J. V. 1958. A study of major congenital defects in Japanese infants. *Am. J. Human Genet.* **10**: 398–445.

Neel, J. V. G., and Schull, W. J. 1956. *The Effect of Exposure to Atomic Bombs on Pregnancy Termination in Hiroshima and Nagasaki.* Atomic Bomb Casualty Commission, Hiroshima, Japan, Washington, D.C. National Academy of Sciences National Research Council Publication No. 461, Washington, D.C.

Neel, J. V. G., and Schull, W. J. 1962. Genetic effects of the atomic bombs. Rejoinder to Dr. de Bellefeuille. *Acta. Radiol.* **58**: 385–399.

Wallace, B., and Dobzhansky, T. 1959. *Radiation, Genes, and Man.* Holt, New York.

III

SOCIAL BEHAVIOR AND

ORGANIZATION

Fighting stags, drawn by Paulus Potter in 1697. Copyright, Rijksmuseum, Amsterdam.

SECTION III

SOCIAL BEHAVIOR AND ORGANIZATION

Populations adapt to their environments not only genetically but also by other means. Experience can cause adaptive change in behavior by the process of learning. When individuals associate and communicate with others, some behavior can spread to others by learning. Passage of behavioral change from generation to generation by learning is cultural inheritance. Cultural transmission of behavior can be much more rapid than the spread of genetically based change, and ability to change in response to the environment has great survival value. Cultural transmission is recognized in animal behavior, but reaches its greatest expression by far in human populations.

Social organization helps populations to withstand onslaughts of the physical and ecological environment, and in some cases is known to help forestall overexploitation of resources.

Many aspects of human social organization have their counterparts in animal populations: the family and care of young, occupation of territory, hierarchies of privilege, and development of social responses in the individual as it grows and learns. Because nonhuman primates are phylogenetically closest to man, study of their behavior may yield particularly valuable results. Man's poor understanding of the origins of his behavior, particularly aberrant behavior and mental illness, is an enormous weakness in medicine and public health; and study of animal behavior provokes new ideas and concepts which must be reevaluated in man himself.

CHAPTER 17

Introduction to Behavior Studies

FREDERIK B. BANG

Animal watching can lead to experimental analyses of behavior, but this is seldom true of people watching, and analysis of human behavior is sorely needed. "Why do we behave the way we do?" is a question reflecting a restless unhappiness prevalent in the world. But this question colors the examination of animal behavior so that, consciously or not, the topics of heredity and environment, instinct and learning, take on personal meanings. In this way studies of animal behavior inevitably reflect the zeitgeist, the spirit of the times, to a far greater extent than do other sciences; and each examination is influenced by the investigator's own culture.

With this initial caution, the chapter will be briefly concerned with some current understanding of genetic bases of behavior and various types of learning; with ontogeny of behavior of the individual; and with some group organization, particularly development of swarming in locusts and schooling in fishes.

THE NATURE, NURTURE CONTROVERSY

The question of how to evaluate for an individual the influence of
    nature *versus* nurture
is an old problem. It can be phrased as
    heredity *versus* environment,
    Mendelian *versus* cultural inheritance.
This is purposely written misleadingly and should really be *times* instead of *versus*. For in each case there is interaction of the two, and it is very difficult to tell how much of each is involved. All that can be done is to set up an experimental situation and attempt to measure how much of an effect may be produced by a change in one or the other of the factors.

GENETICS OF SOCIAL BEHAVIOR

Invertebrates

There are true genetic aspects of social behavior, as in resistance to foul brood, a disease of honeybees, *Apis mellifica* (Rothenbuhler, 1959, 1964). An infected larva gradually shrivels up and dies from a bacterial infection, that can readily spread and involve the whole colony. Resistant and susceptible strains of bees were bred and the resistant strain proved to be resistant primarily because it had certain behavioral characteristics. When an infection occurred in a resistant colony, the bees detected it and cut a hole in the top of the infected cell to remove the diseased larva, and thus kept the infection from spreading.

234

This strain of bees was called hygenic and resistant. The other was unhygienic and susceptible. When the two were crossbred, the F_1 generation had no hygienic behavior and so was susceptible to the infection. The F_1 generation was backcrossed with the resistant parent generation. In the resulting bees there were four patterns of behavior. One-quarter had complete hygienic behavior, one-quarter cut the hole on the surface of the cell but was unable to remove the larva, one-quarter removed the larva if the experimenter cut a hole, and one-quarter had no hygienic behavior at all. Thus one-quarter was resistant and three-quarters were susceptible. The genetics is clear. Ability to cut open the cell was recessive, and another factor determined recessive ability to remove the larva. The hygienic bee was homozygous for the two unlinked recessive characters.

Vertebrates

Whitman (1919) was fascinated by the mating, courting, tumbling, and pouting behavior of pigeons and ring-doves, and he cross-mated various pigeons and doves and analyzed the significance of these patterns. He indicated the need for more careful genetic analysis. Only recently have genetic studies on behavior started, and these are indeed only beginnings.

Scott and Fuller (1959) studied the behavior of various dogs, including crosses between the very excitable basenji and the more placid cocker spaniel. The family environment of the tested dogs had to be standardized, so mothers were used that were the F_1 generation of crossings of the pure basenji and the pure cocker spaniel. These F_1 mothers were then bred with pure basenji fathers or pure cocker spaniel fathers and the offspring were the tested subjects. Avoidance behavior toward a certain person was one of the behaviors scored in standardized tests. A backcross with the basenji produced a dog with a high avoidance score and the excitable behavior of the basenji. The backcross with the cocker spaniel produced dogs with a lower avoidance score, and they were much more easily handled. However, there was not a nice separation of characters. Probably much of the behavior in higher vertebrates will prove to be of the quantitative, overlapping type.

It has been known for a long time that strains of mice vary considerably in their behavior. Young of an inbred strain of mice like the DBA strain will jump suddenly out of the cage at the slightest opportunity. Thompson (1953) studied in 15 inbred strains the rapidity with which they would eat when partially starved; emotionality as shown by the rate and amount of defecation under certain stresses; and exploratory activity. He showed that the strains varied considerably in their scores for these different characteristics.

Southwick and Clark (1968) found differences in 14 strains of inbred mice tested for aggression, social grooming, and activity patterns. Crossbreeding yielded mostly hybrids with aggression scores equal to or higher than the scores of the more aggressive parents.

Lindzey (1951) plotted (Figure 17.1) the percentage of mice of various strains that have audiogenic seizures, against emotionality as measured by defecation

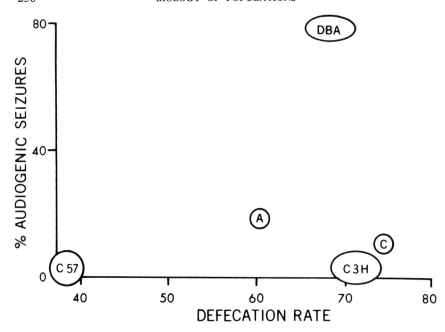

Figure 17.1. The incidence in various strains of mice of defecation during emotional stress, and of seizures when confined in a metal tub with an electric doorbell ringing for 2 minutes (After Lindzey, 1951.)

rate when put in brilliantly lighted new laboratory environment, and showed that these two characters were not controlled by a single genetic factor. Some mice when stimulated with sound of a certain tone have an epileptic seizure that ends in death. Although there was some evidence of a positive relationship between emotionality and susceptibility to audiogenic seizure, the authors concluded that it was likely to be the result of a chance combination in the process of inbreeding.

Rodgers and McClearn (1962) carried the story one step closer to man. They showed that alcohol preference in mice is clearly related to the inbred strain. Some prefer a 10 percent solution of alcohol, like a good table wine, to water. Again, though proved by appropriate experiments to be genetic in origin, lack of segregation of characters in the F_2 and backcrosses indicated a polygenic system.

There are, so far, few data to show whether many of these behavioral characteristics are due to one genetic factor or to two or more, and the degree to which they may be modified by environment. It is difficult to define a behavioral character and to give it a sharp measure. There is also a great deal of variation within a strain. So as breeding crosses are made to test for Mendelian inheritance, the scoring becomes less and less definite. This is typically a problem of quantitatively different characters. In addition, there may be a selective disadvantage in homozygotes for behavioral characters, thus making understanding of the genetic background of behavior very difficult. However, it is reasonable to assume that to varying extents there is some genetic background to much of the behavior of all animals.

ACQUIRED (LEARNED) BEHAVIOR

What parts of behavior are acquired or learned? What influences modify the expression of the genetic material? Thorpe (1963) defined learning as an adaptive change in behavior as a result of experience. Such a broad definition requires some separation of various types of learning. Three types will be discussed—habituation, associative learning, and insight learning.

Habituation

Habituation is a learning not to respond to stimuli that tend to have no significance in the life of the animal. Perhaps the nicest example is the work of Humphrey (cited in Thorpe, 1963), with a snail, *Helix albolabris*, moving up an incline that was shaken at intervals. At first, at each shake the snail withdrew its tentacles, but after a while it ceased to react. It became habituated to the presence of the stimulus, the shaking, and no longer pulled in its tentacles. The animal continued to move up the incline, indicating that it had not become generally fatigued. Perhaps the response of withdrawal of antennae had become fatigued, but this was assumed to not be so. There are numerous examples of habituation in other animals.

Associative Learning—Conditioning Plus Trial and Error; Operant Conditioning

In conditioning, the learned response to one stimulus is conditional upon an original relation of that response to another stimulus. "If two stimuli, S_1 and S_2, are applied in overlapping sequence, S_1 being the antecedent, with repetition of the combination a plastic change in the nervous system is formed, so that S_1 acquires the ability to elicit the response of the same kind as does stimulus S_2" (Konorski, 1948). Pavlov's (1927) original experiment involved a dog that was presented with food and responded with salivation that could be measured. A ringing bell was the antecedent overlapping stimulus. The dog learned to salivate in response to the bell without the presence of food.

This kind of a gradual change in reaction brought about by a new stimulus may be induced in a variety of smaller animals. Yerkes (1912) spent a year studying one worm, *Allobophora foetida*. He put the worm in a T-maze, and when it took one turn it was scraped, and stimulated by an electric current. If it took the other turn it was in a dark and moist area; this was the correct direction, and the earthworm learned to take it with a high frequency. The learning involved conditioning against the new stimulus, S_1, which was associated with the wrong turn, in the trial-and-error situation. This is called associative learning. When the worm had a high score for the correct behavior, Yerkes removed the brain and the worm still made the correct choices. But when the brain regenerated, the worm no longer had the correct behavior. It acquired new nervous tissue and took random turns.

There is a particularly elegant method of improving this kind of trial-and-error learning which has been popularized by Skinner (1961). Thorpe calls the method *instrumental* conditioning; Skinner refers to it as *operant* conditioning. It involves capitalizing on any particular behavior pattern, such as turning the head, by giving a reward every time the behavior occurs. Skinner gave a reward every time a pigeon began to move in required ways, until eventually he trained pigeons to play pingpong by pushing a ball back and forth on a table.

What use for man are such studies? Skinner went on to invent teaching machines which apply operant conditioning to human learning. So, in some cases, it is possible to make a fairly large jump and begin to apply this new behavioral information to human endeavours.

Insight Learning

Insight learning is associated with exploratory behavior—the animal suddenly gains adaptive reorganization of its experience. The animal appears to perceive that something else is involved, and starts a new adaptive behavior. This is often illustrated by the use of tools. A photograph by Dunnett, reproduced in Thorpe's text (1963) shows a European jay, *Garrulus glandarius*, drawing up a string. On the bottom of the string is food that cannot be obtained more directly because it is in a jar. The jay sits on a limb and draws the string up with its beak, then the foot is placed on the string. The next bit of string is pulled up again by the beak. Finally, the jay learns to reach the food by pulling it out of the jar in four or five direct sequential motions. It is presumed that this involves insight into the situation. Surely the bird realizes it must do something particular to bring the food to itself!

Some of the most amusing examples of this kind of learning are furnished by Köhler's (1921) studies of chimpanzees. He first studied their use of tools, such as a stick, to obtain a banana that was out of reach. Frequently the learning occurred only if the chimpanzee saw the stick and the banana at the same time. Sometimes the banana was hanging from the ceiling and the chimpanzee was unable to reach it. A box was placed in the room and the chimpanzee was able to move the box over and then reach up and get the banana. This became more complicated when the banana was placed so high that two boxes were needed and they were at different places in the room. Chimpanzees showed great variation in their capacity to solve this puzzle. A favorite chimpanzee placed the first box under the banana, then ran around the room carrying the second box, banging it against the wall and throwing it around. Finally apparently in a burst of insight it placed the second box on top of the first and quietly climbed up. There seems to be a struggle involved in this sudden and adaptive reorganization of experience.

THE ONTOGENY OF BEHAVIOR

In the embryo the nervous system grows and differentiates as it functions. Hamburger (1964) found that in the latter part of embryonic development of the duck there are rhythmical periods of activity and inactivity. As the heart grows and beats it apparently pushes the head away from the body. As the embryo grows it turns, contracts, and extends. Later, patterns of walking, and paddling in duck embryos, are seen as the foot pushes against the yolk sac. This has also been studied in chickens by Kuo (1967). He insisted that even in the embryonic stage the question of what is inherited and what is acquired cannot be adequately tested, and that a more pragmatic view must be adopted. One must study a series of events and test the subsequent behavior influence with an epigenetic point of view. Furthermore, it has recently been shown that the unhatched chick responds to certain sounds. In addition Gottleib (1968) heard "contentment notes" from unopened duckling eggs as they were turned or warmed, and distress notes when they were cooled.

Kuo (1967) showed how food habits can be controlled by experience. For instance, one group of mynah birds was given a great variety of foods, and another group was given a diet exclusively of soybeans. After 6 months both groups were starved for 24 hours. The first group was then given soybeans, which they all ate. The second was offered a great variety, but soybeans were excluded; only four of these twenty-four took food.

Kuo (1967) has shown extensively that social and antisocial acts can be influenced by preceding events. He produced male dogs that would guard very receptive females in heat, without making any sexual advances. He chose twenty male chow puppies aged 1 week that had good build for fighting. When they were able to feed alone they were left to compete for a food bowl for one, with another weaker male puppy, and the stronger always won and always received praise from the trainer. Twelve of the experimental puppies were kept isolated except for the feeding contests, and in maturity were very aggressive at the entrance of any other dog, including females in heat. Four of them lost their aggression to females in heat, but in 4 months of trials were unable to copulate. The remaining eight experimental puppies were each given the company of a friendly mature female between feeding fights. Any sexual movements to the female were stopped by a "No! No!" that at the beginning of the experiment had been reinforced with an electric shock to the hind leg. When these males were mature, Kuo was able to write "The female acts like a sexually starved, seductive, or even demanding lady-in-waiting of the ancient Chinese imperial palace, where as her guard appears like a dull, uninterested, unresponsive eunuch."

Japanese gray quail are known for their fighting ability, and people prepare them for this by raising them in relatively dark isolated cages, with certain training procedures. Kuo (1967) intensively trained some and kept them in isolation for approximately a year, and they were excellent fighters. However, when he turned a prime fighter free in a flock of twenty quail, it dashed in panic from corner to corner, was chased and pecked free of feathers by all the other quail. For 12 hours the quail crouched in a corner and then it was removed. Such a quail could not be retrained as a fighter and whenever put back with the flock it would become the most submissive bird. This sequence was repeated many times with other individuals.

Kuo influenced social behavior of birds by rearing individuals with species with which they would not naturally associate. A newly hatched bird would be kept entirely with one or more of another species, and at night they would be placed together as closely as possible.

> The Asian song thrush and the masked jay thrush have both been long known in China as predatory birds. In our birdhouse, when they were hungry, they would kill and eat such smaller birds as the South China white eyes and the strawberry finch, and occasionally even Peking robins. However, when they were reared according to the procedure described above, the masked jay thrushes lived peacefully with these birds. Asian song thrushes reared under the same experimental procedure became very friendly with the smaller birds. They ate together, sat together, groomed each other, the large birds letting the smaller ones snatch meat, insects, etc., from their beaks. And on cold nights, they would spread out their wings and let the little birds roost under them. (Kuo, 1967.)

The beginnings of social facilitation in chicks are most difficult to study and yet important. Smith (1957) built a series of straight trays, 9 feet long, 8 inches wide, and 10 inches high. He determined the time that a slightly starved chick would take to run from one end, to food at the other end. With training, the chick increased its speed sevenfold. Speed was also increased by the presence of a feeding chick at the other end. When untrained pairs were introduced the initial speed was twice as fast as with a single chick. It was 14 times as fast after training, thus the factor of improvement by practice seemed the same.

Hogan (1965) studied the reaction of chicks, at 3 to 11 days of age, to the presence of mealworms. Each experiment consisted of careful records of activity during 2 minutes before being given a mealworm and 2 minutes after the mealworm had gone; the behaviors before and after were the same. The presentation caused a new pattern of fixing attention on the mealworm, stretching the neck, turning the head, and depression of total amount of movement. Irrelevant activity increased—head shaking and sleep became more common, shrill calling increased in some cases, and in others decreased. For a variety of reasons Hogan interpreted these results as indicating conflict and fear.

The conflict of approach and withdrawal is a basic issue in the scientific study of behavior. In a recent attempt to synthesize understanding of patterns of behavior, Schneirla (1965) developed a theory of biphasic approach and withdrawal. The key postulate for ontogenesis is that, in early developmental stages, low or decreasing stimulative effects can selectively arouse organic processes of the approach system, and conversely, that high or increasing stimulative effects can selectively arouse organic processes of the withdrawal system. Biphasic mechanisms underlying directionally opposed orientative responses are present in all animals from protozoons to primates, and through natural selection a variety of types of specialization has appeared.

Seemingly opposite behavior may be part of one chain of events. This idea may be more difficult to accept in the western world than in the east. For instance, the gods of India combine opposite principles. One of the fiercer, Kali, brings forth and fosters all creatures, but is simultaneously their common grave. Relentlessly, she swallows back like a monster the beings that she produces. Her garlands are of deadly cobras, not flowers, and yet her hands are lifted in gift-bestowing and fear-dispelling mudras.

Unfortunately, as yet, the many pieces of scientific information do not add up to a grand description of the epigenesis of behavior in definite terms, even when limited to one group of animals, such as the birds, or a narrower group within them.

BEGINNINGS OF SOCIAL ORGANIZATION
Collections of Cells

In order that there be organization within an aggregation of animals there must be communication. Individuality can almost be lost in the organization of what may be considered a superorganism. For comparison, consider an animal composed of various differentiated cells. An embryo can be trypsinized so that its cells are separated and suspended in liquid in a random fashion. If the cells are kept alive they reassociate with the same kinds of cells.

Epithelial cells come together with each other, and cartilage cells form an organoid of cartilage. Embryologists have followed cell movements on glass. In some strains of normal fibroblasts there is inhibition of movement on contact, and this is absent to some degree from many malignant cells. Stoker (1966) suggested there is a stimulus and receptor system, but almost none of this organization is understood.

Barnacles

Some of the mechanisms of gregariousness have been described for the barnacle (Knight-Jones and Crisp, 1953). Antennules of the larvae detect settled members of their own species, or even the remains of cemented bases, and settle down in response. Presence of other barnacles is evidence of success of the species in that location. However, barnacles form colonies rather than societies.

Locusts (Uvarov, 1965)

All the locusts can be modified in their behavior, morphology, color, physiology, and fertility within a lifetime in response to crowding; and maximum changes accumulate through several generations. (Changes in laboratory populations are probably greater than in the wild.) There has been careful, fascinating research on the fact that social contact produces these changes, and on the manner in which the changes are related to habitat and weather. For example, *Schistocera gregaria*, the desert locust, may increase in numbers in dense but very temporary lush herbs. By the time the nearly adult hoppers of the first generation are produced, most of the herbs may have gone and the hoppers may have been forced into greater concentrations around streams and in depressions. By a series of such aggregation, movement to nearby vegetation, reproduction, and aggregation, in a few generations populations of solitary types can be fully transformed into the migratory swarming phase. Flying bands of adult gregarious forms travel with the winds, and wind convergence further increases aggregation and destruction of vegetation. ". . . the locusts came and caterpillars, and that without number, and did eat up all the herbs in their land and devoured the fruit of their ground" (Psalms: 106, verse 34). The desert locust by ability to swarm over enormous distances and produce three generations a year, or settle in solitary phase to pass through one generation in a year, is superbly adapted for rapid exploitation of unstable habitat.

In 1921, Uvarov left an assistant to raise solitary and gregarious forms in separate cages, and when he returned several molts had occurred, and there were gregarious locusts among the solitary ones. He suspected the assistant had mixed them, but repetition produced the same results. The forms had been thought to be separate species because of differences in color, size, and shape of integumental plates, and banding of the eye, but Uvarov showed that they both were *Locusta migratoria*. There are also striking biochemical differences; amino acids, fat, and ascorbic acid are higher in the gregarious form. In both male and female solitary forms the length of the prothoracic gland is about twice the length in the gregarious form, regardless of the weight of the locust itself.   •

Ellis (1964) put *L. migratoria* hoppers in a circular cage without food, and the floor was divided into thirty equal radial sectors. Hoppers of the solitary phase put together for the first time had no more association with each other than would be expected on a random basis, whereas up to 70 per cent of the gregarious form were in groups of two or more. Ellis studied gregarious hoppers marching in concerted bands in cages with a light bulb suspended in the middle. Hoppers marched round and round the floor of the cage in a circle for hours, with some interruptions for rest. Withholding food increased marching. Restriction of social contact reduced marching, so that animals reared in isolation marched very little, even when tested in a crowd. Hoppers in crowded small cages did not march as vigorously as those reared crowded in large cages, and practice made little difference to the individual.

Adult locusts fly, and crowded ones fly around the cage, and have a tendency to aggregate much more than isolated ones. Visual interaction was suggested as the cause of the changes, but continued presence of a mirror had no effect. Ellis demonstrated that keeping a small hopper for 7 hours in a jar in which very fine wires rotated and frequently touched the hopper, produced hoppers which aggregated almost as well as crowded ones.

There is still much to be described to explain, "The locusts have no king, yet go forth all of them by bands" (Proverbs: 30).

SCHOOLING IN FISHES (Keenleyside, 1955)

A school of fish is not just a group but consists of fish of one general size, which may be from 1 inch to 12 feet. When startled, a chance aggregation of fish disperses, whereas a school comes quickly together and moves almost as one organism. There also are "pods", closely packed schools in which the fishes are in physical contact as they swim. Some fish are lifelong schoolers, and others, such as sticklebacks, live in schools except during the breeding season. Two antarctic fish have silvery schooling young but for most of the adult life are dull bottom-dwelling forms. Barracuda may appropriate small schools of prey and watch and herd them. Tuna have been seen closely following mackerel schools, and when one mackerel turned off to the side a tuna would move swiftly forward and the mackerel would get back into line.

Vision plays a large role in this type of aggregation. At night schools become more dispersed, and it usually is fairly easy to demonstrate the interaction across a transparent barrier. Breder and Halpern (1946) studied grouping behavior in goldfish and zebra fish, and the influence of various environments. In general there was a tendency to aggregate in the middle of the aquarium, but if two aquaria were close to each other, then the fish approached the other aquarium unless a card separated them visually. The larger the group of fish, the fewer the strays. Breder (1959) showed that black goldfish tended to aggregate over black squares, gray over gray, and gold over yellow. He hatched some fish separately and raised them without allowing them any experience among other fish. They joined the school of fish as soon as they were introduced into the same container, and this behavior seemed to be independent of previous experience.

Miyadi (1956) studied a Japanese fish, the ayo, *Plecoglossus altivelis*, which has a territory (Figure 17.2). As the population increased, fish tended to leave the area. When the population was dense the whole population left its territorial arrangement and became a school, all moving together.

How does this patterned social behavior develop? What are the stimuli, and how do they alter behavior? Breder (1959) raised two zebra fish as isolated individuals to the age of 6 months. The very raising of single individuals is a difficult task. When they were introduced into the school, they immediately joined the group and swam with it. It was possible to follow one of them because it was slightly crippled and it remained with the others.

Shaw (1960) studied the ontogeny of group behavior in the common silversides, *Menidia menidia* and *M. beryllina* which can be seen in fast-moving schools in the shoal waters of New England. About 1,000 newly hatched fry were transferred to running or stationary sea water. After hatching they aggregated in groups of 15 to 20 individuals near the vessel wall in a layer one fish deep. At length 5 to 7 mm there was no parallel orientation, and when they reached a length of 9 to 9.5 mm they lined up and swam together for 5 to 10 seconds. Schooling was thought to develop as a response initially between 2 fry with numbers increasing as the fish grew. Responses of individuals in glass tubes were measured. All individuals between lengths 13 to 16 mm

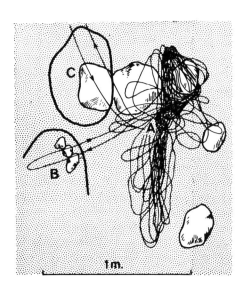 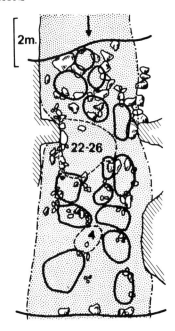

Figure 17.2. Territory in the fish, the ayo, *Plecoglossus altivelis*. *Left:* Tracings of the movements of one fish, A, in its territory, for 3 minutes. Twice it attacked invaders, chasing them into other territories, B and C. *Right:* Territories are shown by the heavy outlines. Figures within the broken lines give the numbers of ayos belonging to nonterritorial aggregations. (Redrawn from Miyadi, 1956.)

oriented to another individual in a tube, rather than to a control tube. Of 400 fish reared in isolation only 4 grew to 15 mm in length. When each of these 4 fish was presented to a school, it joined immediately. However during the first 4 hours they bumped into others and occasionally swam away.

CONCLUSIONS

It has been suggested here that the study of animal behavior, especially social behavior, is greatly influenced by the beliefs of the times, for one finds what one searches for; and field-oriented studies on natural behavior are only now leading to hypotheses that can be tested in the laboratory.

Circumstances can markedly change the social behavior of a species. In locusts, the solitary phase is converted to the swarming plague phase by crowding, a reaction brought about through tactile stimulation. Schooling in fishes is a remarkable group phenomenon whereby the population seems to act as one. It seems more dependent upon visual stimuli, and takes some days to develop to full efficacy in the young fish. Young fry raised alone maintain the capacity to react to the school, if they are able to survive. In other fish, defense of territory is converted into schooling behavior when crowding develops. The study of ontogeny of fear and conflict in the chick illustrated Schneirla's biphasic theory of approach and withdrawal, in which certain contradictions of opposites seem to be resolved into syncretic union.

REFERENCES

Allee,W. C. 1951. *Cooperation Among Animals*. Abelard-Schuman, New York.

Bitterman, M. E. 1965. The evolution of intelligence. *Sci. Am.* **212**: 92–100.

Breder, C. M. 1959. Studies on social groupings in fishes. *Bull. Am. Museum Nat. Hist.* **117**: 393–482.

Breder, C. M., and Halpern, F. 1946. Innate and acquired behavior affecting the aggregation of fishes. *Physiol. Zool.* **19**: 154–190.

Ellis, P. E. 1964. Marching and color in locust hoppers in relation to social factors. *Behaviour* **23**: 177–191.

Fuller, J. L., and Thompson, W. R. 1960. *Behavior Genetics*. Wiley, New York.

Gottlieb, G. 1968. Prenatal behavior of birds. *Quart. Rev. Biol.* **43**: 148–174.

Hamburger, V. 1964. Ontogeny of behavior and its structural basis. In *Comparative Neuro-Chemistry. Proc. 5th Intern. Neurochem. Symp., Austria, 1962*. Pergamon Press, New York.

Hogan, A. J. 1965. An experimental study of conflict and fear: an analysis of behavior of young chicks toward a mealworm. *Behaviour* **25**: 45–97.

Keenleyside, M. H. A. 1955. Some aspects of the schooling behavior of fish. *Behaviour* **8**: 183–248.

Klopfer, P. H., and Hailman, J. P. 1967. *An Introduction to Animal Behavior: Ethology's First Century*. Prentice-Hall, Englewood Cliffs, New Jersey.

Knight-Jones, E. W., and Crisp, D. J. 1953. Gregariousness in barnacles in relation to the fouling of ships and to antifouling research. *Nature* **171**: 1109–1110.

Köhler, W. 1921. *The Mentality of Apes*. Harcourt, Brace, New York.

Konorski, J. 1948. *Conditioned Reflexes and Neuron Organization*. Cambridge, England.

Kuo, Z. 1967. *The Dynamics of Behavior Development: An Epigenetic View*. Random House, New York.

Lindzey, G. 1951. Emotionality and audiogenic seizure susceptibility in five inbred strains of mice. *Jour. Compar. Physiol. Psychol.* **44**: 389–393.

Miyadi, D. 1956. Perspectives of experimental research on social interference among fishes. In *Perspectives in Marine Biology*. A. A. Buzzati-Traverso (ed.). Symposium at Scripps Institution of Oceanography.

Morris, D. 1962. *The Biology of Art*. Knopf, New York.

Pavlov, I. P. 1927. *Conditioned Reflexes: An Investigation of the Activity of the Cerebral Cortex*. Trans. by G. V. Anrep, Oxford Univ. Press, London. Republished 1960, Dover, New York.

Rodgers, D. A., and McClearn, G. E. 1962. Alcohol preference in mice. In *Roots of Behavior*. E. L. Bliss (ed.) Harper, New York.

Rothenbuhler, W. C. 1959. Genetics of a behavior difference in honeybees. *Proc. Intern. Congr. Genet., 10th Montreal, 1958* **2**: 242.

Rothenbuhler, W. C. 1964. Behaviour genetics of nest cleaning in honeybees. *Animal Behaviour* **12**: 578–584.

Schneirla, T. C. 1965. Aspects of stimulation and organization in approach-withdrawal processes underlying vertebrate behavioral development. *Advan. Study of Behavior* **1**: 1–74.

Scott, J. P., and Fuller, J. L. 1959. Heredity and the development of social behavior traits in dogs. *Acta Psychol.* **15**: 554–555.

Shaw, E. 1960. The development of schooling behavior in fishes. *Zoology* **33**: 79–86.

Skinner, B. F. 1961. Teaching machines. *Sci. Am.* **205**: 90–102.

Smith, W. 1957. Social learning in domestic chicks. *Behaviour* **11**: 40–55.

Southwick, C. H., and Clark, L. H. 1968. Interstrain differences in aggressive behavior and exploratory activity of inbred mice. *Comm. Behav. Biol.* **1**: 49–59.

Stoker, M. 1966. Mechanisms of viral carcinogenesis. *Can. Cancer Conf.* **6**: 357–368. Academic Press, New York.

Thompson, W. R. 1953. The inheritance of behavior: behavioral differences in 15 mouse strains. *Can. J. Psychol.* **II**: 145–155.

Thorpe, W. H. 1963. *Learning and Instinct in Animals.* 2nd ed. Methuen, London.

Uvarov, B. P. 1966. *Grasshoppers and Locusts.* Cambridge Univ. Press, London.

Whitman, C. O. 1919. The behavior of pigeons. Posthumous Works of C. O. Whitman **3**: 1–161. A. H. Carr (ed.). *Publ. Carnegie Inst.*, **257**.

Yerkes, R. M. 1912. The intelligence of earthworms. *J. Animal Behavior* **2**: 332–352.

CHAPTER 18

Animal Communication

WILLIAM J. L. SLADEN

Communication is an essential part of social organization. It will be discussed here as a preparation for the following chapters on social organization.

In many respects human behavior is analogous to the behavior of other animals, more than has previously been recognized. Indeed, the study of behavior is probably expanding at this time into the kind of role played by comparative physiology, anatomy, and pharmacology in the study of man. Great caution is required when human behavior is interpreted by comparative studies, but new bases for the understanding of human activities are much needed. Some investigators see in animal behavior analogues of human behavior in land and property possession (Ardrey, 1966), political hustings (Haldane, 1953), the basic mechanisms of morality (Wynne-Edwards, 1962), and perhaps of frustration and boredom in extremely simple animals (Best and Rubinstein, 1962). Darwin thought that all the elements of human communication would eventually be found in some animal or another. Koehler (1956; cited in Thorpe, 1961) believes that no single character infallibly distinguishes the communication of birds by sound from human language, and that human language is unique only in the way that it combines attributes which themselves are not peculiar to man. Of course, such statements need to be carefully defined and documented.

COMMUNICATION INVOLVES AT LEAST A SIGNALER AND A SIGNAL, A "MESSAGE" PASSED, AND A RECEIVER

The signal is described by its form, ideally when it has been reduced to its essentials by experiment. The "message" has to be inferred by the investigator according to his knowledge of the following things: the nature (species, sex, etc.), physiological state, and experience of the participants; the location, season, etc.; the sequence of behavior in which the signal occurs; and the reactions made by the receiver. Such study requires ecological knowledge, and it is part of the discipline *ethology*. Ethology is concerned mostly with animals behaving naturally in their own habitat and also depends on laboratory observation and experiment. It deals not only with units of communication but with how these are built into complex behaviors and social organization, and the relation of all this to survival of individuals and of populations.

CONVENTION IN COMMUNICATION

If signaling systems are to be effective they must be reliable, that is, fairly stable in their form and in the response evoked, so that behaviors are coordin-

247

ated and occur in appropriate situations. As an instance of failure of communication, when penguins from Peru and from Antarctica were kept together in my laboratory, one species failed to appreciate the aggressive threats of the other; there was confusion, and typical patterns of fighting in defense of territory did not appear, at least for some time. Stability of mechanisms for communication within a population can be maintained genetically or by cultural inheritance.

GENETICS

Lorenz (1958) and others analyzed courtship behavior in surface-feeding ducks, particularly in some species that are so closely related that they will interbreed in captivity and even produce fertile hybrids. These species were excellent material for analyzing the movements and postures involved in the courtship display of males. Parts of this display in three closely related species are shown in Figure 18.1. There are ten postures and movements involved in these three sequences of behavior, some accompanied by sounds, but in each species some of the ten are absent and the arrangement is different. In the hybrids there were new combinations, sometimes involving characteristics of both parents; sometimes those of one parent were suppressed, and in some cases patterns were produced that appear only in other species. Further new combinations were produced in the few successfully reared second generation hybrids. All this suggested that the movements from which courtship displays in these species are built "are dependent on comparatively simple constellations of genetic factors" (Lorenz, 1958).

IMPRINTING

Lorenz (1935) divided a clutch of goose eggs into two groups. One group was hatched by the goose and wherever she went the goslings followed her. The other group was hatched in an incubator and saw Lorenz as the first moving object and followed him until they were fully grown. They took no notice of their mother and they recognized Lorenz from other men. It had long been known that nidifugous birds, those that leave the nest shortly after hatching, immediately learn the signal that releases (see later) the response of following. In natural conditions that signal is the mother. Lorenz drew attention to this as a type of learning that occurred only within a short sensitive period. Others have further analyzed the imprinting of young nidifugous birds for their following response. During a period of 10 minutes Hess (1958) imprinted mallard ducklings, *Anas platyrhynchos*, to a moving duck decoy (a model of a male mallard) while he played a recording of a human voice saying "Gock gock gock . . ." Later each duckling was exposed to the same moving decoy and sound and the percentage of times it followed the decoy was recorded (Figure 18.2). Hess showed that in this case the most sensitive period for learning the object to be followed was from 10 to 20 hours after hatching. The highest percentages of following responses in ducklings that had been imprinted at ages 12–17 hours, were obtained for those that had travelled at

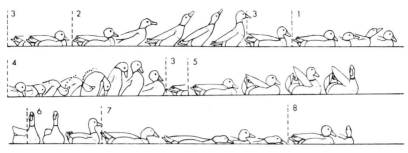

Mallard, Anas platyrhynchos

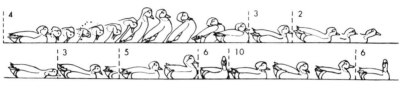

Gadwall, Anas strepera

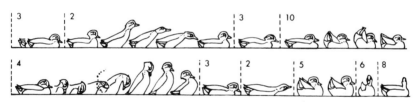

European Teal, Anas crecca

Figure 18.1. Parts of the courtship displays of the male in three closely related species of surface-feeding ducks. Ten identical movements or positions occur in these three species, but their arrangements vary; nine are shown here. Hybrids have different patterns again, and some of the ten elements may not appear. 1 = bill-shake, 2 = head-flick, 3 = tail-shake, 4 = grunt-whistle, 5 = head-up—tail-up, 6 = turn toward female, 7 = nod-swimming, 8 = turning back of head, 10 = down-up. (Redrawn from K. Z. Lorenz, "The evolution of behavior." Copyright ©1958 by *Scientific American*, Inc. All rights reserved.)

least 50 to 100 feet each time they followed the model during the learning-exposures.

Mother-substitutes that have been successfully used with ducks and geese are humans, models of the same species, other species, even moving footballs and boxes. The selective advantage of such learning to respond to a particular stimulus may well be related to the fact that the mother is an invariable part of the environment of the successfully wild-reared bird. For the recognition of

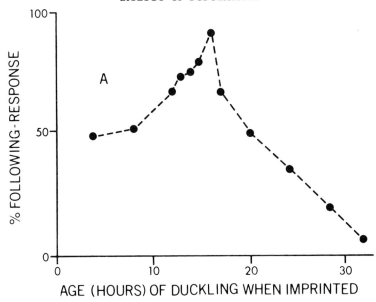

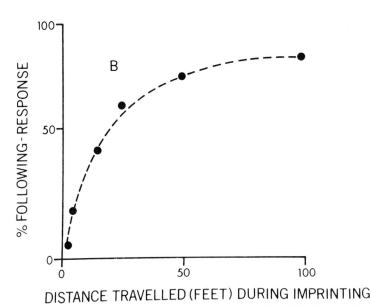

Figure 18.2. Demonstration of the sensitive period (graph A) and the effect of the distance travelled (graph B), for mallard ducklings, *Anas platyrhynchos*, experimentally imprinted to a decoy. Percent following response is the average percentage of times that ducklings followed the decoy when tested some days after they had been imprinted. (After E. H. Hess, " 'Imprinting' in Animals." Copyright ©1958 by *Scientific American*, Inc. All rights reserved.)

the mother-signal to be completely genetically controlled some sophistication of genetic coding and its translation into neural mechanisms for recognition would be required. Surely in these terms imprinting is economical, and also allows the form of the mother to vary. For discussion of imprinting and the extent to which it is known to occur in other animals, including man, Thorpe (1963) and Hinde (1966) are recommended.

OTHER LEARNING OF SIGNALS AND THE DEVELOPMENT OF CULTURAL DIFFERENCES

The song of a common Eurasian bird, the chaffinch *Fringilla coelebs*, has been carefully analyzed to show which parts of it are directly influenced by genetic inheritance and which parts are learned (Marler, 1956). Many small birds have a song, which is a very characteristic series of notes and is used particularly in the establishment and maintenance of territory during the breeding season. It advertises possession of territory, warns off other males, and to some extent is a substitute for fighting, and it attracts females. The song of the chaffinch has ideal locating clues (Figure 18.3 and Table 18.2). It is a series of rapid trills that often change pitch two or three times, and it has a terminal flourish.

A hand-reared chaffinch that has never heard another bird develops a simple song, and this part is considered to be directly genetically determined (Figure 18.3). Species are recognizable from others in bird songs, so it is understandable that some of the song should have a genetically determined base that is characteristic of the species. When chaffinches are raised in an isolated group they develop a more complex song that is indistinguishable from bird to bird, but fairly different from the normal song. In fact, these birds are in the process of developing their own subculture of song. Wild chaffinches apparently learn some details of song from their parents or from other adults during the first few weeks of life. Then the following spring when the young male first sings in territory in competition with other males, he completes his song, learning it from neighbors, some of which are older birds. He may learn several variants. The age of 14 months is the end of the learning period, and thereafter for life the song does not change. The process of learning from several neighbors results in differences in chaffinch song that are so individually characteristic that investigators can recognize a given bird from his song. Surely this is a basis for individual recognition among the birds themselves. Also the practice of learning from other birds results in local dialects. It is interesting that in places such as islands where the chaffinch populations are not large, and individual differences perhaps do not have to be complex, the song is indeed found to be simpler.

So, in the ontogeny of mechanisms for communication, as in other behaviors, there may be rather direct genetic determination and also various types of learning. It should be noted that other bird song does not necessarily develop in the same way as chaffinch song (Thorpe, 1961).

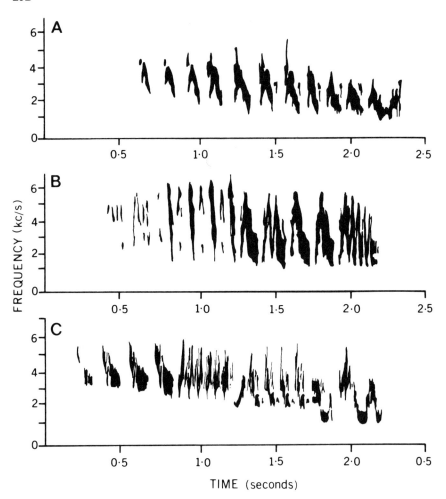

Figure 18.3. Sound spectrograms of the song of the male chaffinch, *Fringilla coelebs.*
A. Single bird hand-raised in auditory isolation.   B. One of two birds raised together but
isolated from all others. C. Full song of normal male. Soundwave frequency is given in
kilocycles per second. (After Thorpe, 1958.)

DISPLAY

Many physical systems are used for signaling—the signaler can take up a
posture or make a movement, sound, smell, light, electrical discharge; or he
can touch the receiver, make vibrations in water, move silken threads, leave
marks in the ground or on trees, and so on. Much of the signaling is called
displaying, according to the following definition. Displays are movements,
postures, and sounds, etc., generally of a conventionalized kind, which have

the capacity to initiate specific responses in other creatures, more particularly in members of the same species. A single display can consist of several signals each with different effects upon the receiver. For instance, the song of the male bird is a display, and females react by approaching, and other males by not approaching.

Many displays and signals are considered to have arisen by exaggeration of normally occurring properties of animals or of incidental effects. These have been enhanced by their survival value into more prominent signals—that is, they have been conventionalized. This can perhaps be understood from consideration of a territorial display of the male beaver. He gives a smack with his broad tail to the surface of his pond as he dives, and this can be heard for a great distance. It serves notice to other beavers that a male is in occupation at a certain locality. Possibly this display began as an inevitable "plop" as he dived. The subject of the origin of such conventions is discussed by Tinbergen (1963).

SOCIAL ORGANIZATION IN TERMS OF CONVENTIONS

Wynne-Edwards (1962) proposed the theory that social organization has the function of substituting conventionalized modes of competition instead of competition for materials of real value. Such competition in accordance with conventional rules for conventional rewards might act as a buffer controlling population density before resources are depleted. This idea gives significance to Wynne-Edwards' category of epideictic behavior, which he defines as behavior concerned with assessment of membership within a population. Sometimes epideictic behavior is part of territorial behavior, and species recognition and individual recognition are important components. According to Wynne-Edwards epideictic displaying may be at regular times such as dawn and dusk early in the breeding season, in particular locations where all participants may gather. Usually only males participate and the animals may otherwise be cryptic. The dawn chorus of birds, and the singing at dusk might function as epideictic displays, as well as the other functions that can be assigned to bird song. Much displaying of groups was previously noted but not understood, such as the displaying of gyrinid beetles on the surface of water and the swarm-dancing of small insects on summer evenings.

THE INFORMATION CONTENT OF SIGNALS:
EXAMPLES, CHAFFINCH SONG AND HONEYBEE DANCES

There is no intention of implying here that *ideas* are communicated among animals. One can go no further than the evidence—that a signal in a certain set of circumstances is followed by a reaction related to certain aspects of the total environment. The reaction can be analyzed and classified according to the aspects of the total environment that are involved.

TABLE 18.1

Some Types of Information Conveyed in Signals[a]

Type of "message"[b]	Kind of evidence	Comments
Information that locates and identifies the signaler (*locator signals*)		
Gives his *species*	The song is sung only by a chaffinch	This part of the signal needs to be stable within the species
Gives his *sex*	The song is sung only by males	
Identifies him as an *individual*	The female responds to an unfamiliar song but, given a choice, there is some preference for a familiar song	This part of the signal needs to be variable within the group
Gives his *exact location*	The female goes directly to him	The physics of song is excellent for identification of its source (see Table 18.2)
Information that directs the reaction of the receiver to certain things (*designator signals*)		
By conveying the *motive* of the signaler		Some of these signals are called intention behavior— when a flock of geese is about to fly, some members make a certain movement which is then taken up by other members, until they all suddenly move off together
a) The signaler is ready to engage in certain activities		
To copulate	Female chaffinch gives a brief signal and copulation occurs	
To feed	Young bird begs food from mother, who then feeds it	
To fight	Very common set of signals from threat to attack, which get agonistic response (i.e. of responses associated with aggression)	
b) The signaler is ready *not* to engage in certain activities		
Appeasement	In many animals submissive behavior stops a fight	
By giving information about the *environment* of the signaler, the receiver is directed to certain activities		
No mate is present	The male chaffinch sings more frequently before he has acquired a mate, than after	

TABLE 18.1 (*continued*)

Type of "message"[b]	Kind of evidence	Comments
The signaler has territory	The male chaffinch does not sing if he does not hold territory	
Food is present	See the section on honeybee communication	
A predator is present	The two alarm calls of the chaffinch alert small birds in the woods and their activity changes accordingly	The alarm calls are not species-specific—many species in the same wood receive and act on the signal (see also Figure 18.6 and Table 18.2)
Information directing the receiver to react with preference to one of several things (appraisor signals)	The male chaffinch sings less frequently when he is mated. The female confronted with two males singing is more likely to go to one singing more persistently	
Information disposing the receiver to perform certain sequences rather than others (prescriptor signals)	When the female chaffinch responds to the male song, she adopts a sequence of courtship activities instead of her previous behavior	

[a] Based on Marler (1961) and Thorpe (1961).

[b] Where not stated otherwise, the examples are from the song of the male chaffinch which is unmated, in reproductive condition, in possession of a territory, and near to a female. The "message" has been inferred from the reactions of the receiver.

Chaffinch Song and Calls

Marler (1961) analyzed the information content of the chaffinch song according to a classification used in human signaling systems. The classification is summarized in Table 18.1, and the chaffinch song is used for most of the examples. The chaffinch also has call notes, some of which are mentioned in the table; the nestlings and fledglings have certain calls, and there are 15 other basic calls used in flight, flocking, feeding, alarm, distress, fighting, etc. Marler has analyzed the information content of these 15 basic calls as follows: 5 give environmental, 9 social, 7 identifying, and 7 locating information. Thus, one call may give several kinds of information.

Honeybee Communication

When a foraging honeybee, *Apis mellifica*, has fed from a rich source and has deposited honey back in the hive, other bees from the hive go to the same source without actually following her, and the numbers of new feeders are regulated according to the sweetness and abundance of the food. The signals have been extensively investigated by von Frisch and his associates (von Frisch and Lindauer, 1956). Scent of the flower species is in the honey that is regurgitated, and also is well retained by the bee's body, which is touched by

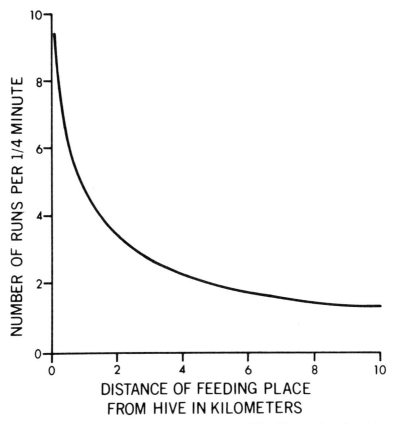

Figure 18.4. The waggle dance of the honeybee, *Apis mellifica*. The number of straight runs per 1/4 minute decreases with increased distance of the food source from the hive. As the rate of runs decreases, the length of the run and the vigor of the waggling increase. Going into the wind and going uphill also decreases the rate of runs per time. The exact relationship of distance and rate is constant. (After von Frisch, 1953.)

the receivers' antennae. Also, the scent of the colony is disseminated into the air by the everted scent gland when she is feeding at a rich source, and flying bees are attracted directly to the food. When the food is within 100–150 meters from the hive the returned forager gives the "round dance" display in the hive. Neighboring bees are excited and follow her in the dance, touching her body, and then depart in all directions and locate the source by its scent. When the food source is more than 100–150 meters away the signaler displays with a "waggling dance," which conveys the direction and distance of the source from the hive. Figure 18.4 shows that the rate of waggling of the abdomen, which occurs during a straight run in the dance, declines with distance. Also, when the bee flies out against a wind or up a hill, the waggling rate declines, so the rate is probably related to energy expended or time involved.

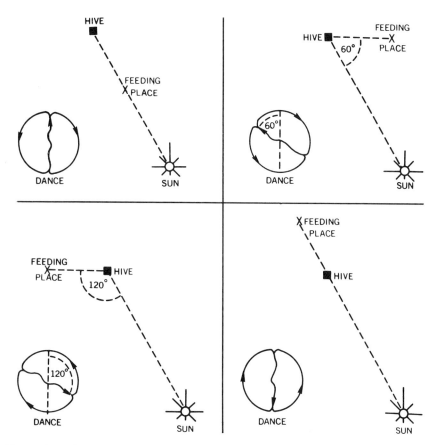

Figure 18.5. The pattern of the waggle dance of the honeybee, *Apis mellifica*, when it is performed in darkness on the vertical surface of the comb. The orientation of the dance to the direction of gravity changes according to the positions of the food source and the sun in relation to the hive. (After von Frisch, 1953.)

These relationships are remarkably stable from hive to hive and from day to day. Figure 18.5 shows how the direction of the food source in relation to the position of the hive, and either the position of the sun or the plane of polarized light in the sky, are related to the orientation of the dance to the sun or to the plane of polarized light. If the dance takes place in darkness on the vertical comb, its orientation is related to the direction of gravity. The accounts by

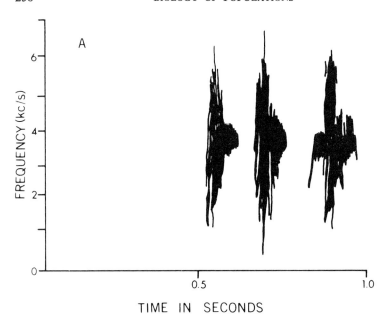

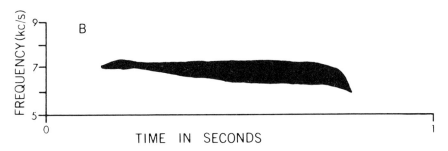

Figure 18.6. Sound spectrograms of two alarm calls of the chaffinch, *Fringilla coelebs*. A. The *chink* provides abundant location clues.   B. The *seet* call, from which it is difficult to locate the signaler (Table 18.2). Soundwave frequency is given in kilocycles per second. (A, after Thorpe, 1958; B, after Marler, 1956.)

von Frisch (1953) and von Frisch and Lindauer (1956) should be consulted for experimental evidence for these statements, and Wenner *et al.* (1967) for a critique of the subject.

The dance displays are used also by honeybees when they are swarming. Scouts that have found a hole in which a new nest could be built return to the cluster and dance on it and by their dance arouse other bees to excitement and to visit the site. Thus different groups of bees on the swarm can be dancing in several different directions at the same time. During several hours or several days the most intensely dancing group gains more and more followers and the lesser groups cease their dancing. A unanimity is reached among dancers and then the swarm moves off to that particular location. Haldane (1953) points out that this is a typical example of intention behavior, which is a mechanism for coordinating group behavior. He interprets the dancing in relation to food, also as intention behavior—the foraging bee communicates to neighboring bees which source of food she intends to visit when she goes out again and arouses them to do likewise.

These two examples of complex communication, chaffinch song and honey-bee dances, are presented to indicate the amount of information involved in some animal communication, as well as to indicate how the experimenter tries to assess the information content.

THE PHYSICS OF COMMUNICATION

Just one example will be taken from this fascinating research field, the means of locating the source of a sound, using for illustration two of the alarm calls of the chaffinch (Figure 18.6 and Table 18.2). If a bird of prey is on a perch or on the ground, small birds tend to mob it. In these circumstances a chaffinch gives the *chink* note which provides abundant clues for locating the signaler.

TABLE 18.2

Methods of Locating the Direction of the Source of a Sound, or of Not Locating it.[a, b]

Direction of the source is determined from comparison of what is received at each ear	The form of the signal if its source is to be	
	Located	Not located
Differences in times of arrival of the sound	Abrupt beginnings and endings; e.g. a series of sharp taps	Fading in and out of the sound
Differences in the phase of the wavelength at the same instant	Wavelength longer than the distance between the two ears; sound sustained	If the wavelength is shorter than the distance between the two ears the information becomes ambiguous
Differences in the intensity of the sound at the two ears, which is a function of the size of the head and the wavelength	Best when the wavelength is very short—high frequency sound	Locating is difficult when the wavelength is long—low frequency sound

[a] Summarized from Marler (1959).

[b] Two of the alarm calls of the chaffinch are used to illustrate this (Figure 18.6) and are explained in the text.

When the bird of prey is in flight, the chaffinch gives the *seet* alarm call, and small birds in the woods dash for cover and hide. This call is a pure tone that fades in and out, at a frequency too high for phase difference detection and too low for ideal use of intensity difference. Man finds it difficult to locate the signaler by this call and presumably so does the predator.

SEQUENCES OF REACTIONS BETWEEN TWO ANIMALS; AND SIGNALS AS "RELEASERS"

The ethologist does not always try to identify the information content of signals. He begins by noting that a particular display or signal in appropriate circumstances is followed by a particular reaction in the receiver. When the events have been well studied he may be able to say that a display or signal *releases* a particular reaction in the receiver and it may not be appropriate to talk of information content. The signal and reaction may be part of a long sequence of behavior that as a whole has great survival value, and the interest is in seeing how the sequence is built from smaller units of signal and reaction. The signal can be termed a releaser only when the reaction of the receiver follows fairly closely after reception of the signal; otherwise at the present level of understanding of behavior the situation becomes too complex for use of the term.

There are two obvious events in the life histories of animals where cooperation between two individuals is essential: in mating and, when there is care of young, between parent(s) and offspring. Examples of these will be given.

The mating behavior of the three-spined stickleback fish, *Gasterosteus aculeatus*, has been analyzed by Tinbergen (1951) into a series of releasers, and reactions that themselves are releasers of the next step in the sequence of behavior. The sequence is as follows, the numbers referring to positions illustrated in Figure 18.7. The male is in nuptial colors and has a territory in which he has built a nest, a tunnel of weeds stuck together over a pit in the sand. When males approach the owner he reacts with threat display which may lead to fighting. When a female approaches he reacts with the zigzag dance (1). Most females flee, but one in spawning condition turns to the male in an almost upright position. Her position and swollen shape releases in the male the behavior of turning and swimming to the nest. The female follows. The male takes up position 5. Then the female enters the nest and stays with head and tail stuck out. The male repeatedly thrusts her tail base with his head. The female lifts her tail and soon she spawns. She passes out of the nest and the male enters and fertilizes the eggs. Several other females may add to the eggs in the next few days. Then the male defends the territory against all comers and raises the eggs and young by himself.

The first action of the male in this series can be released by showing him a fish-shaped model of about the same size as the female. He reacts with the zigzag dance only if the model has a fat body and is held in a slanting position. When the body of the female is no longer swollen because the eggs have been laid, the female releases aggressive behavior in the male and is driven out of the territory. The red color of other males in reproductive condition releases strong aggressive behavior in the males. The importance of the color was demonstrated by Tinbergen's fish whenever red mail trucks went by his

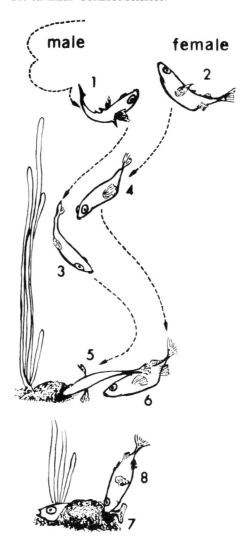

Figure 18.7. Sequence of actions and reactions in the mating behavior of the three-spined stickleback, *Gasterosteus aculeatus.* The numbers are explained in the text. (Redrawn from Tinbergen, 1951.)

laboratory and the males rushed to the window side of the tanks and took up postures of threat.

A male and a female Adelie penguin, *Pygoscelis adeliae* that have a nest site and an established partnership are frequently seen in the mutual display. This display plays a role in maintaining the pair bond and is used in recognition between parents and chicks (Sladen, 1953). Before coition the male scratches out a scoop in the nest of stones and moves to the side of the nest to allow the female to enter the scoop. If the female lies flat in the nest the male will mount on her back. Coition occurs if the female extends her head upwards and they vibrate their chins one against the other (Sladen, 1958). A male will mount a frozen bird if it is lying down with the head raised so that the bill is pointing upwards.

The parent herring gull, *Larus argentatus*, has an orange mark on its lower beak, and the herring gull chick reaches up with its bill and touches the orange mark. The parent then opens its bill and regurgitates food from the crop. The chick feeds from the parent's mouth. In some of this sequence models can be substituted that release just the same behaviors—for instance, an orange mark on the beak of a cardboard head releases the begging behavior (Tinbergen and Perdeck, 1950).

Thus complex adaptive cooperation between individuals can in some cases be seen to be organized as sequences of action-reaction-reaction. The neurophysiology is not known. But such analyses are beginnings toward an understanding of organization of behavior at one of its levels.

CONCLUSIONS

Comparative psychology tends toward analysis of the behavior of single captive animals, more in terms of their physiology and neurophysiology. Ethology describes and analyzes behavior in its adaptive setting of the habitat and the population. This chapter has attempted to show that communication in wild animals can be analyzed in a given social and ecological environment in terms of the behavior that is released in the animal that receives a signal. The action that is released in the receiver can be said to be directed toward certain aspects of the environment, and information content of the signals can be discussed in these terms. Understanding the physics of the signal and its means of reception also clarifies the material. It is particularly necessary that the bridge between ethology and neurophysiology should be strengthened; indeed in most cases this bridge still needs to be created.

REFERENCES

Ardrey, R. 1966. *The Territorial Imperative*. Atheneum, New York.
Best, J. B., and Rubinstein, I. 1962. Environmental familiarity and feeding in the planarian. *Science* **135**: 916–918.
Haldane, J. B. S. 1953. Animal ritual and human language. *Diogenes* **4**: 61–73.
Hess, E. H. 1958. "Imprinting" in animals. *Sci. Am.* **198**: 81–90.
Hinde, R. A. 1966. *Animal Behavior: A Synthesis of Ethology and Comparative Psychology*. McGraw-Hill, New York.
Lorenz, K. Z. 1935. Der Kumpan in der Umwelt des Vogels. *J. f. Ornith.* **83**: 137–213, 289–413.
Lorenz, K. Z. 1958. The evolution of behavior. *Sci. Am.* **199**: 67–78.
Marler, P. 1956. The voice of the chaffinch and its function as language. *Ibis* **98**: 231–261.
Marler, P. 1959. Developments in the study of animal communication. In *Darwin's Biological Work: Some Aspects Reconsidered*. P. R. Bell (ed). Cambridge Univ. Press, London, England.

Marler, P. 1961. The logical analysis of animal communication. *J. Theoret. Biol.* **1**: 295–317.

Sladen, W. J. L. 1953. The Adelie penguin. *Nature* **171**: 952–955.

Sladen W. J. L. 1958. *The Pygoscelid Penguins. I—Methods of Study. II—The Adelie Penguin.* Falkland Islands Dependencies Survey, Sci. Rept. **17**. H.M. Stationery Office, London.

Thorpe, W. H. 1958. The learning of song patterns by birds, with especial reference to the song of the chaffinch. *Ibis* **100**: 535–570.

Thorpe, W. H. 1961. *Bird Song.* Cambridge Monographs in Experimental Biology, 12. Cambridge Univ. Press, London.

Thorpe, W. H. 1963. *Learning and Instinct in Animals.* 2nd ed. Harvard Univ. Press, Cambridge, Mass.

Tinbergen, N. 1951. *The Study of Instinct.* Oxford Univ. Press, Oxford, England.

Tinbergen, N. 1963. The evolution of signalling devices. In *Social Behavior and Organization Among Invertebrates.* W. Etkin (ed.) Univ. of Chicago Press, Chicago, Illinois.

Tinbergen, N., and Perdeck, A. C. 1950. On the stimulus situation releasing the begging response in the newly hatched Herring Gull chick (*Larus a. argentatus* Pontopp.). *Behaviour* **3**: 1–38.

von Frisch, K. 1953. *The Dancing Bees* (translation). Harcourt, Brace, New York.

von Frisch, K., and Lindauer, M. 1956. The "language" and orientation of the honeybee. *Ann. Rev. Entomol.* **1**: 45–58.

Wenner, A., Wells, P. H., and Rohlf, F. J. 1967. An analysis of the waggle dance and recruitment in honeybees. *Physiol. Zool.* **40**: 317–344.

Wynne-Edward, V.C, 1962, *Animal Dispersion in Relation to Social Behavior.* Haffner, New York.

CHAPTER 19

Social Organization in Some Birds and Mammals

WILLIAM J. L. SLADEN

In this chapter there will be brief descriptions of the social structure of populations of the Adelie penguin, *Pygoscelis adeliae*; the Australian magpie, *Gymnorhina tibicen*; the northern fur seal, *Callorhinus ursinus*; the black-tailed prairie dog, *Cynomys ludovicianus*; the European rabbit, *Oryctolagus cuniculus*; and domestic chickens, *Gallus domesticus*. These will be used as examples in discussion of possession of territory, agonistic behavior with reference to territory, hierarchies, pair formation, individual recognition, and differing behavior of different age groups. A few examples will be included of the relation of territory or hierarchy to transmission of disease, control of reproduction, and effects of food shortage.

Marking of Individuals

Unless animals are individually distinguished by scars, color patterns, and other peculiarities, individuals of a group should be marked with a durable label that can be read from some distance so that the animals are not disturbed. Penguins can be identified from a band on the flipper that carries a number large enough to be read through binoculars at a distance of 20 meters (Sladen *et al.*, 1968b; Penney, 1968). For the Australian magpie, Carrick (1963) used a numbered addressed aluminium band on one tarsus and colored bands on the other. For fur seals Peterson (1965) put large numbers or symbols on the backs by shaving off guard hairs and bleaching the under-fur. Mykytowycz (1964) marked rabbits with black dye, and tattooed and fixed disks in the ears.

THE ADELIE PENGUIN (Sladen, 1953 and 1957)

The main events of the breeding season of this species are shown in Figure 19.1. Adelie penguins are very gregarious, remaining together in winter in groups on pack-ice and in summer in breeding rookeries on shores south of the Antartic Convergence. A rookery contains a few, to hundreds of thousands of birds, on nests of stones in closely packed territories* of about 3 feet in diameter. Each pair defends only its own nest site; and the one or two chicks are reared by both parents. Banding has shown that the majority return to the territory and mate of the previous summer; but apparently pairs do not keep together in winter. During the initial occupation of territory and nest building, the breeders do not feed. The female leaves for the sea after egg laying,

* Territory is defined as any defended area.

264

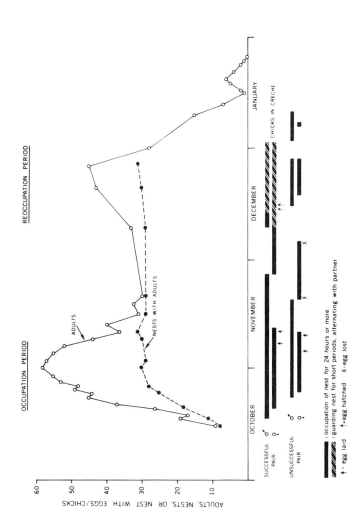

Figure 19.1. A population of Adelie penguins, *Pygoscelis adeliae*, in a small part of a rookery, through the breeding season. Two nest histories are shown; 1—of a successful pair, 2—of an unsuccessful pair. Pairs that have lost their eggs or chicks return to reoccupy territory in December. (After Sladen, 1958.)

while the male incubates the eggs and continues defense of the nest, and the male leaves when the female returns about 15 days later. By then the male has fasted for about 6 weeks and has lost 40 percent of his weight. Other aspects of their social organization will be mentioned in other parts of the chapter.

THE AUSTRALIAN MAGPIE (Carrick, 1963)

These nonmigratory birds live in groups in open savannah, woodland, and pasture, some defending territory all year. A group consists of up to 10 individuals, at least 1 male and 1 female, but never more than 3 breeding birds of each sex. They are usually monogamous, but sometimes a male has 2 females, occasionally 3. The group defends a shared territory of 5 to 20 acres.

The population studied by Carrick (1963) was of about 740 birds, and 200 of them held group territories. The rest had no territory. Carrick classified the whole population according to possession of territory, the nature of the habitat occupied, and breeding success (Table 19.1).

TABLE 19.1

Division of a Population of About 740 Australian Magpies, *Gymnorhina tibicen*, According to Possession of Territory, Breeding Success, and Habitat[a]

	In territory		Not holding territory	
	Number	Percent of total of each sex	Number	Percent of total of each sex
Males	79	21	296	79
Females	107	38	174	62
First year birds[b]	15	—	70	—
	201		540	

[a] Carrick (1963).

[b] Approximate numbers.

FURTHER DIVISION:

1. *Permanent* groups were the main breeding part of the population. A group of 2 to 10 adults defended a group territory of about 10 acres of best habitat of mixed pasture and woodland, that provided adequate food throughout the year.

2. *Marginal* groups occupied territories with an inadequate amount of either cover or feeding area. They were not successful breeders.

3. *Nomadic flocks* of nonbreeders were in treeless pasture.

4. Some groups held *separated breeding and feeding territories* but did not manage to rear young.

Territorial activity was greatest just before and during the breeding season. Birds in poorer habitat were always trying to improve their position by invasion of better areas, and these were fiercely defended by group aggression.

THE NORTHERN FUR SEAL (Baker *et al.*, 1963; Peterson, 1968)

The northern fur seal breeds on beaches of the Pribilof, Kurile, and Commander Islands of the Bering Sea and North Pacific Ocean, and in winter is

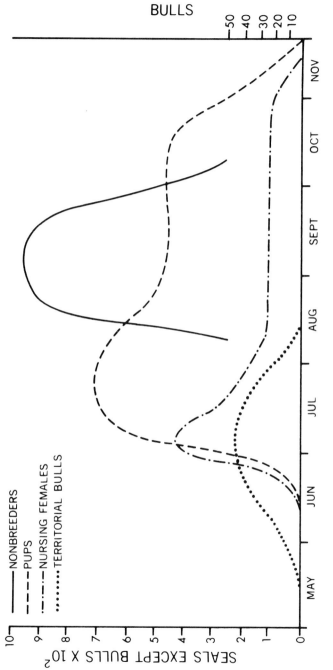

Figure 19.2. The composition of a population of the northern fur seal, *Callorhinus ursinus*, on a breeding beach of the Pribilof Islands, Alaska. (After Peterson, 1965, 1968.)

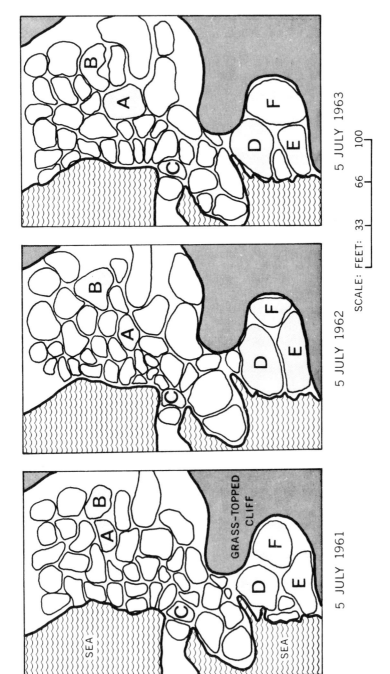

Figure 19.3. Territories of harem bull northern fur seals, *Callorhinus ursinus*, on a beach of the Pribilof Islands, Alaska. A painted grid was made on the stones and rock of the beach, and the territorial boundaries were identified by the boundary displays of the bulls. Six of the bulls whose territories are shown here (A–E) were marked for individual identification, and their territories changed little in 3 years. (After Peterson, 1965.)

widely distributed in the North Pacific. The composition of the populations on the breeding beaches is shown in Figure 19.2. The males are polygamous and very territorial. Bulls aged 10 or more years are big enough to fight for territory and defend it. They appear on the beaches in May and early June and established ones occupy territory of the previous years (Figure 19.3). Territories are close to the shore, and 300 square feet the mean size. Several established males may attack a newcomer, but each in defense of his own space. Bulls with big territories at the shoreline collect harems of up to 100 females.

Peterson (1965) summarized the behavior of females as follows: "They probably have site tenacity, but definitely not territory, and they are served by the male that happens to own the particular place. They are essentially solitary individuals (except briefly during oestrus and in that they suckle the young) that are herded into aggregates by the territorial male. The regularity of female behavior suggests that it is very stereotyped and associated with a good sense of time."

A single pup is born from a half to one day after landing, copulation occurs about 5 days later, and after one further day the female goes to sea to feed. The mean number of females served per territorial male during the season was 16 in Peterson's studies.

For the rest of the season on land, females return for periods of 2 days of nursing at somewhat regular intervals, usually of 8 to 10 days. In the latter half of July, territorial males leave, and the females and young move a little further inland. By the end of November the population has gone from the islands.

THE BLACK-TAILED PRAIRIE DOG (King, 1955 and 1959)

These rodents live in permanent systems of underground burrows, in highly territorial groups. They occur between North Dakota and Texas, where farmers have not removed them, in places where the rainfall is about 20 inches a year and the dominant vegetation is prairie grasses. The permanent, branching burrows have exits in large mounds of bare earth; and inedible plants in the territories are bitten off at the base. Each territory is of about 0.7 acre, and the boundaries rarely change, regardless of ownership. Defense of group territory is abated when females defend nest territories, from gestation until new pups appear above ground.

The territorial group consists typically of 1 to 2 adult males, 3 to 5 adult females and 6 or more young. Very little aggression within a territorial group is seen, but much grooming, touching, and "kissing." All adults kiss and groom the young of the group, and the young may spend nights with other broods within the group, and even suckle from other mothers. The young are rarely rebuffed until they meet adults of other groups at the boundaries of the group territory, and hostile reactions grow more severe as the young grow.

Sexual maturity is at 2 years, and there can be 3 or 4 more years of life. A female has one litter of about 5 pups each year. Gestation and lactation occur from March to May, when females defend individual nest territories. During this time some adult males and 1-year-olds pay daily visits to new boundary areas and construct new burrows there; or they may invade a territory occupied by a very small group. Pups appear above ground in early July. Then territorial groups are reestablished, but those males visiting new sites may begin to spend all day there, then the nights with only occasional returns to the original territory, until females have joined them and new groups are formed. Thus population density is periodically reduced by movement of sexually mature animals and young out into new areas.

The social organization of rabbits and chickens will be discussed later, under "hierarchies."

TERRITORY

Even bacteria may need certain space for their activities. Many animals limit their close contact with other individuals by "staking out" an area for themselves and advertising it as theirs.

Territory is defined as any area or volume that is defended (Howard, 1920). Some territory is retained all year, though it is frequently held only for the breeding season; other species hold territory for feeding and for roosting, and one species may hold different territories with different functions throughout the year or life history. The area may be small, like that of the Adelie penguin or large like that of the Australian magpie. Territory may be maintained by a female or a male alone, as among some birds such as the European robin, *Erithacus rubecula*, in winter; by a male for the female(s) as in the northern fur seal; by the mated pair together as in penguins; or by a group as in the Australian magpie. In a few species, such as the sage grouse, *Centrocercus urophasianus*, European ruff, *Philomachus pugnax*, and Uganda kob, *Adenota kob*, males defend closely packed small territories to which females come only for copulation, and nesting and rearing occur elsewhere in more secluded places. In bower birds, *Ptilanorhynchidae*, male courtship territories are widely scattered and highly elaborated with buildings, "paint," flowers, and various colored objects. The adaptive advantages of such "arena" territories, where males conspicuously compete for females, have not been really proved. Hinde (1956) and Wynne-Edwards (1962) are recommended for classification of types of territory; and for discussion of the use of other areas, such as home ranges. A home range is an area to which an animal or group confines its activities, but which it does not defend.

The male tends to be more responsible for maintenance of territory than is the female. For instance, the Adelie male is usually the first to return after winter and defends the nest site vigorously, while the female often runs away when challenged. In prairie dogs the adult male moves about the territory, traveling about a mile a day, inspecting boundaries, and grooming and kissing other members. Faced with the bared teeth and open mouth of the kiss, a group member reciprocates, but an intruder flees. Two adults at the boundary of their different territories smell the anal glands of the other and attempt to bite the rump. But adults are rarely scarred, and aggression is mostly threat. There is also a great deal of vocal communication within and between prairie dog groups.

An animal in possession of territory appears to have "self-assurance" on the property, greater chance of winning a fight on it, and furtive behavior in the territory of others. When territory has been established, fighting is decreased and there is greater use of threat to repel intruders.

Advertisement of Territory

There are various territorial displays, such as bird song, and most of these displays have large agonistic components—they are associated with conflict. There are three phases of aggressive behavior: threat, attack, and fighting. Each phase can elicit in another animal one of the following agonistic responses:

    reciprocal threat;
    reciprocal threat with attack and fighting;
    reciprocal threat and fleeing;
    reciprocal threat and "freezing" postures, or appeasement behavior;
    avoidance.

Peterson (1965) described the "boundary display" of the northern fur seal.

Boundary display is remarkably parallel to fighting. In both situations, the bulls rush towards each other, stop abruptly, lean far forward and lunge at each other's foreflippers. During fighting contact is repeatedly made, but in the display it is unusual. While lunging at each other, the bulls puff loudly, then rear back and avert their heads, staring obliquely at each other with outstretched necks. During these interchanges, the animals keep their front quarters well back, stretching their necks to the fullest length. During repeated lunges, if one stops suddenly, the other may jerk away quickly, seemingly to avoid contact. There is considerable interplay and mutual stimulation in this conventional display. The boundary display may require 15 seconds or less if a bull is reaffirming its territory with a long-established neighbor. More intense interchanges involving recently established bulls, may continue for one minute. Social status thus seems to influence the intensity of this behavior. Restricted, newly arrived bulls repeatedly initiate display, and evidently attempt to prolong the interchanges once started, while individuals that have been established longer show less reaction.

Territorial behaviors are seen so frequently along the edges of territories that they can be used for mapping the boundaries. Estimation of breeding population size of songbirds in dense woodland is frequently made by plotting positions of singing males.

Many mammals mark their territory from scent glands, and in the prairie dog these are located at the anus. The male European rabbit uses a chin gland and piles of feces to mark its territory with scent (Mykytowycz, 1968). Hinde (1956) discusses other methods of advertising territory.

Functions of Territory

Territorial behavior may have no advantages for some individuals of the population, in that they are excluded and do not reproduce; others by tenaciousness and defense of territory may be more susceptible to predators. On the other hand, various advantages are possible, and six of them are suggested here.

PROTECTION FROM PREDATORS

Protection from predators is enhanced by familiarity with a piece of land providing cover. Prairie dogs keep the vegetation in the territories low and use raised entrances to the burrows as lookout posts from which warnings of the approach of people and predators are given to others. Some studies have demonstrated protective value of territory by comparing mortalities in prime and marginal territory. However it may be difficult to separate other factors, because the occupation of marginal territory may be associated with low social rank and youth. Younger Adelie penguins occupy peripheral nest sites, and their eggs and young are more frequently attacked by the skua, *Catharacta skua* (Sladen, 1958). Carrick (1963) reported that in the Australian magpie the territorial groups in marginal habitat and the nomadic flocks of nonbreeders were more prone to predation than the main breeding territorial groups.

MAINTENANCE OF THE PAIR BOND

The songbird advertises for a mate with his territorial song. In the Adelie penguin the pair is reunited at the territory; this returning to known ground is very likely to aid in their maintenance as a pair.

REDUCTION OF CLOSE CONTACT WITH OTHERS, AND OF INTERFERENCE, PARTICULARLY WITH REPRODUCTION

When territory is well established, owners display at adjacent owners and strangers, and the latter usually react with avoidance. Thus other animals are excluded from interference with pair formation, nest building, copulation, and care of young. In experimentally crowded populations of mice, reproduction can be curtailed by breakdown of nest territory, and also by the smell of strangers (Chapter 8). Carrick (1963) showed that in the Australian magpie reproductive physiology was very sensitive to ownership of territory, and to other aspects of social organization. All adult males came into reproductive condition, but testes were largest in those permanent groups with territory. In females more than 1 year old ovaries became partially developed, but final rapid development of oocytes depended on membership in one of the permanent groups and on the possession of a nest site. Even then, oocyte development was inhibited by intrusion of other birds, domination by another female, or much fighting. In one year failure of 16 females in territory was attributed to intergroup aggression, and failure of 13 more was attributed to dominance by other females within the group. Most of the females that bred in permanent territorial groups raised one young.

The fur seal spends a great deal of energy establishing and defending his territory and harem. But within it he may copulate with up to 100 cows, virtually undisturbed, while younger bulls without territory have to wait for mating until established bulls leave at the end of July.

When Adelie penguin chicks are half grown they are able to get warmth, and protection from the predatory skuas, by grouping together. The exclusive protection by one parent is abandoned, and the nest site territory is no longer defended.

REGULATION OF NUMBERS

Carrick concluded that reproduction in the magpie population was reduced to one-quarter of the potential, and that only those birds in possession of the permanent, good territory were successful breeders. In one year, 292 adult females were present, but 189 were in marginal groups and did not reach reproductive condition. Age may have been involved; and as mentioned previously, 29 females in territory failed to breed, probably because of aggression. Thus increase of the magpies was rather directly and obviously being related to the amount of available optimal habitat.

In most species that have been studied there is an outlying part of the adult or subadult population that is ready to take over breeding territories that become unoccupied. For instance, 25 pairs of Adelies were permanently removed from a block of territories in a rookery of approximately 10,000 nests (Sladen, 1958). The following day new pairs were in occupation and the whole colony looked undisturbed. Two days later, 45 more pairs were removed, the same area being included. The next day the emptied area was occupied again, almost as densely. On this occasion there was a noticeable increase of the territorial ecstatic display, indicating that the new males were young birds previously without territory and mates. In another year, four nests were experimentally emptied, and within 6 hours 4 new birds were in occupation, one a marked 4-year-old male, that had previously been wandering in the rookery with others of his age group.

These and other investigations (see Wynne-Edwards, 1962) support the idea that territorial organization can space individuals in relation to resources, and that by exclusion of some, particularly from breeding, the territorial habit can function to regulate numbers. The habit can allow birth rates, at least, to become density dependent.

FOOD SUPPLY

The territories obviously preferred by the Australian magpie were those with abundant food; and it was in those that the young were successfully reared.

As previously mentioned, prairie dogs "weed" their territory and thereby have an improved source of food nearby that is defended.

Some hummingbirds are examples of species that maintain territory more for food than as a place to breed; their territory is defended most vigorously during feeding periods and abandoned when the supply fails (Pitelka, 1942).

REDUCTION OF DISEASE

Carrick (1963) reported that two epidemics of infection occurred in the Australian magpie population that he studied, one of *Pasteurella pseudo-tuberculosis*, and the other of *Aspergillus*. Both of these infections killed non-territorial birds, whereas no cases were seen in closely adjacent territorial groups.

HIERARCHIES

The European Rabbit in Australia (Mykytowycz, 1960, 1961, 1964)

This study by Myers, Mykytowycz, and others is reported in some detail because it well describes a hierarchical system, in a species that also is territorial, and relates breeding success to social rank. Marked populations in wire enclosures of about 18 acres were studied for more than 4 years (Figure 19.4).

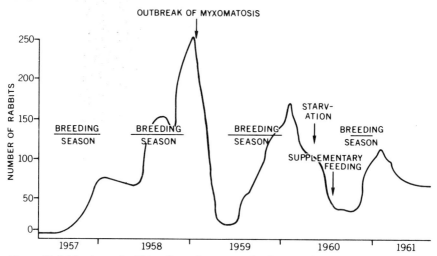

Figure 19.4. Numbers of rabbits, *Oryctolagus cuniculus*, in an enclosure from May 1957 to August 1961. Breeding seasons are shown, and the effects on numbers of a myxomatosis epidemic and of starvation. (After Mykytowycz, 1961.)

In the laboratory, female rabbits are sexually receptive all the year and copulation induces ovulation, but in the wild the breeding season was restricted to the time of good pasture and females were receptive for a short time every 7 days. At the beginning of the breeding season the populations organized themselves into groups of 3 to 8 adults, kept discrete by the exclusive ownership of group territory. Mykytowycz (1964) wrote: "A linear hierarchy of dominance exists among the males, each dominating males below it and being dominated by males above it in the social scale. This hierarchy is established at the commencement of the breeding season by physical contests which are often violent. Among the females, linear hierarchies are not universal; but females can be just as vicious fighters as males, although not so frequently."

Dominant males had the biggest scent glands in the chin and used them most for marking territory, and each group fiercely repelled intruders. Sometimes all members would work together in this, but usually the dominant male was most defensive and mobile. The subordinate males usually confined themselves to certain sectors of the territory. Unattached males ranged through many territories, being attacked in all. A male dominant in his own territory would avoid even subordinate males if he entered another territory. Territorial behavior ceased during the nonbreeding season, but the dominance relationships persisted.

A female moved around little, but was very active in defending her burrow in the breeding season against entry by other females. Some burrows gave better protection than others against flooding and predation. They were in the main burrow system of the territory and were used by top-ranking females. Females of inferior rank tended to dig isolated breeding "stops" as the season progressed. These places gave poor shelter to the young and were conspicuous to man and apparently to other predators. There was much evidence that subordinate females had socially subordinate progeny.

From April to June only high-ranking females bred, occupying good burrows. In a good year a dominant female might rear seven litters, and her early daughters could also raise one litter each the same year. Subordinate females did not start to breed until halfway through the season. Early offspring had the advantages of being oldest and the biggest through the season. Figure 19.5 shows that of the 26 percent of the year's young that survived to January, those born early of the dominant females were represented most.

In wild rabbits in Wales, Brambell (1948) measured high rates of resorption of some embryos within litters, and estimated that usually there was a 50 percent loss of whole litters. In the Australian study, high-ranking females showed little intrauterine loss, whereas in the middle of the breeding season there was much resorption of embryos occurring in lower ranking females. The young of these females also had higher mortality rates at birth and after birth, this being partly due to poor nest sites. In October and November high-ranking females began to lose their interest in nest territory and some subordinate females then occupied the best breeding burrows, and prenatal loss declined. The relationship of lower social rank or territory with embryo resorption is not understood. The authors called it "social pressures."

During a 3-year period there was a trend for population increase to be lower when population size was greater, though the data were insufficient for proof of a direct relationship of the two. Population decline was directly caused by decreased fertility rates, and increased mortality rates of young, and these rates per individual mother varied with her social status.

DISEASE

Of 159 rabbits that became naturally infected with myxomatosis, only 12

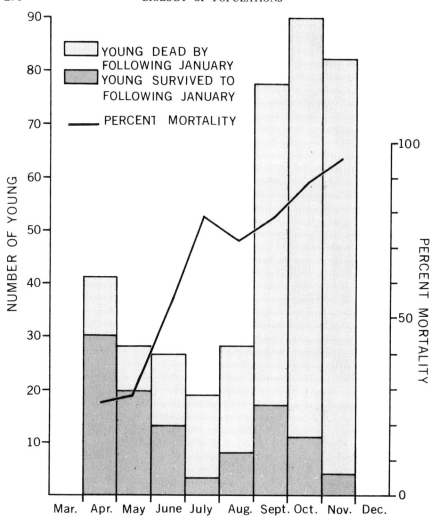

Figure 19.5. Rabbits, *Oryctolagus cuniculus*, born each month, showing numbers that survived until the following January, and numbers dying. Mortality is shown also as a percentage of those born per month. (After Mykytowycz, 1960.)

males and 7 females recovered. The average duration of individual illness was 5 days longer in adults that died in their own territory than in those that died in foreign territory. This shortening was probably due to fighting, difficulty in obtaining food, and perhaps other behavioral effects. In one case a dominant male, which was recovering from a mild form of the disease, died after he was expelled from his territory by the male he had displaced 4 years before.

FOOD SHORTAGE

A food shortage occurred one summer. All young of the year died of starvation, as well as 58 of 100 adults. Dominant males fed in the best places without interruption, while other animals sat watching. The highest survival rate, 77 percent, was for the dominant males aged $1\frac{1}{2}$ to 2 years. Artificial feeding was started, and by June the survivors had fully recovered. Four territorial groups were established, all dominant rabbits having survived and having retained their status.

SUMMARY

The rabbit study demonstrated a social organization whereby some individuals had priority over others in access to food, or space, or in ability to reproduce and were not challenged in fights. Such dominance may be initially established by fighting, and is then maintained by individual recognition and threat. In the rabbits dominance was also influenced by parental rank. Thus dominance, like possession of territory, may be contested and recognized by displays that do not involve fighting.

Visual assessment of size is very frequently involved in the assignment of rank—the size of antlers of deer, the dorsal hump of Indian bison, and height of rearing bull seals.

Flocks of Domestic Chickens (Schjelderup-Ebbe, 1922; Allee, 1951)

There is strong hierarchy in chickens that can be recognised according to which give and which receive pecks, and by the reactions to threats of being pecked (Table 19.2). When two hens are put together for the first time there

TABLE 19.2

Peck Order of a Flock of Thirteen Domestic Hens, Gallus domesticus[a, b]

A	pecks 12 others—	B to M
B	pecks 11 others—	C to M
C	pecks 10 others—	D to M
D	pecks 9 others—	E to M
E	pecks 8 others—	F to M
F	pecks 7 others—	G to M
G	pecks 6 others—	H to M
H	pecks 4 others—	I, K, L, and M
I	pecks 4 others—	J, K, L, and M
J	pecks 4 others—	H, K, L, and M
K	pecks 2 others—	L and M
L	pecks 1 other —	M
M	pecks 0 others	

[a] After W. C. Allee, *Cooperation Among Animals*, Rev. Ed. 1951. By permission of Abelard-Schuman, Ltd. All rights reserved.

[b] The linear hierarchy was interrupted by a triangular relationship of H, I, and J.

is fighting, or one avoids the other from the beginning, and an immature one defers to an older one. The loser or deferrer thereafter submits to being pecked and does not peck in return. Pecking may draw blood from the comb or head, or it may be light, or there may be only threat. In a long-established group of hens the order is firmly fixed. Allee noted that the hens themselves recognized others and their rank. High-ranking hens do not solicit cockerels as much as do lower hens.

Among cockerels the peck order is more likely to be changed by fighting. Those higher in the social order tend to have more social contact than those lower down; and high-ranking ones may prevent lower ones from mating. However, high-ranking cockerels might also allow subordinates to replace them in the copulating position. Allee also compared the condition of hens in a static group with those in a group to which he repeatedly added strange birds. He concluded that hierarchy reduced fighting and that well established groups grew better and bigger than disturbed ones in which relationships were repeatedly being tested and established by fighting.

PAIR FORMATION

Pair formation involves species- and sex-recognition and mechanisms for keeping species apart, is influenced by age, breeding experience, and affinity, and particularly involves the reduction of agonistic behavior between the pair. Much of this was indicated in the previous chapter, where song of birds was discussed and where the sequence of releasers and reactions involved in the brief pairing of the three-spined stickleback was described. Northern fur seal females are herded together by a male occupying territory, although the female is essentially nonsocial in her behavior. However, during estrus the female loses her aggressive behavior toward other females and the male. She excites the male by prolonged nose rubbing and nipping of his neck. Her attitude when ready for copulation induces the bull to mount even if he is starved and exhausted, and has served many others the same day. The female then returns to the maintenance of herself and her newborn pup.

Isolating mechanisms that reduce the possibility of mating between species are usually behavioral, as well as usually being based on morphological differences. The glaucous gull (*Larus hyperboreus*), the herring gull (*L. argentatus*), Thayer's gull (*L. thayeri*), and Kumlein's gull (*L. glaucoides*), all overlap in their Canadian arctic breeding ranges. Smith (1967) showed that colors of the iris, and of the ring around the eye, were effective barriers to cross-mating among these birds that otherwise have almost indistinguishable heads. He changed the colors of the eye ring with paint, and showed that this feature appeared to be a major stimulus to the male in copulatory behavior. He also showed that the female chose a male with an orbital color, and also wing-tip pattern, of her own species. Since the female does not see herself, Smith suggested that the young female is imprinted to the pattern of her parents soon after hatching, and subsequently her choice of mate depends on the imprinting. This interesting hypothesis probably has wide application.

Some vertebrates, such as geese and swans (Anseriformes), mate for life and stay together during all seasons; others such as hamsters are solitary, and pair for a few days of the year for copulation. In most of the long-lived birds that have been studied, the process of pair formation is considerably prolonged in the young that are breeding for the first time, and very short in the older experienced birds that keep the same mates from year to year.

"*Affinity*" is acknowledged in human relationships, but mostly ignored in studies of other vertebrates. Richdale (1957) reported that a female yellow-eyed penguin, *Megadyptes antipodes*, was particularly attracted to a certain male. Though both were paired to other mates, they sometimes "kept company" (see below), and when one mate died they bred together. The prolonged courtship in many social birds and the repeated changing of partners in the early stages of pair formation, indicates that affinity might play an important role in pair formation, though it is difficult to prove.

"*Keeping company*" occurs in Adelie penguins. The term refers to an association of male and female at a nest site that may or may not lead to establishment of a pair bond. Among young Adelies there are frequent changes of partnership. The young male usually holds territory and the young female wanders from nest to nest—one male may have as many as 20 visits from females in a day. A male that has bred in other years usually arrives at the nest site before the female, and if she does not soon follow, another takes her place. As soon as the former mate returns there is immediate recognition between the pair, and the female keeping company is expelled. A female that returns before her mate of the previous year spends some time around the territory before keeping company with an unattached male; and should the partner return soon, she usually joins him. Keeping company starts new pair bonds in preparation for failure of return of the mate and results in better nest building and defense of the property.

INDIVIDUAL RECOGNITION

Studies with marked birds show that parent Adelie penguins feed only their own chicks from the crêche (Sladen, 1953). Pettingill (1960) found this also for rockhopper penguins, *Eudyptes crestatus*; Davies and Carrick (1962) found it for terns, *Sterna bergii*, and Peterson (1965) for northern fur seals, all of which also have young that group in masses. Penney (1968) played records of the voices of parents, and the parents' own chicks were the only ones that responded. Bartholomew (1959) suggested that location, vocalization, and olfaction help the mother northern fur seal find her own pup from among several hundred.

Tinbergen (1964) described recognition in mated red-necked phalaropes, *Libipes lobatus*, as proceeding by a series of steps. A female approached individuals of several species, but next only male and females of her own species; at the third step she approached only males; among these she found, and stayed with, her mate. There is no such chain of separate reactions in the case of an established Adelie penguin returning from winter in the pack-ice.

It returns straight to its mate at the nest and immediately goes into loud vocalizations of the mutual display. When an Adelie penguin keeps company, and then its mate returns, there is instant change of partners even if the female has been in another nest site. If an incubating Adelie is experimentally exchanged with another a few yards away, the returning mate goes to its former nest and vocalizes with the loud mutual display. It is repelled by the new occupant and immediately moves to its mate in the new site and remains there (Penney, 1968).

The development of individual variation in chaffinch song, which may well be the basis for individual recognition in that species, is discussed in Chapter 18.

In a social group recognition of individuals and their status is apparently very significant in reducing aggression. This is particularly striking in a team of husky dogs, where in my experience, a male, after he had lost most of his teeth, remained dominant for many months by threat alone.

AGE GROUPS

Behavior in an Adelie penguin rookery varies with the age group components, as indicated in Table 19.3. Asynchrony in egg laying and hatching in this species can be attributed mainly to age (Sladen *et al.*, 1968a).

TABLE 19.3

Age Categories in an Adelie Penguin Population[a]

a) *Nestlings*—birds of the year.

b) *Nonbreeders in immature plumage*—1 year old; they seldom appear at the rookeries.

c) *Nonbreeding wanderers in adult plumage*—2 to 4 years old. From banding recoveries at the rookeries it appears that there is a 90 percent loss from nestling to this stage, although there is only a 10 percent annual loss of experienced breeders. They appear in and around the rookeries, mostly arriving when chicks are hatching, wander widely, and may stay temporarily in outlying places. During the reoccupation period they take over territory abandoned by older birds and build huge nests from abandoned stones. They cause most of the fights and other disturbances.

d) *Inexperienced unestablished breeders*—mostly 5 years old, though a few females breed at ages 3 and 4. They tend to arrive later than experienced breeders, often change nest site and mate, and have much lower rate of breeding success. They tend to occupy least favorable sites and spend more time in ecstatic display and threat, even fighting.

e) *Experienced breeders*—7 years and older, though a few are 6 years old. They arrive first at the rookery, lay eggs earlier, and are usually faithful to mates and nest sites from year to year.

[a] Based on Sladen (1958) and further observations covering 7 years and 30,000 individuals banded as nestlings (Sladen *et al.*, 1968a).

The fur seal population of the Pribilof Islands has many similarities with the Adelie penguin. The oldest males return first and defend territories, the youngest arrive last in areas away from the harems. In early August territorial males leave, and younger males move in and exhibit quasiterritorial behavior, and some copulations occur, but most appear to be unsuccessful. From early August until September, large numbers of even younger males and females gather in and around the beaches.

CONCLUSIONS

Agonistic Behavior and Social Organization

Ability to fight has great selective value for an individual, in the acquisition of food, mates, and the defense of young and self, and selection to this extent must surely drive toward greater capacity for aggression. On the other hand, social life requires some cooperation. There is a multitude of possible resolutions of these opposites. Thus in social animals there are many mechanisms for overcoming aggressive behavior—appeasement behavior, ritualized contest where fighting is replaced by nonlethal display, and the remarkable influences of social experiences early in life upon subsequent agonistic behavior, such as described in Chapter 17. The value of pacific behavior within a group may be seen in emperor penguins, *Aptenodytes forsteri*, which show almost no antagonisms in their breeding colonies, especially when they huddle for days in tightly packed groups with only backs exposed to Antarctic winter storms. This mechanism for survival in a harsh environment is somewhat paralleled by nonaggressive Eskimo groups, living closely packed together in small dwellings in the Arctic winter.

Although we are aware in general of the effects of aberrant social experiences, particularly of neglect, upon later social facilitation in man, the matter is not fully exploited for the benefit of man. Nor is the plasticity of individual behavior sufficiently examined in experimental animals. The plasticity of social organization in animal species is barely recognized, one exception being Southwick's experiments that are described in Chapter 20.

Competition and Survival

Nicholson distinguished "between two different forms of competition, namely scramble and contest. In a scramble for an inadequate supply of food, all, or almost all, the animals concerned may die. In a contest some perish but others survive, and are little the worse for it. Lack (1954) gives many examples of contests for food between nestlings in which the weakest perish, but the survivors seem fairly normal" (Haldane, 1955). In many species actual physical contest is of limited duration, and its results become established as a hierarchy. Thus the dominant rabbits, organized by a Nicholson-type contest resulting in hierarchy, survived food shortage unharmed, while the subordinate ones died of starvation. Wynne-Edwards (1962) stressed that the function of hierarchy is to identify surplus individuals when population density is high in relation to resources.

Human societies have varying degrees of rigidity of hierarchical organization, but rigid and extreme hierarchies are not the preferred system of organization of most men. If man aims for the good life for all, he should pay some attention to the allegories of animal populations. He should take rapid action to stop deterioration of the land (Chapter 9) and the extreme acceleration of his increase, to prevent all from suffering the effects of something approaching a Nicholson-type scramble for inadequate supplies.

Population Stability and Defended Land

It is of great advantage for a migratory population to return to known places that have previously provided adequate support of life—the winter ranges of cattle, headstreams of salmon, and breeding territories of birds. Defense by the "owners" limits population density, even total size. When young are reared in families they also have the opportunity to learn the territory. Association of groups, particularly of two or more generations in the family and society, enhances the possibilities for transmission of learned behavior. This mode of adaptation to the environment has reached its greatest expression by far in man; and has brought him to a great crisis of "oversuccessful" survival.

REFERENCES

Allee, W. C. 1951. *Cooperation Among Animals.* Abelard-Schuman, New York.
Baker, R. C., Wilke, F., and Baltzo, C. H. 1963. The Northern Fur Seal. *U.S. Fish Wildlife Ser.* Washington, D.C., Circular 169.
Bartholomew, G. A. 1959. Mother-young relations and the maturation of pup behaviour in the Alaska fur seal. *Animal Behavior* 7: 163–171.
Brambell, F. W. R. 1948. Prenatal mortality in mammals. *Biol. Rev.* 23: 370–407.
Carrick, R. 1963. Ecological significance of territory in the Australian Magpie, *Gymnorhina tibicen. Proc. XIIIth Intern. Ornithol. Congr.*, 740–753.
Davies, S. J. J. F., and Carrick, R. 1962. On the ability of crested terns, *Sterna bergii,* to recognize their own chicks. *Australian J. Zool.* 10: 171–177.
Haldane, J. B. S. 1955. Review of David Lack, 1954, *The Natural Regulation of Animal Numbers* (Oxford). *Ibis* 97: 375–376.
Hinde, R. A. 1956. The biological significance of the territories of birds. *Ibis* 98: 340–369.
Howard, H. E. 1920. *Territory in Bird Life.* Murray, London.
King, J. A. 1955. Social behavior, social organization and population dynamics of a black-tailed prairie dog town in the Black Hills of South Dakota. *Contrib. Lab. Vertebrate Biol.* 67: 1–123.
King, J. A. 1959. The social behavior of prairie dogs. *Sci. Am.* 201: 128–140.
Lack, D. 1954. *The Natural Regulation of Animal Numbers.* Clarendon Press, Oxford, England.
Mykytowycz, R. 1960. Social behavior of an experimental colony of wild rabbits, *Oryctolagus cuniculus* (L.). C.S.I.R.O. *Wildlife Res.* 5: 1–20.
Mykytowycz, R. 1961. Social behavior of an experimental colony of wild rabbits, *Oryctolagus cuniculus.* (L.). *C.S.I.R.O. Wildlife Res.* 6: 142–155.
Mykytowycz, R. 1964. Territoriality in rabbit populations. *Australian Nat. Hist.* 14: 326–329.
Mykytowycz, R. 1968. Territorial marking by rabbits. *Sci. Am.* 218: 116–126.
Penney, R. L. 1968. Territorial and social behavior in the Adelie penguin. In *Antarctic Bird Studies.* O. L. Austin, Jr.(ed.), Antarctic Research Series 12: 83–131. Amer. Geophysical Union–National Academy Sciences, Washington, D.C.
Peterson, R. S. 1965. Behavior of the Northern Fur Seal. Thesis for D.Sc. Degree, Johns Hopkins Univ. School of Hygiene and Public Health, Baltimore, Md. Unpublished.
Peterson, R. S. 1968. Social behavior in pinnipeds, with particular reference to the northern fur seal. In *Behavior and Physiology of Pinnipeds.* R. J. Harrison, R. C. Hubbard, R. S. Peterson, and R. J. Schusterman (eds.). Appleton-Century-Crofts, New York.
Pettingill, O. S. 1960. Crèche behavior and individual recognition in a colony of Rock-hopper Penguins. *Wilson Bull.* 72: 213–221.
Pitelka, F. A. 1942. Territoriality and related problems in North American humming-birds. *Condor* 44: 189–204.

Richdale, L. E. 1957. *A Population Study of Penguins.* Clarendon Press, Oxford, England.

Schjelderup-Ebbe, T. 1922. Beitrage zur Sozialpsychologie des Haushuhns. *Zeit. f. Psych.* **88**: 225–252.

Sladen, W. J. L. 1953. The Adelie Penguin. *Nature* **171**: 952–955.

Sladen, W. J. L. 1957. Penguins. *Sci. Am.* **197**: 44–51.

Sladen, W. J. L. 1958. *The Pygoscelid Penguins. I—Methods of Study. II—The Adelie Penguin.* Falkland Islands Dependencies Survey, Sci. Rept. **17**: H. M. Stationery Office, London.

Sladen, W. J. L., LeResche, R. E., and Wood, R. C. 1968a. Antarctic avian population studies, 1967–1968. *Antarctic Journal of the U.S.* **3**: 247–249.

Sladen W. J. L., Wood, R. C., and Monaghan, E. P. 1968b. The USARP bird banding program, 1958–1965. In *Antarctic Bird Studies.* O. L. Austin, Jr. (ed.), Antarctic Research Series, **12**: 213–262. Amer. Geophysical Union–National Academy Sciences, Washington, D.C.

Smith, N. G. 1967. Visual isolation in gulls. *Sci. Am.* **217**: 94–102.

Tinbergen, N. 1964. Recognition. In *A New Dictionary of Birds.* A. L. Thompson (ed.). Nelson, London.

Wynne-Edwards, V. C. 1962. *Animal Dispersion in Relation to Social Behavior.* Hafner, New York.

CHAPTER 20

Population Dynamics and Social Behavior of Domestic Rodents

CHARLES H. SOUTHWICK

The Norway rat, *Rattus norvegicus*, and the house mouse, *Mus musculus*, have been used extensively in studying the interrelations of social behavior and population dynamics. They have been ideal for this because their natural habitats consist of buildings and man-made enclosures. It has been possible, therefore, to construct enclosed habitats for studying demography and social interactions under controlled conditions without undue distortion of behavior. Further, these rodents breed rapidly, they have small home ranges, and their physiology, genetics, immunology, and pathology have been well studied.

SOME ECOLOGICAL AND BEHAVIORAL CHARACTERISTICS OF DOMESTIC RODENTS

Norway rats and house mice have the potential for very rapid population growth because of rapid sexual maturation, short gestation periods, postparturient estrus, and large litter sizes. Theoretically in 12 months one pair of house mice can produce 4,000 mice and one pair of Norway rats can produce 250 progeny. Norway rat populations in English grain ricks often increase at the rate of 35 to 45 percent per month (Leslie *et al.*, 1952). One wild house mouse population in an English wheat stack was found to increase from less than 50 mice to over 2,000 in 6 months, an increase of approximately 100 percent per month (Southwick, 1958).

A typical population growth curve of Norway rats is shown in Figure 20.1 (Davis, 1953). Baltimore rat studies in World War II showed that urban blocks often had from 20 to over 300 rats per block. A rigorous program of rat control in 1943–1944 achieved reductions from 25 percent to nearly 100 percent in these rat populations, but after the poisoning campaigns ended many populations increased rapidly at rates ranging from 4.8 to 30 percent per month. In 9 out of 29 study blocks, the rat populations returned to or exceeded former population levels within 3 to 8 months. This demonstrated the need for environmental sanitation as well as poisoning and trapping in rodent control programs.

The natural social behavior of wild Norway rats involves territorial and hierarchical behavior. Males of both species are territorial if population density is not excessively high, and high-ranking males defend a system of burrows or runways and a number of females living therein (Barnett, 1963; Crowcroft, 1966). The size and location of the territory and the number of females controlled depends upon the dominance of the male and the nature of

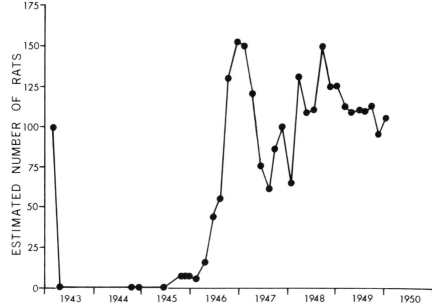

Figure 20.1. Changes in Norway rat populations in a Baltimore residential block. Control procedures were applied only in May 1943. (After Davis, 1953.)

the habitat. Females defend just the nest site and perhaps a small area around it. In both species the social structure is sensitive to individual characteristics of high-ranking members and to various ecological factors, especially population densities in relation to habitat quality.

EXPERIMENTAL STUDIES OF RAT POPULATIONS

Calhoun was one of the first ecologists to illustrate the interaction of social behavior and population dynamics in the Norway rat (Calhoun, 1949, 1950, 1962a). He established a colony in an escape-proof enclosure of 10,000 square feet. An inner area provided easy access to food; a rat in the outer areas had to go through narrow openings to reach the inner food area (Figure 20.2). The population was started with 5 wild pairs. Food and water were continuously and abundantly supplied. Thirty-six nest boxes were provided and the rats were free to burrow in the soft ground of the enclosure. Behavioral observations were made daily and every 6 weeks the entire population was trapped, censused, and marked for individual identification. Within 2 years the population reached a level of only 171 although the potential exceeded 50,000. The population was limited primarily by social interactions which influenced reproductive physiology and behavior.

Calhoun found that the dominant males occupied the more favorable places close to the food supply, established territories there, and dominated

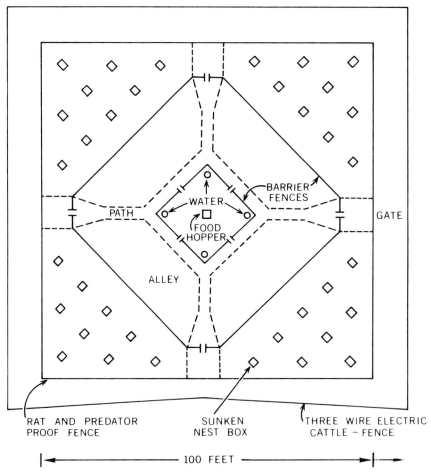

Figure 20.2. Enclosure for an experimental population of Norway rats, *Rattus norvegicus*, started with 5 wild-caught pairs. Areas within the dashed lines were kept free of vegetation. Ground was soft so that rats could easily burrow. Rats in the outer next-box areas had to pass through two sets of narrow openings in the barrier fences to reach the supply of food and water. (After Calhoun, 1962a.)

burrows containing several females. As the population increased, social competition forced lower ranking rats, especially males, into fringe areas and all changes in social rank were downward. These movements increased the social interactions among lower ranking members of the population that accumulated in fringe areas. The colonies there had excess males and unstable social structures. Normal territories did not develop. Estrous females were followed by packs of males which mounted repeatedly in a disorganized way. Calhoun noted: "The behavior of such females indicated they experienced considerable stress. Conceptions were reduced and few young were raised. The normal preparturient patterns of self-isolation, burrow construction and nest building, which are requisite to successful rearing of young, were reduced or entirely inhibited. As the population increased a larger proportion of the colonies were characterized by this low social rank and instability, with concomitant reduction in reproduction."

Eventually the only successful reproduction in the entire population was achieved by the highest ranking animals and was sufficient only to replace mortality. The high-ranking members of the colony maintained normal nocturnal activity rhythms whereas the low-ranking members of the population were able to visit food and water sources only when the dominant members had returned to their burrows. Thus their nocturnal feeding behavior was upset and they were more active in daylight. Calhoun showed that there was inhibition of growth in the socially low-ranking animals. Thus, as the population increased more individuals had abnormal behavior and inhibited growth.

EXPERIMENTAL STUDIES OF MOUSE POPULATIONS

Similar studies on house mice have afforded further environmental refinements and the opportunity of replication of the experiments. House mice require less space and natural habitats can be constructed more easily in laboratories. At the University of Wisconsin such studies were begun in 1947 to investigate the effect of limited food supply on the growth and natural control of confined populations, and the effect of crowding upon confined populations with unlimited food. In both cases, emphasis was placed on the interactions between behavior and demography at various stages of population growth. The mice were reared in large escape-proof enclosures 150 to 500 square feet in area. All physical factors of the environment were controlled, periodic censuses of births and deaths were taken, and observations were made on social interactions.

Mouse Populations on a Limited Food Supply

When confined populations living on a limited food supply reached the stage where the daily allotment of food was consumed, a sudden decline in fecundity occurred. Reproduction in both sexes ceased. The reproductive organs of males and females involuted, so that males no longer produced spermatozoa and females went into a prolonged anestrus. Hence, the population ceased to grow, but the existing mice remained in good physical condition (Strecker and Emlen, 1953).

Many other vertebrates display this phenomenon (Lack, 1954), but some do not. The most notable of the latter are the white-tailed deer, which continue reproduction under severe food deprivation, and man himself, who also remains fecund on semistarvation diets.

Mouse Populations on an Unlimited Food Supply

Six populations which always had a surplus of food showed different and more

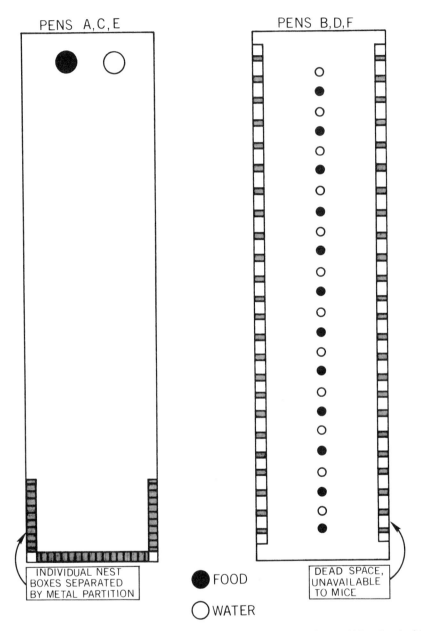

Figure 20.3. House mouse pens, 25 × 6 feet. Shaded areas are nest boxes. (After Southwick, 1955a.)

variable patterns of population behavior. Some reproductive alteration did occur at crucial population levels, but the major mechanism of population control was increased mortality rates of infant mice (Southwick, 1955a).

In three of these populations, food and housing arrangements were placed to allow maximal population dispersal and thereby reduce social contact at nesting and feeding locations (Figure 20.3). In three other populations, food and nesting facilities were concentrated in order to force a high degree of crowding and social contact. Both types of pens had the same amounts of food, water, space, and nest boxes. The purpose was to determine whether the additional crowding in the concentrated pens would alter the population growth, densities achieved, control mechanisms and the behavior, physiology, and health of the mice.

Each population was started with 4 pairs of wild-trapped house mice and the populations were permitted to grow for 2 years. An overabundance of food and water was always present. Within 18 months all the populations had reached their maximum sizes and were apparently controlled by social phenomena related to crowding. However, the density at which crowding phenomena occurred was highly variable, as was also the way in which crowding affected the population.

Some populations attained levels of 140 adult mice, while others did not grow above 25 mice and this variability was not related to the habitat differences. There were no consistent correlations of reproductive, behavioral, or demographic characteristics with the concentrated or dispersed food and cover arrangement. Each population responded with individual characteristics of behavior and reproductive rates. It was surprising that populations living in identical environments in terms of space, temperature, light, humidity, food, and cover should show such pronounced individuality. The reaction to, and utilization of, these identical physical resources differed widely and, in the final analysis, depended primarily upon the group behavior and social organization of each population.

Mechanisms of Limitation

The only consistent similarity among the six populations was that limitation was primarily due to high infant mortality, not lack of births (Figure 20.4). In some cases infant mortality was due to cannibalism by adults, but in other cases it was due to desertion and milder forms of improper parental care. In all there were signs of social disturbance: aggression, gregariousness at the nest site, intermixing of the sexes, nest destruction, and abnormal communal nesting. All of these are normal types of behavior, but when distorted they became harmful to individuals.

Aggression

Within all populations the amount of aggression increased as the population density increased (Figure 20.5). When the intensity of fighting reached a level

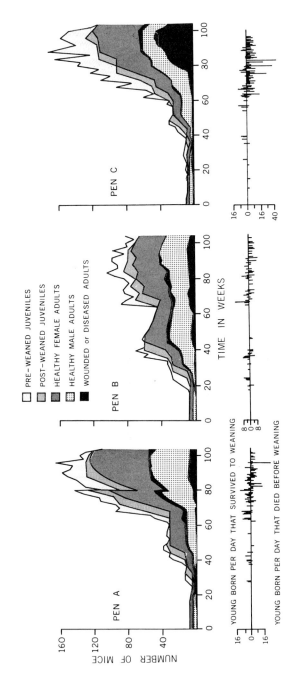

Figure 20.4. Confined populations of house mice. *Upper graphs:* Population growth, showing the sexes, age classes of juveniles (before and after weaning), and adults, and numbers of wounded or diseased adults (males shown below healthy males, females below healthy females). *Bar graphs:* Young born per day with survivors (above the axis) and those dying before weaning (below the axis.) (After Southwick, 1955a).

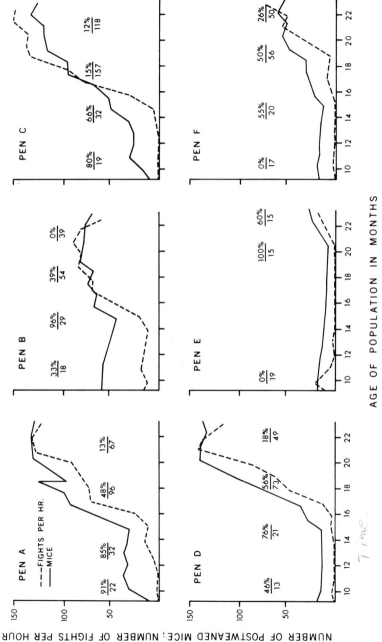

Figure 20.5. Confined house mouse populations: population growth, litter survival, and fighting. Explanation of ratios: 91 %/22 = percent of survival to weaning of young born in 2 months/number born in 2 months. (After Southwick, 1955a.)

of about one fight per mouse per hour, population growth was retarded. Some populations reached this intensity of fighting at lower densities than others. Pen E exhibited 20 fights per hour when it had a population of 17 mice, and it ceased growth at this point; whereas Pen C did not cease growth until it had reached a level of 130 mice and was exhibiting 140 fights per hour. In both pens the intensity of fighting per mouse was about the same at the time of population limitation.

One or just a few individuals greatly influenced the amount of aggression within a population and fighting often spread rapidly. Scott and Fredericson (1951) found that pain, sudden movement, noise, the presence of the male hormone, and previous social experience are all important stimulants for fighting. Hence, one fight tends to stimulate another. The mechanism by which fighting may have influenced litter survival is not known, but it apparently disrupts normal maternal behavior.

Experimental Studies on Aggression

Vessey (1967) also observed increased aggression with increased population growth in confined populations of wild house mice, and found that maximum aggression was correlated with the cessation of population growth. The primary mechanism of population limitation was again high infant mortality. When chlorpromazine was administered to two populations at the rate of 0.75 milligrams per gram of food, aggression rates were markedly reduced, infant survival increased, and population growth was renewed. When the drug was removed, aggression rates increased and the population declined. By comparing seven different populations, Vessey found an inverse relationship between aggression and ultimate population size. The mechanism by which aggression influenced infant survival was still not known, and Vessey concluded "Future studies should focus on relationships between aggressive behavior and infant survival." Thus it is possible that both high aggression scores and high infant mortality are symptoms of social disorganization in rodent populations.

Brown (1967) recently completed a long-term study of the bioenergetics of confined house mouse populations. At lower stages of population growth, major amounts of energy went into reproduction and parental care and relatively little into agonistic behavior (all responses associated with aggression: threat, attack, fighting, and submission.) As the population increased more energy went into agonistic behavior and less into reproduction and parental care. Initially, the energy expenditure increased greatly in threat behavior, but later in direct fighting. The population was limited when the energy expenditure in threat and fighting became excessively great and that available for reproduction and parental care became progressively less.

It should be emphasized again that fighting and other aggressive activities are normal social activities in mice and are beneficial to the population within certain bounds, but in excess may be detrimental. In the Wisconsin populations intense fighting directly affected the health of the population by producing wounds of the rump and scrotal area. Tropical rat mites, *Bdellonyssus bacoti*, attacked mice there and produced extensive scabbiness. Although both sexes were living in the same mite-infested boxes, only the males that fought developed the acarine dermatitis.

Nest Box Crowding (Gregariousness)

In the Wisconsin populations litter survival also decreased as adult crowding in nest boxes increased (Table 20.1). If only 1 or 2 adults attended a litter the probability of survival was nearly 60 percent. If 3 or more adults were present the probability of survival fell significantly to 37 and 27 percent.

TABLE 20.1
Survival of Litters of House Mice and Numbers and Sexes of Attendant Adults[a]

Adults present	Number of Litters	Total young	Total young surviving to weaning	Percent survival
Total adults present				
1	55	248	144	58
2	72	376	216	57
3 or 4	52	253	93	37
5 or more	57	267	72	27
Only adult females present				
1	45	205	136	66
2	52	281	196	70
3 or 4	25	143	60	42
5 or more	7	42	3	7
Adult males present alone or with females there also				
1	10	43	8	19
2	20	95	20	21
3 or 4	27	110	33	30
5 or more	50	225	69	31

[a] Southwick (1955b).

Intermixing of Sexes

The presence of one adult male in a nest box reduced the chances of the survival of a litter as much as the presence of several adult females. Table 20.1 shows that the highest survival occurred when only 1 or 2 adult females attended a litter, and that if 1 adult male was present, the survival was significantly lower. These studies did not demonstrate whether adult males were directly responsible for litter mortality or whether their presence elicited unfavorable female behavior.

Crowcroft (1966) and Anderson (1961) have shown that wild house mice, particularly lactating females with young, are territorial under normal un-crowded conditions. This territoriality is complicated and variable. Two lactating females may share a nest, but normally most other adults are ex-cluded, particularly adult males. The isolation of the lactating female is apparently very important to normal reproductive physiology and maternal behavior.

Dominant males also show territorial behavior and may exclusively domin-ate small groups of breeding females. Thus a population can be divided into small breeding units and all the paternal genes come from a few individuals (Reimer and Petras, 1966). If the population is structurally stable this may persist for several generations.

Nest Destruction

Another behavioral characteristic of the Wisconsin populations which was correlated with increased nestling mortality was the failure to build and maintain good nests. Normally a good house mouse nest is completely en-closed by nesting material. However, sometimes the nest may be bowl-shaped and not covered on top. The presence of many adults in a nest box often reduced the nest to a flat platform. Despite the presence of abundant nesting material, many nests in crowded populations degenerated to platform nests. Table 20.2 relates nestling survival to nest type and adult composition. In covered or bowl-shaped nests survival was about 60 percent regardless of adults present. In platform nests survival was good if only 1 or 2 adult

TABLE 20.2

Survival of Litters of House Mice in Type of Nest[a]

A. *Favorable adult attendance 1 or 2 females*

Type of nest	Number of litters	Total young	Total young surviving	Percent survival
No nest	3	18	0	0
Platform	46	226	133	59
Bowl	24	123	77	63
Covered	10	43	37	86

B. *Unfavorable adult attendance at least 1 male and/or 3 or more females*

Type of nest	Number of litters	Total young	Total young surviving	Percent survival
No nest	6	24	0	0
Platform	64	341	63	18
Bowl	9	44	25	57
Covered	1	4	4	100

[a] Southwick (1955b).

females were present but poor if adult males and/or additional females were present.

Communal Nesting

Communal nesting of 2 parturient females frequently occurred and survival was good if other factors were favorable. Even 3 parturient females might nest together successfully. However, at high population densities in several pens large numbers of young representing several litters were deposited in a single nest box and then deserted. In Pen C, 55 young, from 1 to 3 days of age were found in one nest box and were attended by only 2 females. The other females contributing to this large number, probably 8 or 9, had deserted. None of these young survived.

There were slight increases in the mortality rates of postweaned subadult mice in 5 populations but this made only minor contributions to population control. The mortality rates of adults were unaffected by population density in these studies.

Reproductive Responses to Crowding

Significant differences in fecundity and fertility rates occurred in different populations. In population B birth rates continued unabated after population limitation and none of the reproductive traits other than infant survival was affected by crowding. However, in populations A, C, D, and F birth rates gradually declined with continued crowding. In A and D this was apparently due to declining fecundity associated with a reduction in food consumption despite an abundance of food. Several highly dominant mice in these two populations prevented subordinate mice from feeding freely. A system of "rotating monarchy" developed. Different dominant males dominated the feeding areas throughout the day and night and prevented subordinate males from feeding freely so that there was a decrease in average per capita food consumption. In populations C and F food consumption did not decline and fecundity remained high yet fertility declined. Behavior observations on these pens, particularly pen C, indicated the probable existence of what Calhoun (1949) called "copulation pressure." Many males simultaneously attempted to mount an estrous female. These populations had an unstable social structure in which no individuals were dominant, a state of anarchy.

Thus populations A and C provided a particularly interesting comparison of behavior and demography in homogeneous environments. Both populations reached levels of 130 to 140 mice at the same time and displayed the same growth curve, had similar patterns of fertility, and had high infant mortality as the main mechanism of population control. Nonetheless, they showed major differences in physiology, behavior, and physical condition, as outlined in Table 20.3.

In these populations, as in others, the prime determinant of fecundity and fertility patterns was the social organization of the population. Although

TABLE 20.3

Comparison of Confined House Mouse Populations in Identical Environments[a]

Population A (Fecundity less than 75%)	Population C (Fecundity over 90%)
Declining and subnormal per capita food consumption	High per capita food consumption
Low to moderate aggression	Very high aggression
Little or no wounding	Most males wounded
No dermatitis	Severe acarine dermatitis
Rigid dominance structure—rotating monarchy	No clear dominance structure—anarchy

[a] Southwick (1955b).

crowding was present in all of these populations the responses to crowding were by no means consistent.

BEHAVIORAL SINKS

Calhoun (1962b) described a self-stimulating aggregation of Norway rats which appeared in certain crowded populations. In some populations excessive aggregations developed in very limited parts of the habitat. Calhoun felt that the animals became conditioned to a high level of social contact at certain levels of crowding and then they continued to seek these high levels even though this could have been avoided. Thus an extreme state of sustained aggregation developed which Calhoun called a "behavioral sink" or a form of "pathological togetherness."

As the behavioral sink developed many abnormal behaviors and reproductive disturbances appeared. Calhoun described these as follows:

Females experienced difficulty in carrying fetuses to term, and if they carried to term they were sometimes unable to deliver young. Death frequently occurred at this time. If they survived, one region of the uterus enlarged until it was sometimes as large as the former size of the rat. Such affected rats always died. Females developed a mortality rate 3.5 times that for males.

On the behavioral side, males developed a pansexuality in which they would mount other rats regardless of their age, sex, or receptivity. Infliction of wounds during mounting developed. An abnormal response of biting the tails of other rats also developed. Nest-building behavior became completely disrupted. Transport of young by lactating rats became so disorganized that young became so scattered that they were no longer nursed.

A behavioral sink can become a self-stimulating vortex of destructive behavioral and physiological responses.

STUDIES ON OTHER RODENT SPECIES

Recent studies of *Peromyscus* populations by Terman (1965), Sadleir (1965), and Healey (1967), and other rodent groups by Bronson (1963) and Warnock (1965) have also emphasized the key role of social behavior in determining patterns of population growth and regulation. Terman (1965) concluded from experimental studies of *Peromyscus* that "... controlling mechanisms were activated in the presence of surplus food and water and at varying population densities under identical conditions of the physical environment. The data presented above agree with those of other studies and suggest that intrinsic regulatory mechanisms are basically behavioral and operate *via* physiological mechanisms to regulate populations."

Sadleir (1965) proposed that the survival rate of juvenile deermice (*Peromyscus maniculatus*) was determined by adult aggression. He felt that territoriality reduced the survival of juveniles dispersing throughout the habitat, and thus aggression served a direct role in regulating population density. These theories have been confirmed experimentally in both field and laboratory studies by Healey (1967). He concluded that a small group was a unit within which aggression was reduced to conserve energy, but that each animal was aggressive toward any strange animal entering his territory. Aggression led to reduced growth in the nonresident mouse or, in many cases, to direct mortality.

CONCLUSIONS

Population growth and control in rodent populations are variable and complex. Reproductive responses to behavioral stress are also complex and not adequately explained by simple theories of density-dependent stress.

Many animals with highly evolved and adaptive social behavior show territorial and hierarchical patterns which act as intrinsic self-regulatory population mechanisms. Members of the genus *Peromyscus* are good examples, and these animals rarely if ever develop eruptive population growth and excessive densities. However, many other species, such as the white-tailed deer, do not have sensitive behavioral checks on excessive population growth and these animals often exhibit disastrously high population densities. In Norway rats and house mice normal patterns of territory and hierarchy are easily upset by various ecological circumstances such as rapid population growth. This creates social instability and crowding, which result in a long series of behavioral and physiological abnormalities.

Social organization plays a critical role in determining the effect which crowding has upon reproduction, health, and welfare of the population. The most deleterious effect of crowding may be to disrupt normal behavior patterns and to create pathological social conditions.

REFERENCES

Anderson, P. K. 1961. Density, social structure, and nonsocial environment in house-mouse populations and implications for the regulation of numbers. *Trans. N.Y. Acad. Sci.,* Ser. II. **23**: 447–451.

Barnett, S. A. 1963. *The Rat: A Study in Behavior.* Aldine, Chicago, Illinois.

Bronson, F. 1963. Some correlates of interaction rate in natural populations of wood chucks. *Ecology* **44**: 637–643.

Brown, R. A. 1967. Bioenergetics of house mouse populations. Unpublished ms.

Calhoun, J. B. 1949. A method for self-control of population growth among mammals living in the wild. *Science* **109**: 333–335.

Calhoun, J. B. 1950. The study of wild animals under controlled conditions. *Ann. N.Y. Acad. of Sci.* **51**: 1113–1122.

Calhoun, J. B. 1962a. The ecology and sociology of the Norway rat. *U.S. Public Health Serv.* Publ. No. 1008. 288 pp.

Calhoun, J. B. 1962b. A "behavioral sink." In *Roots of Behavior* E. L. Bliss (ed.), Harper, New York, pp. 295–315.

Crowcroft, P. 1966. *Mice All Over.* Dufour Press, Chester, Penn.

Crowcroft, P., and Rowe, F. P. 1957. The growth of confined colonies of wild house mouse (*Mus musculus* L.). *Proc. Zool. Soc. London* **129**: 359–370.

Davis, D. E. 1953. The characteristics of rat populations. *Quart. Rev. Biol.* **28**: 373–401.

Healey, M. C. 1967. Aggression and self-regulation of population size in deermice. *Ecology* **48**: 377–392.

Lack, D. 1954. *The Natural Regulation of Animal Numbers.* Oxford Univ. Press, Oxford, England.

Leslie, P. H., Venables, U. M., and Venables, L. S. V. 1952. The fertility and population structure of the brown rat (*Rattus norvegicus*). *Proc. Zool. Soc. Lond.* **122**: 187–238.

Reimer, J. D., and Petras, M. L. 1967. Breeding structure of the house mouse, *Mus musculus*, in a population cage. *J. Mammal.* **48**: 88–99.

Sadleir, R. M. S. F. 1965. The relationship between agonistic behaviour and population changes in the deermouse (*Peromyscus maniculatus* Wagner). *J. Animal Ecol.* **34**: 331–352.

Scott, J. P., and Fredericson, E. 1951. The cause of fighting in mice and rats. *Physiol. Zool.* **24**: 273–309.

Southwick, C. H. 1955a. The population dynamics of confined house mice supplied with unlimited food. *Ecology* **36**: 212–225.

Southwick, C. H. 1955b. Regulatory mechanisms of house mouse populations; social behavior affecting litter survival. *Ecology* **36**: 627–634.

Southwick, C. H. 1958. Population characteristics of house mice living in English corn ricks: density relationships. *Proc. Zool. Soc. Lond.* **131**: 163–175.

Strecker, R. L., and Emlen, J. T. 1953. Regulatory mechanisms in house mouse populations: The effect of limited food supply on a confined population. *Ecology* **34**: 375–385.

Terman, C. R. 1965. A study of population growth and control exhibited in the laboratory by prairie deermice. *Ecology* **46**: 890–895.

Vessey, S. 1967. Effects of chlorpromazine on aggression in laboratory populations of wild house mice. *Ecology* **48**: 367–376.

Warnock, J. E. 1965. The effects of crowding on the survival of meadow voles (*Microtus pennsylvanicus*) deprived of cover and water. *Ecology* **46**: 647–664.

CHAPTER 21

Social Behavior of Nonhuman Primates

CHARLES H. SOUTHWICK

In recent years field studies of the ecology and behavior of primates have been rapidly expanding. Before the mid-1950's there had been fewer than 10 major expeditions and knowledge was available on only 6 or 7 species.* Since 1955 there have been more than 100 major expeditions to study primate behavior and ecology in natural habitats and sound field data are now available on about 30 species.

Monkeys and apes have been studied in captivity for many years and several prominent misconceptions about their behavior arose. For example many primates, especially baboons, *Papio* spp., gorillas, *Gorilla gorilla*, and chimpanzees, *Pan troglodytes*, were thought to be highly aggressive. Recent field work in natural environments has shown that most primates have peaceful well-integrated social lives with relatively rare aggressive encounters. Most have social hierarchies and/or complex display and signaling systems which reduce and control overt aggressive behavior. However, some of the macaques, particularly the rhesus macaque, may be pugnacious in natural habitats. In these cases the amount of fighting is often related to crowding and habitat conditions.

Most primates were thought to breed throughout the year and this continual sexual activity was considered to be the primary basis of primate social life. In the laboratory the Indian rhesus, *Macaca mulatta*, breeds in all months, but natural populations in India show sharply seasonal reproductive behavior with sexual activity being most frequent from September to December and 95 percent of the births from late March to early June. During the months of sexual inactivity there is no diminution of group life and no weakening of social bonds. During most of her adult life a female monkey is either pregnant or lactating and is not sexually active, but her attachment to the group remains consistently close. Sexual bonds are important in primate groups, but other social ties are even stronger. Among these are the maternal–infant bond, peer-group bonds of juveniles expressed in play, feeding, and investigative behavior, and social bonds among adults of the same sex.

Another major misconception of primate social behavior has been that most primates are territorial. Some are, notably the howler *Alouatta palliata*, vervet, *Cercopithecus aethiops*, lutong, *Presbytis cristatus*, and *Callicebus*

* The Order Primata consists of more than 200 species in two suborders and approximately 12 families. The Suborder Prosimii includes the shrews and the tree shrews (Family Tupaiidae), lorises (Lorisidae), lemurs (Lemuridae), and the tarsiers (Tarsiidae). The Suborder Anthropoidae, includes the marmosets (Family Hapalidae), monkeys (Cebidae and Cercopithecidae), great apes (Pongidae), and man (Hominidae).

299

monkeys, but many, including the macaques, baboons, chimpanzees, and gorillas are not strictly territorial.

Primate field studies of the last 10 years, in addition to dispelling these prominent misconceptions, have tended to emphasize four major topics: the variability, adaptability, complexity, and social basis of primate behavior.

Variability within and between species has been shown in group organization, social relations, and behavioral capabilities. Many of these patterns of variability have been shown to be ecologically significant, permitting primates to adapt to a wide range of environmental circumstances and permitting them to live within complex and often hostile communities. For example, variations in food habits enable rhesus monkeys to survive in an extraordinary range of habitats in India, from tropical monsoon forests to high mountain ranges.

Complexity in primate social behavior can be seen in terms of group structure, communicative networks, and kinship relationships. Until recent field studies in Japan and Puerto Rico, the true complexities of primate groups were not apparent. Kinship relations extending into adulthood and social transmission of learned behavior, often thought to be uniquely human traits, have now been documented in nonhuman primate groups.

The final point of emphasis of recent research is the importance of primate social life. Field and laboratory studies have demonstrated the great impact of social forces on normal behavioral development in primates. Harlow and Mason (Harlow, 1959) have shown that an infant rhesus without satisfactory maternal and peer relationships develops many serious behavioral anomalies.

Washburn and DeVore (DeVore, 1965) have demonstrated in field studies of baboons, *Papio doguera*, in East Africa, the ecological reasons for primate social life. They have shown that the primary adaptation of baboons to their environment is a social adaptation—the only way of life possible for a baboon is a closely integrated group life. Solitary individuals or seasonal herds cannot survive. Whereas many other mammals have achieved adaptive success through various morphological changes, the primates have remained anatomically generalized while they have achieved ecological success by social means.

To document some of these generalizations this chapter will discuss primate grouping patterns, intragroup organization, intergroup relationships, patterns of socialization, and patterns of social change.

PRIMATE GROUPING PATTERNS

With few exceptions primates live in well-organized, permanent, heterosexual groups throughout the year. All young are given prolonged care within a structured society. Table 21.1 shows some typical grouping patterns in several species. Ecological factors such as population density and habitat characteristics may have major influences on group sizes and structures, but in typical habitats many species have a tendency to form groups of certain sizes. For example, gibbon groups are characteristically small with only 3 to

TABLE 21.1
Examples of Primate Grouping Patterns

Species	Total group size		Group composition			
	Average	Typical range	Adult males	Adult females	Infants	Juveniles
Rhesus in villages	17.4	3–60	3.5	7.2	4.6	2.0
Rhesus in						
temples	41.9	5–80	7.9	15.2	9.5	9.3
forests	49.8	5–120	5.6	19.2	11.4	13.6
Howlers	18.5	3–45	3.3	9.1	3.0	3.1
Langurs	—	5–30	—	—	—	—
Baboons	—	20–120	—	—	—	—
Gibbons	4.4	3–6	1.0	1.0	0.5	1.9
Gorillas	16.9	5–30	3.2	6.2	4.6	2.9
Chimpanzees—no typical or consistent patterns of group formation						

6 members, whereas baboon groups are usually large with 40 to 80 members.

Primate groups also vary in the extent to which strangers are excluded. Some species usually have closed social groups from which strangers are completely excluded. This is probably true for howlers, gibbons, marmosets, *Hapale*, *Tamarinus*, *Oedipomidas*, *Mico*, *Callimico* spp., and titi monkeys, *Callicebus moloch*. Other species, Indian rhesus, Japanese macaque, *Macaca fuscata*, some baboons, and the gorilla, have semiclosed groups which have stable members, but frequently 10 to 20 percent of the members may change groups annually. There are other species with open social groups which have free exchange of individuals between groups. In the African redtailed monkey of Uganda, *Cercopithecus ascanius*, nighttime groups of 8 to 10 merge during the day into large feeding aggregations of 40 to 50 monkeys, and then disperse into smaller groups in the evening.

The chimpanzee is one of the few species of primate known to have no consistent grouping pattern. Chimpanzees may occur together in groups of 2, 10, or 30, and the associations change from day to day. In other ways chimpanzees show the most variable and complex behavior of all non-human primates.

INTRAGROUP ORGANIZATION

Rhesus groups display several features of organization which are characteristic of many monkeys, particularly of the macaques and baboons. The basic social structure depends on the dominance patterns of adult males. For example, in a temple group 40 to 50 individuals would typically be divided into three subgroups (Figure 21.1): the largest, forming a central core for the entire group, would consist of a central adult male, one or two subdominant males, and a group of adult females, infants, and juveniles generally associated with the central male; a second subgroup would consist of a second dominant male and his accompanying females, infants, and juveniles; a third subgroup

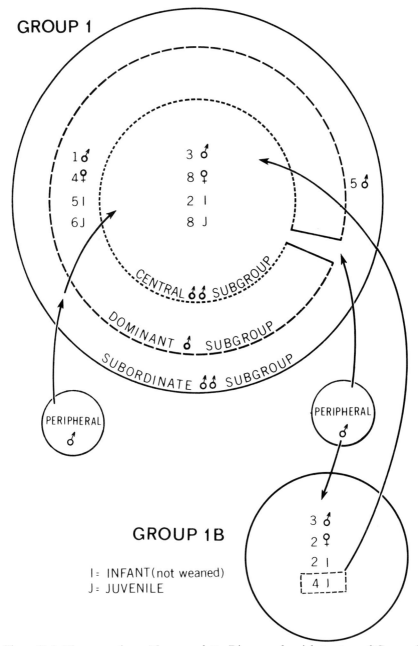

Figure 21.1. Rhesus monkeys, *Macaca mulatta*. Diagram of social structure of Groups 1 and 1B of Figure 21.3. (After Southwick, in DeVore, 1965.)

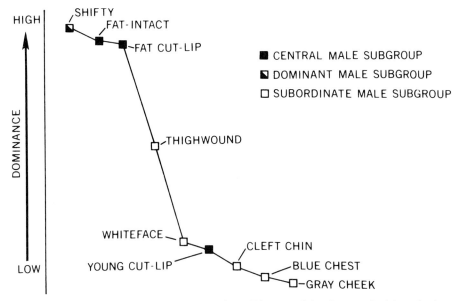

Figure 21.2. Rhesus monkeys, *Macaca mulatta.* Diagram of dominance of adult males in Group 1 of Figure 21.3. (After Southwick, in DeVore, 1965.)

would consist of young adult subordinate males which roamed the periphery of the main subgroups and were usually repelled by the dominant males if they attempted to enter the main subgroups. These subordinate males were extremely aggressive, particularly to other groups.

The compositions of these subgroups would not be rigid and constant. The greatest frequencies of favorable social interaction—grooming, feeding, sleeping, and play—occurred within subgroups, whereas the greatest frequencies of strife or aggression usually occurred between subgroups. There were some favorable interchanges between subgroups. This was particularly true for juveniles, who often played together but it was also true for adult females who occasionally changed subgroups during estrus and mated briefly with adult males of other subgroups.

Outside the main group structure there were often one or two peripheral males. These were fully adult males, usually in good physical condition, who led solitary lives. They oriented to the group but retained a distance of 100 to 300 yards. They moved when the group moved and often fed and rested when it fed and rested. If they attempted to approach the main group they were usually threatened and kept away by one of the group males. The origin of this ostracism is not known. Occasionally a female in estrus would come out and mate with a peripheral male during a consort period of a few hours.

The key to the subgroupings seems to be the dominance relationships of the males. Males may be ranked according to status and spatial position in

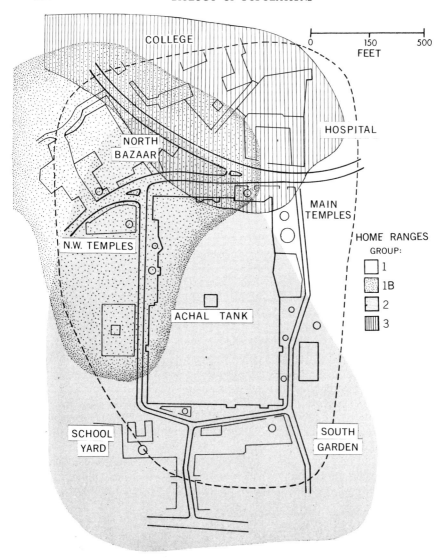

Figure 21.3. Home ranges of four groups of rhesus monkeys, *Macaca mulatta*, in a temple
habitat in Aligarh, India. (After Southwick, 1962.)

relation to the central core of the group (centrality). This is shown for the
Aligarh temple Group 1 in Fig. 21.2.

Many primate groups do not show patterns of subgrouping. In langurs,
howler monkeys, Barbary macaques, and probably many others, there is
greater homogeneity of group structure without evident social boundaries

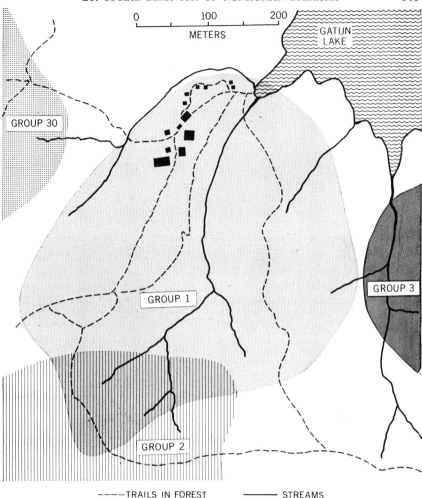

Figure 21.4. Home ranges of four groups of howler monkeys, *Alouatta palliata*, around main laboratories on Barro Colorado Island, Panama Canal Zone, in 1951. (After Collias and Southwick, 1952; Southwick, 1962.)

between members of the same group. The possible origin of subgrouping will be discussed later under socialization.

INTERGROUP RELATIONSHIPS

Social interactions among social groups of the same species may take a variety of forms. In some species groups are usually antagonistic in the proximity of other social groups. In rhesus this intergroup antagonism is often expressed by direct fighting, whereas in howler monkeys and gibbons it is expressed in a complicated pattern of threats and vocal displays rather than direct fighting.

Other primate species show passive or indifferent intergroup relations. In the langur, *Presbytis entellus,* and some baboons different social groups may come into close contact without mixing and without conspicuous interaction. The groups largely ignore each other, each passing on its way.

In other primate groups slight or complete mixing of individuals may occur with little or no aggressive interactions. Gorilla groups often meet and interchange a few individuals without aggressive activity, and African redtail monkeys have a complete intermingling of many groups into large feeding aggregations almost every day.

Although certain patterns of intergroup relationships are generally characteristic of various species these patterns are often modified by ecological conditions. For example, both langurs and baboons which normally have passive intergroup relations may also show considerable aggression under crowded conditions. Population density, availability of food, water, and cover are some of the factors known to influence the pattern and quality of intergroup interactions.

Rhesus and howler monkeys provide a particularly interesting comparison of intergroup relationships. Intergroup antagonism exists in both species but in rhesus it is in the form of damaging aggressive behavior, whereas in howlers it is in the form of ritualized threat without direct fighting. One prominent difference underlying this seems to be that the rhesus has no effective means of intergroup communication other than visual contact. They have no vocalizations characteristic of movement and no well defined territories. Home ranges overlap extensively (Figure 21.3) and groups may meet suddenly and unexpectedly. When this occurs fighting erupts spontaneously. Howler monkeys, on the other hand, have both effective intergroup communication at the vocal level and more accurately defined territories with little overlap of home ranges (Figure 21.4). Howlers begin each day with pre-dawn vocalization, which apparently serves to orient the groups to each other within the forest. Then each movement of the group is accompanied by characteristic vocalizations which can be heard through the forest foliage. Thus groups remain in communication and fully aware of other groups in the vicinity even though they cannot see each other. Sudden and unexpected meetings of groups are rare, occurring only during thunderstorms or some other distraction which confuses their normal communication system. The rhesus has not achieved an effective system of intergroup communication comparable to that of the howler monkey and gibbon.

PATTERNS OF SOCIALIZATION

The term socialization is used here as the process by which an individual develops behavioral associations with other individuals during infancy. In primates socialization is prolonged and complex. It consists of parental care, peer-group associations, and occasionally total group relations.

In rhesus care of a newborn infant is usually provided by the mother alone for several weeks. Other adults do not hold, carry, or groom the infant until it is 7 weeks of age. Adult male rhesus usually ignore infants. There is no semblance of paternal care in most rhesus and only rarely do they show positive behavior toward infants. It is, in fact, generally impossible to recognize, on a behavioral basis, the father of any one infant in a group that has more than one adult male. Thus the mother meets all the needs of the infant rhesus for several weeks and only after 2 or more months do social relations with peers and other adults become established.

Satisfactory maternal care and subsequent peer relationships are absolutely essential for normal behavioral development in the rhesus. Harlow and Mason (Harlow, 1959) have shown in the laboratory the drastic effects of rearing infant rhesus under maternal and peer-group deprivation. Such individuals develop asocial and aberrant behavior, including self-clasping, self-mutilation, catatonia, and idiopathic stereotypes. As adults such individuals cannot establish satisfactory social relations with other individuals. Laboratory research of this type should be extended to other species which have different patterns of socialization. In langurs maternal care is much less restrictive and from the first day of the infant langur's life the mother permits other females and juveniles to carry and groom the newborn. Thus the infant langur receives a very broad social experience with most members of the group within its first few weeks.

The Barbary macaque, *Macaca sylvana*, also has a broad pattern of socialization which even includes the adult males. In captive groups infant Barbary macaques are carried, groomed, protected, and fondled by all members of the groups, particularly the adult males, within the first week of life.

Neither langur nor Barbary macaque groups show the pattern of subgrouping that the rhesus does. It is an interesting hypothesis that the social barriers characteristic of rhesus groups have their social origin in restrictive socialization, whereas the group homogeneity of langurs and Barbary macaques has its origins in extensive socialization. This type of hypothesis could be tested in long-term fostering studies in captive or natural groups.

True paternal care is seen in only a few primates. In some of the marmosets, where a permanent bond develops between a mated adult pair, the male parent assumes an active role in infant care.

It would be interesting to undertake deprivation experiments on different species of primates with various patterns of socialization. Such work could make important contributions to understanding social factors in behavioral development. The effects of maternal and peer-group deprivation observed in the rhesus may or may not prevail in other species.

PATTERNS OF SOCIAL CHANGE

Field studies on Japanese macaques have clarified some of the mechanisms of social evolution and cultural changes in nonhuman primates (Imanishi and Altmann, 1965). Isolated groups show considerable variation in food habits and social behavior. For example, some groups eat wheat whereas others living in wheat growing areas do not. In some groups adult males take an active role in infant care, whereas this has never been observed in other groups. Japanese primatologists have undertaken careful studies of these differences in an attempt to understand their origin and development.

Some of the differences in food habits among groups of Japanese macaques originated under artificial feeding. The Japanese started artificially feeding wild groups in the early 1950's. One near Koshima was given cut-up sweet potatoes and these soon became a major portion of the diet. Early in 1952 one young juvenile took some of these potatoes into shallow sea water and played with them, turning them over in the water before eating. Other juveniles saw this behavior and copied it. Perhaps they liked the salt or it may have been just play. In a matter of months the mothers started to do this and then the trait passed from mothers to all infants and juveniles to other adult females and finally to adult males. Within 2 years most monkeys in the group were routinely washing their sweet potatoes before eating them. The group has consistently done this since 1954. Dr. Kawamura calls this the "washing-sweet-potato subculture of Koshima." This is a nonsense trait unless it has some adaptive value in the acquisition of salt. It provides an example of how a behavior trait started and spread slowly through an entire group of monkeys from individual to individual and from generation to generation becoming an established pattern.

Another group at Takasakiyama was given small bits of candy. The infants were the first to eat it. They liked it and the candy eating habit spread from infant to infant, then to older siblings and finally to adult females. In 18 months 51 percent of 500 monkeys had the candy-eating habit. Some have continued to eat candy but about half the group has not developed the habit. The habit is passed from mother to offspring but many of the adult females and adult males have not acquired it.

A third group, known as the Minoo-B group, was given wheat. Wheat and rice were both in their environment but they had not been eating wheat. One of the first to do so was a high-ranking adult male. Within 4 hours all members of the group were eating wheat and they have continued to do so for several generations.

Japanese primatologists have analyzed other examples of behavioral change and they have found that if a new trait is first expressed by an infant, juvenile, or other low-ranking individual, it spreads very slowly or not at all. If it is first expressed by a high-ranking individual it may spread very quickly. Monkeys tend to mimic high-status individuals more readily than low. The passage of a new behavior downward in the dominance hierarchy proceeds quickly, whereas it proceeds upward with difficulty.

The fact that the social behavior of nonhuman primates is as fixed and consistent as it is, the Japanese feel, depends largely on the fact that adult behavior is fairly stereotyped. Infants and juveniles are the prime innovators within a monkey society. They are continually trying new ways of doing things and very often the new behavior does not harmonize with established behavior. Hence an infant or juvenile playing near an adult and exhibiting some new behavior may get a severe rebuff. Many times I have seen a young rhesus get attacked, bitten, or hit for new play or investigative behavior around an adult. This suddenly ends the social transmission of that new behavior. Itani calls this "cultural inhibition."

CONCLUSIONS

Most species of primates live in permanent heterosexual groups which influence and control the behavior of individuals. The groups are highly organized, with complex patterns of social bonds, communicative networks, and a differentiation of individuals into various social roles.

There is wide variability within species and among species of primates in food habits, activity patterns, home range and territorial patterns, parental care, and basic social structures. There are no simple, clear-cut correlations between taxonomic position of various primate groups and their patterns of social structure.

The infancy of primates is long compared with other mammals and the infant is born into a structured society with established patterns of social behavior. These established patterns are imposed upon the infant primate during its long period of dependency. Capacities of young primates to innovate are not usually given full expression. Social change is usually rare and slow in most nonhuman primate groups, but it does occur.

The social group provides the primary protection and environmental adaptation for the individual, whereas these functions in many other mammals are provided by burrows, nests, cryptic coloration, nocturnal behavior, or morphological adaptations.

Primitive human groups differ in several basic ways from nonhuman primate groups. There are distinctly human patterns of language, kinship, exogamous mating, ownership, and federation of different social groups into pyramids of social institutions. Furthermore, man is distinct from all other primates in his dependence on tools and other artefacts, use of fire, and great elaboration of religion and philosophy. These features do not refute the fact that man shares a profound and extensive biological heritage with nonhuman primates.

A central problem in the study of comparative behavior is to develop criteria for true behavioral homologies between primate taxa, as in the more stable behavioral features such as facial gestures, and basic vocalization and motor patterns. Many aspects of primate social behavior are much more labile, being in some sense ecological adaptations.

REFERENCES

Altmann, S. A. (ed.) 1967. *Social Communication among Primates*. Univ. of Chicago Press, Chicago, Illinois.

Collias, N. E., and Southwick, C. H. 1952. A study of population density and social organization in howling monkeys. *Proc. Am. Phil. Soc.* **96:** 144–156.

DeVore, I. (ed.) 1965. *Primate Behavior: Field Studies of Monkeys and Apes*. Holt, Rinehart and Winston, New York.

Harlow, H. F. 1959. The development of affectional patterns in infant monkeys. In *Determinants in Infant Behavior*. B. M. Foss (ed.) pp. 75–97. Wiley, New York.

Imanishi, K., and Altmann, S. A. 1965. *Japanese Monkeys: A Collection of Translations*. Yerkes Regional Primate Center, Atlanta, Georgia.

Jay, P. (ed.) 1968. *Primates: Studies in Adaptation and Variability*. Holt, Rinehart and Winston, New York.

Morris, D. (ed.) 1967. *Primate Ecology*. Aldine, Chicago, Illinois.

Southwick, C. H. 1962. Patterns of intergroup social behavior in primates, with special reference to rhesus and howling monkeys. *Ann. N. Y. Acad. Sci.* **102**: 436–454.

Southwick, C. H. (ed.) 1963. *Primate Social Behavior*. Van Nostrand, Princeton, New Jersey.

CHAPTER 22-A

Human Social Behavior

ANTHONY J. READING

Man is indeed a social animal. He lives out his days immersed in diverse and complicated social matrices which he produces and which, in many ways, produce him. The fabric of human existence is so interwoven with threads of social organization that man cannot be defined independently of his social context. All human social systems possess language and cultural traditions, stable pair bonds with no special breeding season, division of labor and occupational specialization, emphasis on the family, and sets of relationships based on kinship patterns. There are taboos against incest and murder, but not against the killing of members of another tribe or nation, and rules and rituals regulating adult status, marriage, and property rights. In addition, each human society has elaborated a complex sociocultural superstructure of its own in response to its particular needs. The resulting variations among the multitude of known human social systems so greatly outweigh their similarities that it is not possible to distill a set of universal characteristics which adequately portrays human social behavior. While general characteristics indeed serve to distinguish human social behavior from that of other animals, they fail to capture its essence completely. Variability itself is the outstanding human social characteristic; the ability to respond in a profusion of ways indeed being the basis for much of man's adaptiveness, both individually and as a species.

EVOLUTION

The origins of human social behavior are hidden in the origins of man himself. Although the story of human evolution is still far from complete and much of it is still controversial, modern paleoanthropology is excitingly beginning to piece together the limited and fragmentary fossil record of early man (Washburn, 1961; Dobzhansky, 1962; Hoagland and Burhoe, 1962; Clark, 1964; Howell, 1968). It seems that man is a relatively recent species whose earliest ancestors probably separated from those of the current great apes in the Miocene, some 12,000,000 to 25,000,000 years ago. The Australopithecines, who first appeared some 2,000,000 years ago at the beginning of the Pleistocene are generally considered to be the first true hominids (more manlike than apelike). They were ground-living, fully bipedal animals that used their hands for manipulation and for fashioning simple chopping tools. Their canine teeth were small, like those of contemporary man, but their brains of 600 cubic centimeters were scarcely larger than those of present-day apes. Small bands of various races of *Australopithecus* probably wandered the

311

tropics for hundreds of millennia, hunting small game, scavenging, gathering fruits, defending themselves from predators, and evolving only slowly. About 500,000 years ago they had either evolved into or were replaced by a distinctly more human species, *Homo erectus*, which became widely distributed throughout Europe, Asia, and Africa. Their brain cases of 900 to 1,100 cubic centimeters were considerably larger, and they made large stone choppers, hunted big game, sheltered in caves, and had the use of fire. Neanderthal man, the next distinct form, appeared particularly in Europe about 120,000 years ago and seems to have evolved from *Homo erectus*. The cranial capacities of the Neanderthals were comparable to our own of 1,400 cubic centimeters. They hunted with spears and pitfalls, made fires, improvised shelters, and apparently engaged in some forms of ritual observance. Although seemingly successful, the classic Neanderthalers of Europe were fairly suddenly replaced about 40,000 years ago by modern man's immediate predecessor, Cro-Magnon man. Although the origins of Cro-Magnon are still uncertain, there is some evidence that they were the descendants of non-European Neanderthaloid races.

The Cro-Magnon were hunters and gatherers who wore clothes, had elaborate shelters, and made many highly specialized tools. They also had religious ceremonials and elaborate burial customs, and they adorned their bodies and painted scenes on rocks, some of which still survive. They were the wanderers and colonizers who first spread man to the four corners of the earth. Cultures such as that of the Australian aboriginals were later isolated when the land bridges between the continents disappeared, and the aboriginal remains today as a somewhat archaic remnant of these earlier times. The transition to Modern Man is generally placed about 12,000 years ago, close to the beginnings of animal and crop domestication that occurred at the end of the last great glaciation. Societies began to change their nomadic ways for more settled agricultural and pastoral ones, and from these settlements the first villages began to appear in Asia about 8,000 years ago. Villages and towns rapidly prospered and coalesced into states and nations. Opportunities for economic diversification, occupational specialization, and trade also grew, and from these roots the great civilizations of early recorded history started to form. By 4,000 B.C. the Egyptians had invented the plow, had constructed a calendar, and were writing; and by 2,000 B.C. the pyramids were already 500 years old. Civilizations rose and fell in Near and Far East, in Africa, in Europe, and in the New World, and in each of them man struggled in different ways to master the increasingly complex patterns of social organization that were involved, much as he still seems to be doing.

In the course of evolution, man has become increasingly adaptable and able to modify his environment, rather than himself become narrowly adapted to particular environments. He has also become increasingly efficient at the tasks of simple survival, thereby freeing himself for the wide range of non-subsistence behaviors that now characterize many of his sociocultural systems. Whereas some primate populations spend some 80 percent of their waking hours gathering and consuming necessary foodstuffs, these pursuits occupy only about 10 to 20 percent of the energies of modern human societies. Man's manipulation of his physical and biological environment have, however, often brought about changes in the human sector of his environment to which he has had to adapt with changing patterns of social behavior. Along the long way from stone axe to agriculture, printing press, penicillin, and H-bomb, technological change has been followed by social change, such social changes themselves often providing the impetus for further technological development. As present man struggles to develop new social patterns to enable him to survive his new technologies, several authors have begun to contend that not only may man's capacity for social adaptation be limited, but that it may be quite finitely constrained by the genetic legacy from his evolutionary ancestors. Ardrey (1966) has extensively documented the case for man's being a territorial animal and for much of his social behavior being inescapably bound by this. Morris (1967) has marshaled evidence which he feels indicates that such fixed and innate components pervade an even larger range of human social interactions. There is no general agreement, however, about the extent and the manner in which human social behavior is genetically determined.

CULTURE AND CULTURAL EVOLUTION

Quite early in his history man evolved the capacity to partake in a second great evolutionary system which came to supplement and interact with the genetic one (Spuhler, 1959). This second system of cultural evolution although present in other animals in rudimentary forms, is almost unique to man and has come to represent an exceedingly potent mechanism of adaptation. Culture is best viewed in this context as a process of extra-genetic information transmission which allows, as one of its consequences, each generation to profit from the experiences of preceding generations. Like the genetic evolutionary system with which it interacts, the cultural system is able to produce innovations whose survival depends largely on their adaptive value but, unlike the genetic system, information transfer is accomplished through language and other forms of social communication. While genetic information is transferred exclusively from the individual's parents, cultural systems have available various other information sources which permit far greater short-term adaptability and dissemination. All human cultural systems depend heavily on the use of symbolic language for the transmission of complicated bits of information and differ qualitatively and quantitatively from the "proto-cultural" systems of other animals which appear to rely exclusively on imitation. Lenneberg (1967) is strongly of the opinion that there is no continuity between the communication systems of other animals and the language of man, but as spoken language leaves no fossils, its origins are unknown. It seems likely, however, that isolated and nomadic groups might not have been large enough to generate and sustain the elaborate syntactical structure on which modern languages depend. The great cultural efflorescence of the past 6,000 years may indeed be related to increase in abstract and symbolic language functions. Contemporary man, in addition to being capable of these complex language functions, possesses elaborate means for language storage and transmission which potentially unite the entire species for the first time within a single social matrix.

Man evolved against a turbulent background which included four major periods of glacial shift, each accompanied by significant changes in the fauna and flora. It seems that surviving human populations became increasingly reliant on cultural methods of adaptation. For instance, populations that innovated and utilized axe or spear, fire or shelter, undoubtedly had advantage over ones that did not, and as these populations prospered use of these implements became more widespread. Adaptive cultural traditions become increasingly represented in succeeding generations.

Cultural innovations can, however, spread rapidly through a population independently of their adaptive value. Modern market research and advertising are largely directed towards inducing widespread changes in culturally determined behavior regardless of their adaptive value—"use Brand X soap." Much of public health practice, on the other hand, deals with attempts to introduce hopefully adaptive customs such as family planning, or to extinguish presumed maladaptive ones such as cigarette smoking. Not enough is known about the factors that govern the diffusion of cultural innovations to guarantee the success of such programs, although several descriptive studies of the process have been attempted (Rogers, 1962). These studies have mainly been limited to the spread of technical innovation among peers, for instance the introduction of hybrid corn in Ohio, and have shown that successful diffusion has depended on the apparent utility of the new practice and on lack of conflict with preexisting values. Although people may receive information from an impersonal source, it seemed that they were most likely to adopt an innovation only after personal contact with someone who had already adopted it. Because of this, it seemed that once 20 to 30 percent of a population had made the change, further diffusion occurred relatively rapidly. The spread of new ideas or behaviors among peers would thus seem to follow an epidemiological model, with certain portions of the population being considered susceptible and others resistant, and with the rate of diffusion dependent on contact between "infected" and "uninfected" susceptibles. The dancing manias of thirteenth-century Europe, the witch hunts of Salem, and the "Invasion from Mars" and other mass hysterias (Linton, 1956) probably represent extreme cases of cultural "contagion." On a smaller scale, brain-washing and some psychiatric and religious conversions (Sargant, 1957) attest to man's extreme susceptibility to cultural influence under certain special conditions.

Some of the most pertinent data on diffusion across generation barriers come from the studies of the Japanese macaque populations (Chapter 21). Innovation was most prevalent in the adolescents while dissemination was most thorough from the dominant adults. Human parallels are abundant, both in the amount of influence but relative lack of innovation of dominant individuals, such as Presidents and their ladies, and in the amount of innovation but relative lack of influence of adolescents such as "hippies" and college students. It is almost as if most cultural innovations have to mature with each generation before passage to the next. The culture serially passes through each young generation to be modified, if need be, before they have the opportunity, as adults, to pass it on to the next generation.

COOPERATION AND COMPETITION

Cooperation and competition are the two major themes that dominate the lives of all social animals, man included. Social living always implies cooperation of some sort among the members of a group and such cooperation bestows many selective advantages on the group. Both among and within cooperative social groups there is also some degree of competition, especially for essentials such as food, mates, and space. Each social species has made its own balance between these two opposing modes of social interaction and Chapter 19 discussed the manner in which many species have balanced these forces by means of hierarchies and territorial systems. Many vertebrate animals developed ritualistic ways of dealing with the aggressive potential that exists between and within discrete social groups. Such mechanisms may break down in unusual environments such as those leading to Calhoun's "behavioral sinks" (Chapter 20); and can lead to severe intraspecific fighting (Chapter 8). *Within* normally functioning human groups, competition also tends to be ritualized or expressed indirectly, but competition *between* such groups appears to lack stabilizing mechanisms and man has spilled much human blood as a result. Even within a well-functioning human family unit there is competition in the form of sibling and oedipal rivalries, the resolution of which plays a central role in psychoanalytic theories of personality development. In discussing competition in man it should be remembered that the word "aggression" is used to describe a number of behaviors which range widely from necessary self-assertion to extremes of brutality and sadism and, while these may all have features in common, they are by no means entirely similar in origin.

Freeman (1964) reviewed the anthropological and historical evidence of man's animosity and destructiveness to other men and suggested that the ubiquity of human intraspecific cruelty and bloodshed indicate that these processes served some evolutionary advantage. It is conceivable that the nomadic bands that wandered the earth during most of man's evolution began to compete with each other during times of relative scarcity of food and shelter and to use their new found weapons in the process. The more competitive groups may thus have permanently removed the less successful ones, and their genes, from further competition by annihilating them. Modern man may well be the descendant of small groups that were selected for their ability to kill and maim members of other small groups. Such behavior would be less and less adaptive with the rise of civilization and the consequent merging and diffusion of mankind into larger, more powerful groups. While Montagu (1957) asserts that human aggressiveness is learned, it seems, on the contrary, that civilized man has had to learn to control his innate destructive aggression, and much of our social and cultural education is directed toward this end. Unfortunately the educational attempts are not always successful, for our young still learn by imitation as well as by instruction (Frank, 1968).

Man evolved as the only primate to include significant amounts of flesh in his diet. Like most other carnivores, human adults share food equally within their social units rather than having the dominant animal take the best portion, as generally occurs in the other primates. Even in contemporary primitive societies emphasis is placed on sharing and generosity within the group, and prestige is often based on who gives away the most (Mead, 1961). Like many other carnivores, man has always placed a premium on cooperation within the primary social group of the family. However, with the development of more settled communities cooperation was increasingly extended to larger, secondary social groups, and eventually into states and nations. The ties that unite the family have a long evolutionary history. The infant's smile reflex and the mother's almost reflex response to it have undoubtedly played an essential role in establishing the "maternal bond" in human families. Morris (1967) has also emphasized the importance, for stabilizing the monogamous adult pair relationship, of the fact that man has the most highly developed secondary sexual characteristics and the most elaborate and active sexual life of all the primates. The ties that link man in the more extended secondary social groups are more recent and are probably much less securely founded. In these groups, identity is as much maintained by those the group excludes as by those it includes, and cohesion is as much a function of something external to unite against as it is of something internal to unite for. As a result, cooperation within secondary social groups may have been achieved at the expense of increased intergroup competition, and the consequences of this arrangement may have mounted as these groups have become ever extended. To minimize competition and enhance cooperation, stable secondary groups tend to evolve a structure in which different members occupy different roles. The simplest structure involves a dichotomy into "leader" and "followers" (Berne, 1963), but large, stable groups such as armies or univeristies have much more complicated hierarchies than this. Such hierarchies probably serve much the same function as they do in the previously described animal groups. One of the main challenges of contemporary society is whether it will be able to channel the competition among large antagonistic intranational and international groups into nonlethal, even mutually advantageous, ends. The only effective alternative would seem to be to decrease their competitiveness by subsuming them into even larger groups through sufficiently increased communication. Attempts in these directions include the Olympic Games and Free Trade in the one mode and Profit Sharing and the United Nations in the other. The same general principles can be applied to deal with the competitive forces among smaller groups and often offer an effective answer to the jealousies, rivalries, and factions that undermine the cooperative efforts of many organizations.

SOCIALIZATION

The human infant is born helplessly immature and remains so far longer than any other animal. His behavioral repertoire is undeveloped and his brain has not even attained a quarter of its eventual mass. Rhesus and gibbon brains at birth, by contrast, are 70 percent of the adult size. Brain size at birth is limited by the size of the female pelvic outlet which has not changed appreciably since the time of *Australopithecus*. The evolutionary increases in human cranial capacity have mainly been accomplished through increased postnatal development. Together with this came evolution of effective life-supporting services within the social network in order to enable the young to survive the long period of immaturity that is involved. During the first year of life the human infant's brain more than doubles its size to reach about 750 cubic centimeters and there are accompanying increases in his social behavior. The protective matrix that surrounds him gradually begins to require more active participation on his part, and he begins to differentiate himself from his environment and form an identity of his own. Socialization occurs primarily within the family unit and much of the individual's later behavior is modeled on the patterns of his interactions within it. Such fundamentals of cooperative living as trust, generosity, and consideration for others are learned to greater or lesser degree within the microcosm of the family, as also are some more specific orientations to his fellow men. Future relations with the opposite sex, for example, are considerably moulded by his experiences of the parent of that sex, future relations with peers by his relationship with his siblings, and future attitudes toward authority by his interactions with parental controls. Toman (1961) has shown that marital success can be significantly predicted on the basis of the individual's family constellation. He has shown that individuals that have opposite-sexed siblings have statistically greater marital success than those that do not. Success tends to be even greater in marriages composed of partners with complementary sibships— where a boy with only younger sisters marries a girl with only older brothers. Much of the individual's future relationship with himself, his self-concept, is also fashioned by interactions within the family. For instance, studies of hermaphrodites as reported by Money (1965) have demonstrated that later gender identity is largely determined by the sex assigned such individuals by their parents rather than by the predominant biological assignment.

As children grow older and their arena of social contact enlarges they gradually become less dependent on their parents and begin to acquire new social skills from other sources. Eventually a fairly permanent repertoire of social skills becomes established which characterizes the individual. As the varied and complex experiences that determine social development inevitably differ from individual to individual, even within the same family, there results a wide range of normal variation in adult behavior. Much of what society labels as mental illness undoubtedly reflects the more aberrant variations in these formative social processes (Opler, 1967).

ENCULTURATION

Enculturation is a process distinct from socialization, although for the most part the two proceed inseparably. From the moment of birth the infant is immersed in the same cultural milieu that immerses his family, and their cultural heritage becomes his. The language he first speaks is that of his parents and in much the same way he becomes heir also to their values and attitudes, their religion, their ethnic traditions, and their social and economic station. The parents, in addition, have the responsibility of transmitting the various sanctions and prohibitions that the culture demands. Most of what each society calls manners involves learning to inhibit previous infantile behaviors: in our society the child learns that he is no longer able to run around naked, burp in public, or scream at the top of his voice if his needs are not immediately gratified. Cultures vary little in the more serious sets of interdictions that the young have to learn, almost all having proscriptions against stealing, murder, adultery, and the like. Enculturation also involves the instilling of new behaviors into the individual, and these vary greatly according to the needs of the society. Depending on his culture, the individual may be required to learn how to hunt his own food or negotiate a super-market, how to make simple pottery or parts for automotive engines. Especi-ally in literate societies, parents delegate much of the responsibility for such positive education to others, a practice that has become institutionalized in the formal school systems.

In complex societies, individuals come to have varying degrees of member-ships in many subgroups in addition to their common membership in the larger social system. As well as their membership in the subgroup of their family and its extensions, they may belong in different degrees to different occupational, residential, social, recreational, educational, religious, econ-omic, political, and ethnic subgroups, to name a few, and they may have different roles in each of these. Within the general culture of the society, each of these subgroups has a specific culture of its own which its members share and which they transmit to succeeding generations of members. Cultural inheritance, like genetic, transmits to each individual within a society his own unique heritage in addition to the standard heritage which he shares with his fellows at large. The greater part of man's social identity and individuality is a function of the degree and nature of these past and present group affiliations. If the values and attitudes of the various subcultural groups to which an individual belongs are not reasonably compatible, he may be forced to with-draw or become alienated, or else suffer anxieties due to the unresolved conflicts.

COMMUNICATION

The degree of an individual's membership in a cultural or subcultural group can be defined by his ability to communicate with other members. Indeed, one of the chief functions of a cultural system is to furnish mutually agreed upon rules which allow communication to be interpreted and understood (Hall, 1959). However, even within the same general cultural system meanings of words and actions may differ among persons of significantly different subcultural backgrounds, and communication may be fraught with misunderstandings. Despite the fact that much emphasis in our society is placed on the verbal language that links individuals in their secondarily extended social groups (Cherry, 1957), the primary language of close interpersonal transactions remains essentially nonverbal. Strictly verbal language is limited in its ability to convey the emotional meanings that are often involved in these transactions. Human nonverbal communication can be classified into four fairly distinct modes: (a) action language, probably the most primitive, which includes approach and withdrawal and the various forms of physical contact; (b) body language, which includes the range of facial expressions and body tone and posture; (c) para-verbal language, which includes the intensity, pitch, timing, inflection, etc., of verbal language as well as general vocalizations such as "ooh" and "ah"; and (d) a heterogeneous group of largely inanimate signs and signals that include traffic lights, flags, barbers' poles, and music, sculpture, and painting. Many nonverbal communicatory behaviors, such as smiling and crying, cowering in submission and shaking one's fist in rage, are ancient and well-nigh universal, and probably have a genetic basis. Although nonverbal communication, by virtue of its generally being more primitive, is more likely to have its meanings shared across cultural barriers than is the verbal, it is by no means free of cultural influence. The polite social distance for conversation in some South American countries, for example, is about 6 inches closer than it is in the United States; to stand further away is a sign of unfriendliness. A South American may thus seem rudely intrusive to a North American as he actively tries to establish his accustomed conversational distance while probably wondering what is causing his unfriendly companion to keep backing away. A great deal of our social behavior is governed by feedback from such nonverbal cues. However, we tend to be less consciously aware of such nonverbal messages even though we remain quite bound in our reactions to them, and a great deal of confusion and anguish, and possibly some mental illness, results from conflictingly mixed verbal and nonverbal messages. To some extent we are all prisoners of our own cultures, forever limited to interpreting the verbal and nonverbal behaviors of others in our own idioms and capable of communicating most effectively only with those who are most like ourselves.

When people assemble in large, heterogeneous crowds, all that is available for shared communication and group membership are their lowest common nonverbal denominators of action and emotion. When such assemblages move in the direction of action, they are, as mobs, capable of behaving as if the social and cultural heritages that have just been described had not yet evolved. Despite the fact that mob and crowd behavior has such relevance to the contemporary scene, their formal study has received scant recent attention and analyses such as those of LeBon (1895) and McDougall (1920) remain little improved. The former classically described how participation in a crowd degraded the individual to a lower level of operation, and the latter of how such fortuitous assemblies lacked the formal structure necessary to pursue their goals in an orderly fashion. The intensified emotional reaction and the decrease in inhibitions that such large and heterogeneous groups can force upon their members may indeed produce reactions other than mob violence. The intensely shared emotional experience immediately after President Kennedy's assassination also speaks of their power.

CONCLUSIONS

Although anthropology and sociology provide much descriptive information about human societies, there is, as yet, no systematic body of biological knowledge about the behavior of human populations. Human behavioral sciences have almost exclusively devoted their attention to individuals and to small groups, and yet there is a growing realization that the models that have been formulated from these are not always applicable to the broader perspective of populations. Animal studies, as this book demonstrates, have focused a great deal of attention at the population level of organization and serve well, at present, both as generators of hypotheses for human population behaviors and as reminders of the general constraints of such biological systems. Man and his closer animal relatives have had to solve many of the same evolutionary problems, and study of the latter help to remind us that what we call "human" is simply one of a number of alternative evolutionary "strategies." These studies also serve as reminders that behavior, and especially social behavior, to be fully understood must be interpreted as a function of its total ecology, both past and present. Human behavior must, however, eventually be understood in its own terms, not just in terms of its similarities with and differences from other species.

An attempt has been made here to examine the social behaviour of human populations as a biological system. The genetic and cultural evolution of human social behavior has been reviewed, particularly with reference to cooperation and competition and the processes of socialization, enculturation, and communication. The ubiquitous role that cultural processes play in human social behavior has been emphasized throughout the chapter. Some of the implications of this approach for contemporary social problems have been touched upon.

REFERENCES
Ardrey, R. 1966. *The Territorial Imperative.* Atheneum, New York.
Berne, E. 1963. *The Structure and Dynamics of Organization and Groups.* Grove Press, New York.
Cherry, C. 1957. *On Human Communication.* Wiley, New York.

Clark, W. E. L. 1964. *The Fossil Evidence for Human Evolution.* 2nd ed. Univ. of Chicago Press, Chicago, Illinois.

Dobzhansky, T. 1962. *Mankind Evolving.* Yale Univ. Press, New Haven, Conn.

Freeman, D. 1964. Human aggression in anthropological perspective. In *The Natural History of Aggression,* pp. 109–119. J. D. Carthy, and F. J. Ebling (eds.) Academic Press, New York.

Frank, J. D. 1968. *Sanity and Survival.* Vintage Books, Random House, New York.

Hall, E. T. 1959. *The Silent Language.* Doubleday, New York.

Hoagland, H., and Burhoe, R. W. (eds.) 1962. *Evolution and Man's Progress.* Columbia Univ. Press, New York.

Howell, F. C. 1968. *Early Man.* Time-Life Books, New York.

LeBon, G. 1895. *The Crowd.* Compass Edition, 1960; Viking Press, New York.

Lenneberg, E. H. 1967. *Biological Foundations of Language.* Wiley, New York.

Linton, R. 1956. *Culture and Mental Disorders.* Thomas, Springfield, Illinois.

McDougall, W. 1920. *The Group Mind.* Putnam's Sons, New York.

Mead, M. (ed.) 1961. *Cooperation and Competition Among Primitive Peoples.* Beacon Press, Boston, Mass.

Money, J. 1965. *Sex Research: New Developments.* Holt, Rinehart and Winston, New York.

Montagu, A. 1957. *Anthropology and Human Nature.* McGraw-Hill, New York.

Morris, D. 1967. *The Naked Ape.* McGraw-Hill, New York.

Opler, M. K. 1967. *Culture and Social Psychiatry.* Atherton Press, New York.

Rogers, E. M. 1962. *Diffusion of Innovations.* The Free Press of Glencoe, New York.

Sargant, W. 1957. *Battle for the Mind.* Heineman, London.

Spuhler, J. N. (ed.) 1959. *The Evolution of Man's Capacity for Culture.* Wayne State Univ. Press, Detroit, Michigan.

Toman, W. 1961. *Family Constellation.* Springer Pub. Co., New York.

Washburn, S. L. (ed.) 1961. *Social Life of Early Man.* Aldine Pub. Co, Chicago, Illinois.

Human Group Aggression

JEROME D. FRANK

From the public health standpoint, new weapons of mass destruction have come to represent the greatest threat to human life, far outdistancing such natural disasters as epidemics and famines. Under some conditions humans lose all restraints against killing each other and resort to the most powerful engines of destruction at their disposal. All that has saved mankind from destruction by this means so far has been the inefficiency of even the most powerful weapons. When a single depot of nerve gas contains 100 billion lethal doses, and existing stockpiles of nuclear weapons could kill the world's population several times over, this safeguard has been removed.

As a result, the danger of race suicide becomes a reality and the problem of controlling mass violence requires solution if humanity is to survive. To this end, this presentation reviews some of the psychosocial factors that disinhibit group violence, both domestically and internationally, and some of the ways that existing restraints on it can be made more effective, or new ones can be created.

Role of Threat within Groups

Group conflict, the chief instigator of violence, is the law of life, and the threat of violence, with occasional resort to it to keep the threat credible, plays an important role in maintaining social order. Through it the weak gain concessions from the powerful before they are driven to desperation and the rulers maintain order. In healthy societies the actual display of violence is rare, and threats are largely symbolic. The vote, for example, can be viewed as a substitute for a fight in that it is a test of strength determining which side would probably win in a battle. Moreover, order is maintained by institutions, in which all members of society have confidence, for adjudicating power struggles and protecting the loser, operating under a code of laws.

Altruism in Defense of the Group

Humans are group creatures and the group, not the individual, is the survival unit. When it is threatened, individual members are expected to sacrifice their lives in its defense and are motivated to do so. In this humans differ not at all from baboons, and are not too different from ants. Thus, paradoxical as it may seem, the greatest threat to human survival is not aggression but altruism. It is the threat to the group of which one is a member, not to oneself, which evokes the most powerful aggressive response. Advocates of nonviolence can easily accept the notion of accepting death for themselves rather than

322

attacking their opponent, but find it much harder to defend the nonviolent position when the threatened victim is a wife or daughter. In this too, humans resemble animals, who fight especially viciously to defend their young.

But humans add a unique dimension to their allegiance to the group, through their capacity for what Arthur Koestler has termed self-transcendance. The human group gains the allegiance of its members not only because it is the biological survival unit, but because it embodies and preserves certain ideals, values, and symbols. When Kamikaze pilots committed suicide for their emperor, they had more in mind than the little man sitting on the throne of Japan; and when men offer up their lives for the Flag or the Cross, it is for the concepts these bits of cloth or wood represent. Even an abstraction as vague as a better world for our grandchildren may suffice to call forth the supreme sacrifice.

SOURCES OF GROUP VIOLENCE

Goals That Appear Mutually Exclusive

Group conflict arises when each group perceives its goal as achievable only at the other's expense. Domestically this type of conflict becomes violent when groups feel intolerably frustrated or threatened, and have lost faith in the institutions of society to satisfy their claims or to protect them. Thus in the United States today, many Negroes have lost faith in the ability or desire of the white community to satisfy their legitimate aspirations, and many whites fear that their jobs, homes, and lives are jeopardized by militant Negroes and that law enforcement agencies can no longer protect them. As a result both sides take to arms. A main source of resistance to gun registration is the fear of all groups that they will be unilaterally disarmed and thereby left at the mercy of their enemies. Thus in his presidential campaign Governor Wallace spoke of the good people giving up their guns and the bad people keeping theirs, and Negroes express fears that if they register their guns, they will be disarmed and the whites will not.

Analogously, on the international scene violence erupts when nations find themselves in conflict and cannot appeal to any supranational institution to resolve the issue and protect the weaker (not because such institutions have broken down, but because they do not yet exist), and each perceives yielding to the other as a threat to its survival.

Image of the Enemy

Groups in conflict for any length of time regularly form images of each other as enemies, and the enemy image is remarkably similar, no matter who the conflicting parties are.

This is illustrated by the findings of repeated surveys of Americans concerning their characterizations of people of other countries. In 1942 and again in 1966 respondents were asked to choose from a list of adjectives those that best decribed the people of Russia, Germany, and Japan. In 1942 the first five adjectives chosen to characterize both Germans and Japanese (enemies) included warlike, treacherous, and cruel, none of which appeared among the first five describing the Russians (allies); in 1966 all three had disappeared from American characterizations of the Germans and Japanese (allies) but now the Russians (no longer allies, although more rivals than enemies) were warlike and treacherous. Data were reported for the Mainland Chinese only in 1966, and predictably, they were seen as warlike, treacherous, and sly (Gallup Poll, 1966).

The adjectives applied to the Japanese and Germans as wartime enemies no doubt accurately described their behavior—as they do that of all nations at war, including the United States—but it is noteworthy that Americans did not apply these adjectives to the Russians when they were allies, although the Germans undoubtedly saw them as warlike and treacherous, and that Americans now use these terms for the Russians and the Chinese although there is no direct evidence that they apply. The Russians have been talking and acting with great restraint and Chinese bellicosity is restricted to words; neither nation has shown any particular signs of treachery.

The characteristics of the enemy image have been most extensively studied with respect to the United States and the Soviet Union. A detailed content analysis of selected mass and elite publications in these nations a few years ago found that virtually 100 per cent of the relevant items described the national goal of the other as military expansion, and their military doctrine as including a preemptive strike. About two-thirds of each believed that negotiations were possible and could be successful only if our side were stronger. A majority of each saw foreign aid given by their country as motivated by altruism, that offered by the other as in the service of their expansion (Angell, Dunham, and Singer, 1964).

This is supported by many reports of personal interviews, some systematic and some informal. One of the latter by an American scientist who had an opportunity for long, informal conversation with his Russian counterparts is worth quoting: "The Westerner regards the Russians as controlled, for the most part without their knowledge, by an oligarchy of rapacious and malevolent men who seek constantly to foment world revolution. The Russian is equally convinced that the West (which means really America, for in Russian eyes all other Western countries are American satellites) is being victimized by a small group of profit-mad 'monopolists' who pull the strings that control government, press, and radio and who try to instigate wars in order to sell munitions. On the level of informal conversations such as ours it was impossible to resolve this difference in viewpoint. Each of us was repeating what he had read in his own newspapers, and each was suspicious of the other's sources" (Krauskopf, 1961).

The enemy image leads to the application of a double standard of morality which in turn reinforces it. Since the enemy's motives are always presumed to be bad and those of our side are good, the identical act is viewed as good if performed by our side and bad if performed by the enemy. A psychologist showed some American fifth- and sixth-graders photographs of Russian roads lined with young trees. When he asked why the Russians had trees along the road, two answers were: "So that people won't be able to see what is going on beyond the road," and "It's to make work for the prisoners"; but when he asked why some American roads have trees planted along the side, the children said "for shade" or "to keep the dust down" (Bronfenbrenner, 1963).

The double standard of evaluation was illustrated by a formal study conducted in 1965 in which a large number of college freshmen were presented with fifty statements concerning belligerent and conciliatory actions that had been taken by both the United States and Russia; for half the students the acts were attributed to the United States, for the other half to the Soviet Union. They were asked to indicate their feelings about each statement by marking it from $+3$ (most favorable) to -3 (most unfavorable), and as might be expected, an action was scored more favorably when attributed to the United States than when attributed to Russia. For purposes of statistical analysis the scores were transformed into a 0-to-6 scale, so that scores above 3 would be favorable and those below 3 unfavorable. For example, the average score for "The U.S. (Russia) has established rocket bases close to the borders of Russia (the U.S.)" was 4.7 for the United States version and 0.5 for the Russian one. "The U.S. (Russia) has stated that it was compelled to resume nuclear testing by the action of Russia (the U.S.)" was scored 4.2 in the United States form and 1.0 in the Russian one (Oskamp, 1965).

The enemy image that nations form of each other, of course, more or less corresponds to reality. Nations that failed to recognize that their enemies might be treacherous and warlike would not long survive.

However, the enemy image impedes resolution of the conflict by causing both sides to overemphasize confirming information and filter out information that does not fit. It also creates a self-fulfilling prophecy, leading enemies to acquire the evil characteristics they attribute to each other. In combating what they perceive to be the other's cruelty and treachery, each side becomes more cruel and treacherous itself.

REMOVERS OF RESTRAINTS ON GROUP VIOLENCE
Formulating the Conflict in Terms of Value Systems
The mutual enemy image is one example of how symbolic components aggravate human conflicts. Even more troublesome is the tendency to formulate both domestic and international conflicts in terms of values rather than tangible goods. Negroes fight for better housing and better jobs, but these are subsumed under such concepts as freedom, equality, and justice; whereas whites appeal to self-determination and law and order.

Although domestic struggles may be bitter and destructive, they do not seriously threaten the existence of the social system as long as both sides profess the same values. The life of the society is at stake when one side challenges its basic assumptions as when Blacks demand a separate state, which attacks the concept of integration of peoples of all races, or militant students of the New Left argue that the democratic system is no longer viable and must be overthrown.

International conflicts are caused by conflicts of national interests. As long as these interests are viewed primarily as tangible, such as control of territory, population, or resources, and neither side strives for the complete occupation of the other, they too remain within bounds. They become threatening to the survival of the contestants, and today to the survival of mankind, when the interests at stake are couched in terms of which world view shall prevail, and when the promulgators of each world view believe that it cannot survive as long as the other exists. Unfortunately, today many people view international struggles in this way—they see all as skirmishes in an apocalyptic struggle between Communism and Free Enterprise to determine which will eventually secure the allegiance of the whole world, and proponents of each ideology see the other as a never-ending threat that can be eliminated only by the destruction of its advocates.

Ideological components of conflict remove constraints on violence for two reasons. In contrast to fights over tangible aims, which end when the aim is achieved, ideological struggles have no natural end. A fight over property stops when one side has firmly secured it, but an idea ceases to be a danger only with the death of the last survivor who holds it, and even then it may crop up again. Hence religious wars tend to be especially bitter, stopping only when both sides are exhausted, with the survivors still clinging to their beliefs.

Furthermore, ideologies are often more important sources of psychological security than possessions, so a challenge to them is a greater threat. Because of their power to conceptualize, humans are forced to recognize the insignificance of their individual lives, which appear to be nothing more than brief, tiny flashes of experience in a universe that does not seem to care. This is intolerable to many people, and to counteract it they create ideologies which give meaning to existence. For them the loss of their ideology, as might follow defeat by a group that maintains an incompatible ideology, may be worse than biological death, so they prefer to die.

Impersonality of Modern Methods of Killing

Traditional international wars were waged by the military of each side, that is by institutions of a nation whose special function it is to defend or promote its interests, and the civilian populations were regarded as more or less innocent bystanders who might get trampled underfoot to be sure, but who were not directly involved. In this, war differed from domestic strife in which the whole population is on the firing line. Today as a nation's military strength has come to include its human and industrial resources as well as the size and efficiency of its army, and methods of killing have become more and more massive and indiscriminate, this distinction between international and domestic strife has largely broken down. A further reason for the disappearance of this distinction is the rise of guerilla warfare as the main defense of weak nations against stronger ones' attempts to subdue them. Since the

success of guerillas is known to depend on the active support of the population, everyone becomes a combatant.

In any case, international conflicts show a strong tendency to escalate;
that is, once nations become engaged in war, whatever inhibitions humans
have against killing their own kind—and like other predators, we probably
have some—are abolished.

One reason for this is that so much of the killing is remote. If ordered to
douse a child with burning napalm, even the most hardened soldier might
hesitate a bit, but he can do it on a massive scale from an airplane. That is,
mass methods of killing reduce the enemy to a statistic.

Television, which puts everyone on the firing line, as it were, reduces this
impersonality, and so may have an inhibiting effect—perhaps partly responsible for unprecedented opposition to the war in Vietnam and a decline in the
sale of war toys.

Obedience

An often overlooked disinhibitor of killing in humans, which plays into the
newer methods of mass killing, is sheer obedience. All social structure depends
on a hierarchy of power, and members of even the most democratic societies
readily obey legitimate authority. If they did not, social living would be impossible. The power of obedience was elegantly if somewhat horrifyingly
demonstrated by a study in which normal American adults were told by the
experimenter that an experiment for which they had volunteered required them
to deliver very painful, possibly lethal, shocks to an inoffensive stranger. About
two-thirds of the subjects carried out these orders. The closer the subjects were
to the victim psychologically, the more likely they were to disobey, but even
when they forcibly had to hold the struggling, screaming victim's hand on the
shock plate, about a third complied. (The victim, of course, was an accom-
plice—he received no shocks.) Perhaps the most disquieting finding was that if
the subject only had to throw a master switch which enabled someone else to
give the shock, over 90 per cent complied (Milgram, 1968). That is, we are all
too ready to hand over responsibility for our acts to a legitimate authority. So
it is not surprising that when a Polaris submarine commander was asked how
he felt about the destructive power under his control, he replied: "I've never
given it any thought. But, if we ever have to hit, we'll hit. And there won't be a
second's hesitation" (Cary, undated).

Dehumanization of the Enemy

The power of obedience probably also explains in part why enemy soldiers
slaughtered each other just as ruthlessly in hand-to-hand combat as with
shells and bombs, but other factors are undoubtedly also involved. Sheer
self-defense is undoubtedly one, but an additional factor is the "dehumanization" of the enemy. Denial of common humanity to the enemy is one effective
way of removing any inhibitions humans may have against self-slaughter.

One source of dehumanization, already mentioned, is that the enemy holds an antipathetical ideology. He is no longer a human but the embodiment of a hated abstraction, such as Communism, Imperialism, Islam, and the like. As a believer in false gods he partakes of the demonic, and so must be destroyed.

Another way in which the enemy is dehumanized is to make him into a beast, on the grounds that he commits atrocities. Whether or not a method of killing is viewed as an atrocity depends on whether it is considered legitimate, not on the amount of pain or suffering it inflicts. Who can say whether it hurts more to be disemboweled or to be roasted to death by napalm? Yet in the Vietnam war, the United States regards the former as an atrocity while the National Liberation Front refers to napalm and crop poisoning as "the most cruel and barbaric means of annihilating people." (Malinowsky, 1966). Each side continually invokes the atrocities of the other to arouse indignation and increase the urge to destroy the enemy.

The Psychology of Commitment

Another important disinhibitor of the human propensity to kill may be related to the psychology of commitment. A very powerful motive for human behavior is the protection of one's image of oneself. Since the military man's self-image rests on his courage and willingness to die for his country, his self-image is apt to become involved in the fight. This creates the danger that if a conflict is sufficiently desperate and prolonged, the primary goal of both antagonists may shift from gaining its initial goals to proving that it is more determined or more courageous than the enemy and the best way to do this is to show that it is willing to keep on fighting longer. History is replete with examples, a recent one being the Battle of Verdun, which, as may be remembered, cost the lives of about a million soldiers in World War I. According to a military historian, it continued long after its military significance had passed away because: "it had somehow achieved a demonic existence of its own, far beyond the control of generals of either nation. Honor had become involved to an extent which made disengagement impossible" (Horne 1966). Perhaps the state of mind of the generals (the soldiers, who did most of the dying, had no choice but to obey orders) was analogous to that of persons who suicide rather than face intolerable loss of self-esteem.

SOCIO-PHYSIOLOGICAL MEASURES TO CONTROL INTERNATIONAL CONFLICT

The best hope of getting group violence under control internationally lies in reducing the sense of frustration of the peoples of the underdeveloped world and strengthening the sense of community which is a prerequisite for the development of effective international peacekeeping institutions.

Reduction of Frustration

It is a truism that a major instigator of violence in humans is anger, and a powerful source of anger is frustration. The important point is that the sense of frustration depends less on the absolute level of deprivation than on the size of the gap between what a person has and what he expects to have, or believes himself to be entitled to.

Recent urban riots cannot be explained by the low standard of living of the rioters—in absolute terms even the poorest dweller in the Negro ghetto is better off than millions of Asians. Yet the denizens of the Calcutta slums quietly starve to death, while the Negroes in our cities explode with rage and frustration. Their bitterness arises from the gap between what they have and what they have been led to expect—and, thanks primarily to the war in Vietnam, the gap has been increasing. The same phenomenon, which has been called "relative deprivation," explains the rising tide of violence in most of the world's newly independent countries. The citizens of these lands expected independence to bring a rise in the standard of living—instead it has often been accompanied by a fall.

The sense of relative deprivation has been aggravated by television and, especially, the transistor radio, which makes the poor of these nations aware, as never before, of what other people have. To overcome their frustration requires rapidly raising their living standard, but as with attempts to improve the lot of lower-class American Negroes, aid to them must be offered in such a way as to preserve the self-respect of the recipients and permit them to share in decisions regarding their welfare. The point was well stated by a Negro Vista worker: "Black people want black control of their lives and activities more than anything else. If they make mistakes, let them be black mistakes— we're tired of white mistakes in our lives."

Failure to observe this principle largely explains why American economic aid has created more resentment than good will and why the Peace Corps has been relatively successful.

The central principle, as described by one of its senior psychological consultants, is that "the Peace Corps must be seen as a program that is compellingly relevant to the recipients' well being, as a form of assistance that can be accepted without compromise of autonomy or loss of personal dignity". And to this end it operates on "revolutionary" principles: "The volunteer lives simply, with the people. . . . He is assigned to work under the supervision of host nationals within existing administrative structures" (Hobbs, 1963).

The Peace Corps' aid projects are defined as collaborative efforts, in which the volunteer learns from as well as teaches the host, and the goal is for the host to take over the job himself. The volunteer works as a peer among peers, not as a superior with inferiors, using the host's language and operating in general on the principle that other people are more receptive to your views if you show that you understand and respect theirs.

Creation of a Sense of World Community

Just as domestic tranquility in the United States depends on restoring the sense of community of all Americans, so world peace requires the creation of a sense of community of all the world's people, transcending their national allegiances. In the past, there was no prospect of achieving this goal, but now for the first time tremendous advances in electronic communication and mass transportation may be bringing it within reach. We have not even begun to use the potentialities of international communication satellites, for example, to increase international understanding.

With potential enemies, these means offer new opportunities for constant communication (without the distorting effects of intermediaries) such as the hot line, as well as direct surveillance by satellites, both of which should yield more accurate information as to the opponents' intentions and capabilities. In itself, this would impose restraints on preparations for hostilities by both sides, and would also help to reduce any distortions arising from the enemy image.

An important inhibiter of violence within communities is public opinion, and one can discern—if one looks sharply—the beginnings of a world opinion, whose increased power can perhaps be seen in the striking differences between Russia's behavior toward Czechoslovakia this year and its attack on Hungary a decade ago. Its motives are probably the same, but the subjugation of Czechoslovakia is not easy to achieve with the whole world, and especially the other Communist parties, looking on.

More importantly, nations whose interest conflict also have many interests in common which are jeopardized by an intensification of hostility, and these common interests loom increasingly large as the world shrinks and opportunities for fruitful international cooperation increase. Today the United States and the U.S.S.R. have a strong, short-term common interest in preventing the spread of nuclear weapons to nonnuclear powers, and a long-range one in promoting international stability, since local flare-ups always carry the potential of forcing a major confrontation. This creates incentives for joint undertakings to raise the economic level of the impoverished nations.

Social psychologists have shown that the most powerful antidote to enmity among groups is cooperation toward a goal that both groups want but neither can achieve alone. At first glance survival would seem to be such a goal since all people desire it and its achievement requires international cooperation. Under some circumstances, however, survival takes a back seat compared with the urge to destroy the enemy. Moreover, the long-term measures required for national survival, such as general disarmament, appear to increase the short-term risks of destruction by an enemy so mobilizing the urge to survive works both ways.

Modern science, however, has created many opportunities for cooperative activities among nations to attain goals that all of them want but none can achieve alone. We know from the experience of one such activity which is already in operation, the International Geophysical Year, that this fosters habits and attitudes of cooperation which gradually become embodied in institutions. Scientists have devised dozens of such projects which can be activated as soon as the world's leaders are willing.

In addition to building habits of cooperation, some of these activities afford constructive outlets for man's aggressive competitiveness, which must be rechanneled if it is not to find periodic release in war. The conquest of outer space and the undersea world meet this need since they demand the fullest exercise of courage, determination, and all the other manly virtues and are highly competitive. At the same time, people everywhere experience these ventures as projects of mankind, not the monopoly of any one nation. Thus both Americans and Russians can sincerely mourn when an astronaut or a cosmonaut is lost, and they can sincerely congratulate each other on a new space triumph, which they could hardly do on the invention of an improved nuclear missile.

A sense of community among peoples of different nations would lay the essential groundwork for the creation on an international scale of the domestic model for controlling aggression, namely the rule of law administered by specialized institutions.

The job of devising these new international institutions with machinery for enforcing their decisions under international law is primarily one for political scientists and jurists. Sociopsychological analyses can, however, make significant contributions. On the one hand, they can identify the psychological factors that intensify the mutual image of the enemy and aggravate armed conflict, as the first step toward effectively combating them. On the other, they can suggest how certain psychological principles might be mobilized to foster the feeling of world community on which the new international institutions must rely for acceptance. Perhaps these contributions, when integrated with the more important ones from other disciplines, may yet be enough to tip the balance in favor of survival.

REFERENCES

Angell, R. C., Dunham, V. S., and Singer, J. D. 1964. Social values and foreign policy attitudes of Soviet and American elites. *J. of Conflict Resolution* **8**: 329–491.

Bronfenbrenner, U. Jan. 5, 1963. Why do Russians plant trees along the road? *Saturday Review*, p. 96.

Cary, W. H., Jr. Undated. *Madmen at Work, The Polaris Story*. Philadelphia: The American Friends Service Committee.

Gallup Poll. June 26, 1966. "Image" of Red Powers. *Santa Barbara News-Press*.

Hobbs, N. 1963. A psychologist in the Peace Corps. *Amer. Psychologist* **18**: 47–55; p. 53.

Horne, A. Feb. 20, 1966. Verdun—the reason why. *New York Times Magazine*, p. 42.

Krauskopf, K. 1961. Report on Russia: geochemistry and politics. *Science* **134**: 539–542.

Malinowsky, R. V. May 2, 1966. *New York Times*, p. 1.

Milgram, S. 1968. Some conditions of obedience and disobedience to authority. *International J. of Psychiatry* **6**: 259–276.

Oskamp, S. 1965. Attitudes toward U.S. and Russian Actions: A Double Standard. *Psychol. Reports* **16**: 43–56.

IV

DISEASE IN POPULATIONS

SECTION IV

DISEASE IN POPULATIONS

Populations of all living organisms except viruses may maintain parasitic populations. Damage to the individual host by the parasite and overreaction by the host are disease. Interacting populations of the host (the herd) and the parasite, produce many patterns of disease manifestation. Both the host population and the parasite population change their numbers, distribution, genetics, and behavior; and their complex interactions enhance these variables. Disease is part of the total ecology, and disease itself evolves. Experimental epidemiology, an attempt to study herd-parasite interactions, although now largely neglected, served as a transition between the study of individual illness and the study of epidemics of man and animals.

The ecological setting of human disease is most obvious where the parasite passes part of its life cycle in other hosts, where vertebrate populations are reservoirs, amplifiers, and indicators of the agent, and where invertebrates are involved in transmission (are vectors). Life-table approaches to parasite, reservoir, amplifier, and vector populations are needed in order to understand mass phenomena of infection.

Length of life and generation time of a species are reflections of its niche, its place in the total ecological complex of populations. It is not known whether there is a basic mechanism for the decease of man at the end of his life span that is separable from death due to disease, and what such a mechanism might be. Starvation, environmental damage, overpopulation, disease, damaging aggression, and other behavioral ills must be controlled if most people are to live and enjoy the full length of life, which surely is the aim of public health workers.

333

CHAPTER 23

Variation in Disease Susceptibility

FREDERIK B. BANG

The interaction of populations of disease organisms with populations of hosts, subject to continuous change with time, inevitably becomes a complex of variables. Under one set of circumstances the host may be favored, under another the parasite. The ideas of population genetics, of ecology, indeed of animal behavior can be applied to these situations.

USE OF GENETICALLY PURE LINES

Many factors influence susceptibility to disease, but the genetic aspects will be emphasized here. Much of the information comes from studies of genetically homozygous animals, particularly mice. Strains of animals are made homozygous by repeated inbreeding. In each generation brothers and sisters are mated. It is believed that about half the genetic dissimilarity is lost. After some 20 generations of such inbreeding, without having purposely selected for any particular characters, the animals are considered to be homozygous.* A homozygous colony nevertheless shows some phenotypic variation because of environmental effects and variable penetrance.

PLANTS AND INSECTS

Strains of tobacco and tomato plants that are resistant to particular viruses can be selected. The fact that this can be done means that there is a genetic basis for such resistance. It is well known that some wheat plants may show resistance to a particular rust, and from them resistant populations can be grown. Then the rust may change its characteristics and become infectious again, and so the relationship evolves. Insects vary considerably in their resistance to various infectious agents. This becomes a problem when insecticide viruses are tried for control of insect numbers.

DEGENERATIVE DISEASE AND DIET IN RATS

Diet can modify genetically based variation in disease susceptibility. Caries of the teeth in rats is apparently similar to the caries that occurs in various other animals. On a coarse diet susceptible strains of rats developed caries in about 22 days, and a resistant strain did so in about 392 days (Hutt, 1958). When the diet was changed to one which caused less trauma to teeth, then both strains

---

* These estimations do not take into account possible homeostatic mechanisms that could maintain genetic variability in a closely inbred population. A character could be controlled by polygenes, with extremes of the character being completely unfit (associated with lethal effects or sterility), and all intermediates could thus continue to occur.

334

of rats had the development of caries delayed for a long time. Thus the environment changed the expression of the genetically determined susceptibility to the disease.

COMPLEMENT-DEFICIENT GUINEA PIGS

A strain of guinea pigs with a genetically based deficiency of complement used to exist. Complement is that heat labile complex of substances in the blood, which acting with antibody causes lysis of various invading organisms. If complement were absent from blood, it is likely that the individual would have decreased capacity to destroy various invading agents. In one study of this strain, 77 percent of the guinea pigs died when infected with *Salmonella cholera suis*, while of normal infected ones only 20 percent died. Perhaps it was lack of complement that caused this laboratory stock to die out, with the result that these studies could not be continued.

VARIATION IN ANTIBODY RESPONSE

Variation in antibody response has been shown in inbred mice with *Ectromelia* infections (Schell, 1960). This is a viral pox disease, frequently called mouse pox, which in many ways mimics human smallpox. The virus enters through the lungs and disseminates throughout the body. It causes peripheral lesions on the feet and the skin and kills the host. One strain of mice, the C57 black, is highly resistant to mouse pox. When the virus is inoculated into the footpad of a C57 mouse the disease takes a day or two longer to develop than it does in susceptible mice, and the mortality rate is low. Multiplication of virus in the footpad is of the same order in the resistant and the susceptible mice, but the resistant mouse develops a rapid antibody response that prevents the usual extensive dissemination of the virus.

However, if the virus is put into the lung of the resistant mouse the immediate reaction to its presence is so great that the mouse dies. If the virus is put into the brain of the resistant mouse, the virus multiplies rapidly there, the antibody response is too late, and the mouse dies. These experiments indicate, then, that resistance and susceptibility depend upon the mode of inoculation and upon other circumstances in this experimental model.

YELLOW FEVER IN MICE

The first clear study of genetics of susceptibility of animals to virus diseases was done by Sabin (1952). He was interested in the susceptibility of mice to neurotropic viruses, particularly the arboviruses of Group B, which include yellow fever and various encephalitides. He showed that the Princeton strain of mice was resistant to yellow fever virus, and that the Webster strain was susceptible. When these two strains were crossbred the F_1 generation was resistant (Figure 23.1). When F_1s were backcrossed with the resistant Princeton strain, all offspring were resistant. When F_1s were backcrossed with the

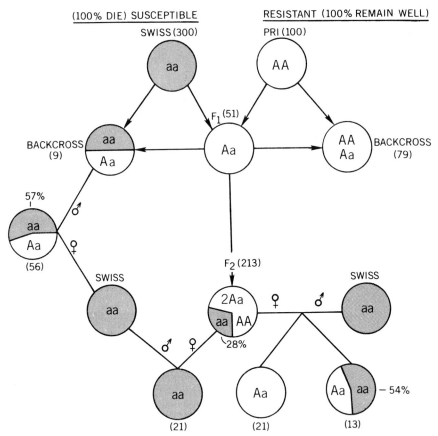

Figure 23.1. Inheritance of resistance to 17D yellow fever virus in PRI and Swiss strains of laboratory mice. Letters within the circles are the genetic interpretation of the data: *A* represents an allele for dominant resistance, and *a* an allele for recessive susceptibility; and proportions of susceptible mice are shown by shading. Figures in parentheses are numbers of mice tested for resistance. (After Sabin, 1952.)

susceptible Webster strain, half the offspring were susceptible and half resistant. The F₂ generation was 72 percent resistant and 28 percent susceptible. These data, supported by further breeding experiments, showed that resistance and susceptibility depended upon alleles at a single locus, with resistance being dominant.

This was an artificial arrangement in the sense that one particular strain of virus was used and that the mice were tested at only one age. Another strain of virus, a French neurotropic strain, killed some of the resistant Princeton mice. The ratios of resistant to susceptible mice changed again when the virus was inoculated intracerebrally instead of intraperitoneally, and again when younger mice were used. So the genetic manifestation of resistance and susceptibility clearly sorted out in simple Mendelian ratios, but only if certain age groups and strains of mice were chosen.

Furthermore, when tests were made with Group A arboviruses such as western and eastern equine encephalitis viruses, or with other viruses such as polio, rabies, and lymphocytic choriomeningitis, genetic differences in these strains of mice were not demonstrable. Thus an important second point was demonstrated—the genetic basis of susceptibility to one agent does not necessarily determine susceptibility to other agents. This study did not reveal the presence of a general characteristic enabling the host to overcome many sorts of infection.

ROUS SARCOMA VIRUS IN CHICKENS

There are some less comprehensive but equally interesting data about susceptibility of chickens to Rous sarcoma virus involving adults, day-old chicks, embryos, and tissue culture cells derived from embryos. Two strains of chickens were used, the Fayoumi and the White Leghorn, and they were not fully homozygous. For instance, the White Leghorns that were used by Crittenden et al. (1964) had been closely inbred since 1939 but were still heterozygous at the loci controlling histocompatibility.

Prince (1958) showed that Fayoumi adults were relatively resistant to the infection and did not develop antibodies, and that White Leghorns were susceptible and did develop antibodies (Figure 23.2). This seemed a contradiction but probably meant that infection failed to be established in the Fayoumi strain and so there was no stimulation of production of antibodies. When virus was added to the chorioallantoic membrane of embryos, lesions were developed by a low proportion of Fayoumis and a high proportion of White Leghorns. When the strains were cross-mated the resulting F_1 embryos were resistant to degrees intermediate to the resistance of the parents (Figure 23.3).

Crittenden et al. (1964) used susceptible and resistant strains of White Leghorns. Cells cultured from resistant embryos were exposed to Rous sarcoma virus, and the results suggested that virus was not released into the culture fluid. Resistance therefore appeared to be at the cellular level, and has since been shown to reside in fibroblast cells. In the last few years the virus has been demonstrated to be a complex of perhaps three different types. Crittenden et al. (1967) extended this original work and showed that there are two different loci for resistance to the Rous complex, one for each of the two main types of virus. The nature of resistance to the third type of virus is not yet elucidated.

SUSCEPTIBILITY TO MOUSE HEPATITIS

Mouse hepatitis has been studied in our laboratory. In susceptible mice aged 3 to 4 weeks this viral infection caused rapid destruction of the liver, and death

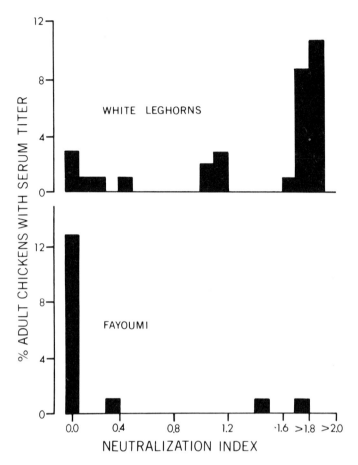

Figure 23.2. Distribution of neutralizing antibody to Rous sarcoma virus in sera of adult White Leghorn and Fayoumi chickens. (After Prince, 1958).

in 2 or 3 days. The genetic basis of resistance followed a pattern which is a mirror image of Sabin's results with neurotropic viruses. Princeton mice were susceptible to hepatitis and susceptibility was dominant and due to a single allele. The C3H strain of mice was resistant, and resistance was recessive. The F_1 mice were all susceptible (Figure 23.4A). The F_2 generation was 75 percent susceptible and 25 percent resistant. A backcross of susceptible F_1 mice with the resistant parent strain of C3H mice at first produced mice that had few susceptibles. However with continued backcrossing a 50 percent susceptibility and resistance was consistently found.

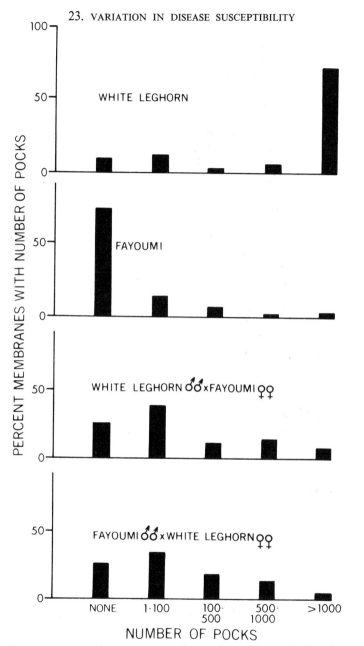

Figure 23.3. Response of chicken embryos to inoculation on the chorioallantoic membrane with Rous sarcoma virus, scored as number of pocks of infection produced. White Leghorn had many pocks, Fayoumi had very few, and hybrid embryos had intermediate numbers. (After Prince, 1958.)

The Role of Macrophages in Susceptibility

There are three basic cell types—macrophage, epithelial, and fibroblast. Cells were washed from the peritoneum of young mice and macrophages were grown in tissue culture. Those from mice resistant to hepatitis virus were resistant in culture, those from susceptible mice were susceptible in culture (Bang and Warwick, 1960; Figure 23.4B). The resistance or susceptibility of these cells matched that of the donor animal. Perhaps the resistance of the whole animal to the disease resided in these cells. Thus the route of dissemination in the host may not be fibroblasts or epithelial cells, but may be somewhere in the reticuloendothelial system.

The Ontogeny of Resistance in Mice

Sabin (1952) showed that genetically based resistance developed in its manifestation as the mouse grew. There was ontogeny of resistance as there is ontogeny of other phenotypic characters. Gallily *et al.* (1967) examined this in the C3H strain of mice which is resistant to mouse hepatitis virus. When mice 1 day old were infected by the virus, even very small amounts of it, they all died in 8 days. This was 5 days longer than it took genetically susceptible older mice to die. When C3H mice were 14 days old large amounts of virus failed to produce disease. Resistance had developed in 2 weeks. Macrophages cultured from 1-day-old mice succumbed to infection within 12 days. When the mice were 8 days old they yielded macrophages that were resistant. It is not known what was involved in this development of resistance within the animal and within the cell.

Change in the Virus

When mouse hepatitis virus was transferred from young resistant mice to others, sometimes the characteristics of the virus changed. Thus during an experiment designed to show that the genetic factor of a host influenced resistance, resistance could be changed not only by the age of the host and the strain of the virus, but the virus itself could alter its characteristics.

GENETIC RESISTANCE IN THE COMPLEX SITUATION OF MOUSE TYPHOID

The Effect of Diet on Disease Resistance in Inbred Mice

This work was begun by Webster (1933) at the Rockefeller Institute. He selected strains of mice that varied in their susceptibility to several diseases. He showed, for instance, that he could select a virus-resistant strain of mice that was not resistant to bacterial infections; and that strains selected for resistance to bacterial infections did not show resistance to viruses. The two phenomena of resistance were separate. He studied the effect of diet on resistance in outbred strains of mice. He inoculated 50,000 *Salmonella enteritidis* bacteria by stomach tube into each of one group of mice, and then 500,000 into others of another group. With one diet he got 66 percent and 37 percent survival in the

A. MOUSE

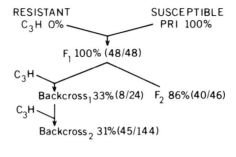

RESISTANT   SUSCEPTIBLE
C_3H 0%   PRI 100%

F_1 100% (48/48)

C_3H

Backcross$_1$ 33% (8/24) F_2 86% (40/46)

C_3H

Backcross$_2$ 31% (45/144)

B. LIVER MACROPHAGE

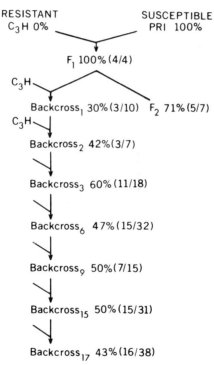

RESISTANT   SUSCEPTIBLE
C_3H 0%   PRI 100%

F_1 100% (4/4)

C_3H

Backcross$_1$ 30% (3/10) F_2 71% (5/7)

C_3H

Backcross$_2$ 42% (3/7)

Backcross$_3$ 60% (11/18)

Backcross$_6$ 47% (15/32)

Backcross$_9$ 50% (7/15)

Backcross$_{15}$ 50% (15/31)

Backcross$_{17}$ 43% (16/38)

Figure 23.4. Laboratory mice (A), and liver macrophages in tissue culture derived from those mice (B), were tested for susceptibility to mouse hepatitis virus. The C3H strain of mice was resistant to the infection, and the PRI strain was susceptible, and the various genetic crosses shown here demonstrated that susceptibility was dominant and conferred by one allele. The numbers of susceptibles are shown over the numbers tested in parentheses, and susceptibility is given also as percentages of those tested. Macrophages were tested from twenty generations of backcrosses with the resistant parental C3H group. Only a sample of these results is shown.

two treatments, and with another diet 4 percent and 17 percent survival. This was repeated over a long period. Thus it was shown that under the circumstances of these experiments, resistance and susceptibility in part depended on diet.

Schneider (1948), a former student of Webster's, continued the investigation by using inbred mice and pure strains of bacteria. He selected bacteria that were more virulent or less virulent in mice and removed genetic variability in the bacteria by culturing from single clones. He found he could no longer demonstrate that different diets had different effects on susceptibility of the mice to the bacterial infection.

The Need for Bacterial Populations with Genetic Variety in Studies of Dietary Effects

Schneider then mixed together pure cultures of more virulent and less virulent organisms, and inoculated them into mice that were not inbred. He showed consistently that under these conditions diet did affect resistance. Table 23.1 gives his results and Figure 23.5 summarizes what was shown in these experiments.

The Bacteria within the Host

Schneider put genetic markers on avirulent and on virulent organisms in *Salmonella* populations. The avirulent organism entered the host and multi-

TABLE 23.1

The Effect of Diet on Resistance of Outbred Laboratory Mice to Infection with *Salmonella typhimurium*[a, b]

Percent survivors on diet "#100"	Percent survivors on diet "#191"	Difference in percent survivorship
75	10	65
75	5	70
75	10	65
75	10	65
45	5	40
40	0	40
75	20	55
60	0	60
60	20	40
75	25	50
	Average	55

[a] Schneider (1948).

[b] Twenty outbred mice were on each diet, in each of 10 tests. An infection with avirulent *Salmonella* was followed 24 hours later by a dose of virulent *Salmonella*. Survival was measured at 30 days.

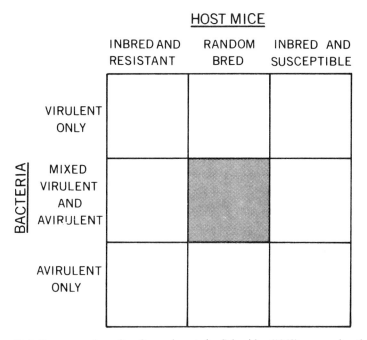

Figure 23.5. Summary of results of experiments by Schneider (1948) concerning the effect of diet on susceptibility of laboratory mice to *Salmonella typhimurium* infection. The differential effect of diet was demonstrable only when conditions represented by the shaded area occurred.

plied in such a way that it partially blocked the development of the virulent organism. So there was competition within the host between populations of the virulent and the avirulent organisms. When this competition occurred within heterozygous hosts it could be influenced by dietary changes. If selection narrowed variation to such an extent that there was little phenotypic variation in the host, then the nutritional factor being studied could not be demonstrated. This is a lesson in terms of the value of experimental work, and of the difficulty of applying it directly from the laboratory into larger, more heterozygous populations. Schneider (1967) has suggested that other nutritional factors, not known to be vitamins, may play important roles in disease resistance.

CONCLUSIONS

It has been shown in this chapter that various genetic factors are known to influence susceptibility and resistance to diseases. This is to be expected if the adaptation of an organism to its host is seen as the ecological situation of the adaptation of an organism to its niche. The niche is not exactly the same in every case. The degree of adaptation of a parasite to a host varies tremendously. When a particular case is studied, it is sometimes easy to show that there is an inherited variation on the part of the host. Despite the fact that clear-cut inheritance of resistance of the host may be shown for one infectious agent, such as a virus, this may have no bearing on resistance of that host to another agent, even another virus. Perhaps the macrophage system plays a role in variation in resistance to those agents which disseminate throughout the host, but perhaps it has no role at all for other agents.

It has been indicated that there is variation in the capacity of the infectious organism to invade its host. This, and variation in virulence, will be discussed further in the next chapter.

Resistance to infection can be studied with controlled genetic constitution of hosts and parasites. In such a case it is possible to lose what may have been the intended subject, the variation caused by changes of the environment.

REFERENCES

Bang, F. B., and Warwick, A. 1960. Mouse macrophages as host cells for the mouse hepatitis virus, and the genetic basis of their susceptibility. *Proc. Natl. Acad. Sci. U.S.* **46**: 1065–1075.

Crittenden, L. B., Okazaki, W., and Reamer, R. H. 1964. Genetic control of response to Rous sarcoma and strain RPL12 viruses in the cells, embryos and chickens of two inbred lines. *Natl. Cancer Inst. Monograph* **17**: 161–177.

Crittenden, L. B., Stone, H. A., Reamer, R. H., and Okazaki, W. 1967. Two loci controlling genetic cellular resistance to avian leukosis-sarcoma viruses. *Am. J. Virology* **1**: 898–904.

Gallily, R., Bang, F. B., and Warwick, A. 1967. Ontogeny of macrophage resistance to mouse hepatitis *in vivo* and *in vitro*. *J. Exptl. Med.* **125**: 537–548.

Hutt, F. B. 1958. *Genetic Resistance to Disease in Domestic Animals.* Comstock Publishing Assoc., Cornell Univ. Press, Ithaca, New York.

Kantoch, M., Warwick, A., and Bang, F. B. 1963. The cellular nature of genetic susceptibility to a virus. *J. Exptl. Med.* **117**: 781–798.

Prince, A. M. 1958. Quantitative studies on Rous sarcoma virus. Mechanism of resistance to Rous virus in chick embryos. *J. Natl. Cancer Inst.* **20**: 843–850.

Sabin, A. B. 1952. Nature of inherited resistance to viruses affecting the nervous system. *Proc. Natl. Acad. Sci. U.S.* **38**: 540–546.

Schell, K. 1960. Studies on the innate resistance of mice to infection with mouse-pox, II. *Australian J. Exptl. Biol. Med. Sci.* **38**: 289–300.

Schneider, H. A. 1948. Nutrition of the host and natural resistance to infection. *J. Exptl. Med.* **87**: 103–118.

Schneider, H. A. 1967. Ecological ectocrines in experimental epidemiology. *Science* **158**: 597–603.

Webster, T. 1933. Inherited and acquired factors in resistance to infection. *J. Exptl. Med.* **57**: 793–843.

CHAPTER 24

The Evolution of Disease

FREDERIK B. BANG

A population consists of a variety of individuals or genetic types. Environmental factors acting on a population tend to select certain individuals that are the contributors of the genes of the next generation; thus the gene pool is changed, and the changes are called evolution. These principles apply also to the interaction of host and parasite populations. The latter may experience large increases and declines in number. Group behavior reflects many of the components of the environment, including crowding. From recognition of these generalities are derived the four main points of this chapter. The first is that aberrancy in host selection, that is, spread of a parasite to a new host species, gives rise to a new imbalance that is likely to cause disease of the new host. Second, virulence of disease is a spectrum of characteristics. Third, evolution of disease may tend toward development of symbiosis.* Fourth, introduction of a second parasite may so alter the situation that new imbalance is produced, with the imbalance dependent upon the presence of both agents.

As to the nature of the evidence, it should perhaps be remembered that Darwin's thesis of 100 years ago on the origin of species by means of natural selection was based largely on close comparisons of the forms that were present at that time, not on observations of evolution in action.

Thus reconstruction of events from existing forms may indeed be valid. With the exception of myxomatosis in rabbits, there are no data on the actual evolution of disease as it occurs. It might be said that the history of scarlet fever or of measles, which have become much less virulent in the last hundred years, or the spectacular history of syphilis from the virulent disease of the 1500's through its more and more mild forms of subsequent centuries, are evidence of the evolution of disease. But the accounts are from historical documents and are difficult to evaluate. In measles, for instance, it is not known how much of the virulence was due to poor medical care, to dietary differences of the hosts, and to associated organisms.

ABERRANCY, THE SPREAD OF PARASITE TO NEW HOSTS, GIVES RISE TO NEW DISEASE

In the balanced situation like the symbiosis of fungi and algae in the lichen,

* Symbiosis is used by the ecologist to mean living together, and includes parasitism, commensalism, and mutualism. It is used here in the usual sense of the parasitologist, only for such living together as is mutually beneficial, which is the mutualism of the ecologist. The uses of the term are reviewed by Allee *et al.* (1949, pp. 243–245).

345

or the rather neutral situation of the pinworm in man, the association is relatively harmless for the host. The aberrant parasite is that which occurs in a new host, a new ecological niche. According to Theobald Smith (1934), who was a pretty straight-laced gentleman himself but must have had a secret admiration for these escapades of nature, aberrancy is the adventurous element in the life of the parasite which leads either to death or to new conquests. Establishment in the new host paves the way for the formation of new races of the parasite and new diseases. One example of this might be schistosomiasis of man. There are three species of this parasite, but the discussion will be restricted to *Schistosoma japonicum*, which is a relatively common parasite in the Orient. In most of the areas of the Orient where it is present, such as the Yangtse valley of China and some of the islands of the Philippines, it is a disease that readily infects man. However in Taiwan humans are in contact with the cercariae which emerge from infected snails there, but man does not develop a full well-developed infection and the reason for this is not understood. Hsü (1960) did some very careful studies on this infection, including experimental infections of monkeys, and showed that the strain on Taiwan is different in its biological capacity to develop its full cycle of development in man. Thus on Taiwan transmission does not occur from man to snail and he rarely if ever becomes infected with a recognizable clinical form of the disease. Surely this parasite must have had the same evolutionary origin as the schistosome of the Philippines and the mainland of China. A separation in schistosome distribution then presumably occurred. The separation continued long enough for divergence of biological properties to have become established. In one locality the worm survived by its capacity to infect animals, including some domestic ones such as the pig, but not man, while in other localities it infects animals and man. Possibly the capacity for a full development in man was lost because the method of feces disposal did not allow for infection of the snail from human feces.

MECHANISM OF ADAPTATION TO A NEW HOST—EXAMPLE, NIPPOSTRONGYLUS IN RATS AND HAMSTERS

Aberrancy in parasites has been referred to here as an adventurous element. This poetic description does not lead to analysis of the forces involved in the transmission of a parasite from one host to another. For the student, helminthic infections have one large advantage over various other infections such as protozoal, bacterial, or viral infections. The dose of the infecting form, the larva, plays a tremendously significant role, since the worm does not produce more individuals of the same form within the host. Thus a dose of 100 larvae of the helminth, *Nippostrongylus*, cannot produce more than 100 adult worms within the intestine of the rat. This facilitates analysis of the transmission of a population of parasites from one host to another.

Haley (1958) took a rat-adapted strain of *Nippostrongylus muris*, a common parasite of the rat in nature. He compared its capacity to infect the laboratory rat with its capacity to infect hamsters, which are not naturally infected. The life cycle is similar to that of the hookworm. The larvae enter through the skin and migrate into the circulatory system and through the lungs. They pass through the capillaries of the lungs into the alveolar spaces, climb up the trachea and are swallowed, and become localized in the intestine where they become sexually mature adults. This takes a few days from the time of the original infection. Haley showed that with a dose of 1,200 infectious larvae put intracutaneously in the rat, it was possible to recover about 54 percent of them from the intestinal tract about a week later. In contrast to this, with the same dose of larvae to hamsters only 4.3 percent were recovered, 8 percent in males, 0.6 percent in females.

Table 24.1 shows that time played a role in determining the numbers recovered from the two infected host species. Both hosts were susceptible to

TABLE 24.1

Recovery of *Nippostrongylus muris* (rat strain) from Laboratory Rats and Laboratory Hamsters After Intracutaneous Injection of Larvae[a]

Host	Number injected intracutaneously	Site of recovery of worms in host	At day	Percent recovered
Rat	1,020	Lungs (larvae recovered)	2	23.1
			6	0.34
			10	0.05
	430	Intestine (adults recovered)	6	67.4
			10	51.8
			13	26.1
Hamster	1,020	Lungs (larvae recovered)	2	56.2
			6	41.2
			10	48.9
	1,640	Intestine (adults recovered)	6	9.2
			10	8.5
			14	4.7
			18	4.6

[a] Haley (1958).

the initiation of infection and the larvae reached the lungs of both, but in the hamster most of the larvae stayed in the lungs. In the rat the larvae passed on into the intestine. The worm was not adapted to its new host to the extent that development could be regularly completed. A small percentage of the parasites did go all the way to the intestinal tract in the insusceptible hamster. If larvae were cultivated from the eggs of these successful worms in hamsters, and were inoculated back into hamsters a higher percentage of migration to the intestine was observed. After a series of 24 passages, Haley had a hamster strain that had a 33 percent recovery rate in the intestine, that is, a third of the injected larvae were able to get all the way through. In one of the two series of adaptation to the hamster the strain lost its capacity to grow in the rat. In the other case the strain maintained its capacity to produce infection in the rat. This showed, perhaps, that the acquisition of virulence for one host did not necessarily depend upon the loss of virulence for the original host. It may be assumed from this that one of the major factors in transfer from one host to another is selection of the potentially successful aberrant parasite. But if the parasite is successful in establishing itself in the second host, it may produce the imbalance which is disease.

MYXOVIRUSES AS RESPIRATORY PARASITES OF MAN AND ANIMALS

The evolutionary relationships of some of the viruses resident in the upper respiratory tract of man and various animals will now be considered. The conclusions cannot be historically documented, but depend on indirect evidence

Influenza, mumps, Newcastle disease, and a variety of others called para-influenza viruses belong to the group myxoviruses (Figure 24.1). The relationships are based upon antigenic similarities, a variety of morphological characteristics, and behavior of the viruses in chick embryos. Under influenza A is a series of animal strains producing disease in pigs, swine, horses, ducks,

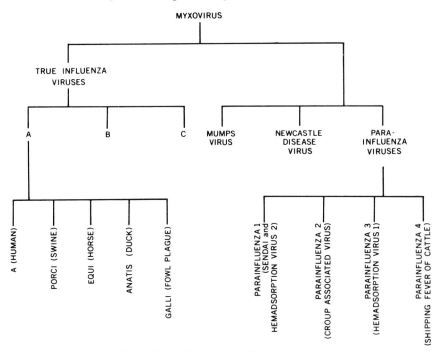

Figure 24.1. Relationships of the myxoviruses.

and finally there is fowl plague virus. All of these have been shown to be closely related to human influenza A. Indeed they are more closely related to influenza A than influenza A is to influenza B or C. A, B, and C cause influenza in man. It is not known whether any of the animal influenzas are transmitted to man. The separation of the animal influenzas from the human influenza A by antigenic means is small, but reproducible. The degree of relationship must indicate something in terms of evolutionary relationships. This complex of agents must have arisen from a single stock in the past. The question really is not whether they did arise from one strain, but when they arose. Shope (1944) believed that influenza A and swine influenza became separated in recent times.

The group of myxoviruses called parainfluenzas are separated into parainfluenza 1, 2, 3, and 4. Parainfluenza 1 is related to a virus called Sendai virus. Parainfluenza 2, which was isolated from cases of croup in children, is related to an infectious agent of monkeys, SV5 virus. When a myxovirus was isolated from shipping fever in cattle it was at first thought that this was identical with parainfluenza 3, an infection of man. The relationship was so close that some cows which were infected with shipping fever developed more of an antibody response to the human infection than did humans themselves. When antibodies to the two agents were made in rabbits, it was not possible to distinguish them. Only by careful antigenic comparisons with sera from actively infected animals was it possible to distinguish them. So again it must be assumed that these two had a common origin.

It might be argued that respiratory disease of man and animals is one continuum of slightly differing agents which are going back and forth among various susceptible hosts. The need for antigenic variation to keep epidemic respiratory infection going in man may be also related to a continued antigenic variation in different hosts. A human host population may become free of the infectious agent at a time when the agent is infecting an animal species. Perhaps the agent changes its antigenic properties in response to the changed host. The human population in time loses individuals with immunity to the original agent and becomes again a very susceptible herd to the slightly changed animal disease. Antigenic change of the human virus might also occur by genetic recombination in a doubly infected host (Chapter 10).

VIRULENCE IS AN ECOLOGICAL TERM DESCRIBING A DEGREE OF ADAPTATION TO THE HOST

That virulence is an ecological term describing a degree of adaptation to the host is rather obvious. Virulence in terms of the parasite cannot be separated from susceptibility in terms of the host. This point will be examined for the case of Newcastle disease.

Newcastle Disease Virus

Table 24.2 compares the relative virulence of three strains of Newcastle disease virus in various hosts. One strain kills chickens readily, another is

TABLE 24.2

Virulence of Three Strains of Newcastle Disease Virus (NDV) for Different Hosts[a]

| Host | NDV Strain[b] | | |
	Velogenic (virulent)	Mesogenic	Lentogenic (vaccine)
Cat	+	0	0
Mouse	+ +	+	0
Adult chicken	+ + +	+	+
Chick	+ + + +	+ + +	+
Chick embryo	+ + + +	+ + + +	+ +
Tissue culture	+ + + +	+ + + +	+ +

[a] Bang and Luttrell (1961).

[b] The terms velogenic, mesogenic, and lentogenic were introduced by Hanson and Brandly (1955) to describe three degrees of virulence of NDV.

used for vaccination of chicks. The hosts range from the cat, which is least susceptible, to tissue culture cells, which are most susceptible. There is gradual acquisition of virulence on the part of the parasite as the host becomes more and more susceptible. All three strains of virus show the same change of relationship so that if a diagonal line is drawn across the chart, the more virulent are separated from the less virulent. The relationship is always an interaction of host and parasite.

At the cellular level, in this case, the virulent strain multiplies readily in the chick embryo. There is logarithmic increase of numbers during the first 24 hours of infection, then retardation of increase based presumably on depletion of cells. Surprisingly, the rate of growth of the avirulent strain in the chick allantoic sac is just as great as the growth of the virulent strain. Yet the avirulent form kills the chick embryo only very slowly, while the virulent one kills within 2–3 days. What then is the basis of this particular virulence? With the electron microscope it can be seen that the virulent form destroys the cells and, in contrast to this, the avirulent form produces only major surface alterations (Figure 24.2). A series of projections develops from the surface of the cell and produces masses of infectious material which are liberated into the fluid without apparent destruction of the cell. Indeed the avirulent form may be able to produce more virus. If the amount of virus in other tissues of the chick embryo is measured, again there are differences. The virulent form is soon found in large amounts in the blood. The avirulent form is never found in more than very small amounts in the blood.

In older chick embryos the lung tissue becomes susceptible to the avirulent strain. At this stage of maturation, the chick embryo may die from infection with avirulent strain. In contrast, all tissues of the embryo are destroyed by the virulent strain.

The virulent form is not able to multiply more rapidly or to a greater degree

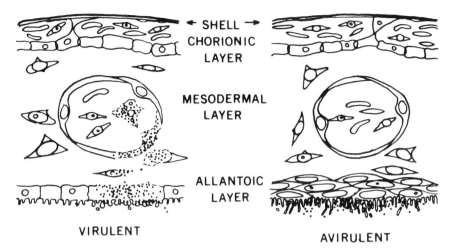

Figure 24.2. Effect of virulent and mild strains of Newcastle disease virus grown in the chorioallantoic sac of the chick embryo. (Redrawn from Bang and Luttrell, 1961.)

than the avirulent form, but is able to progress from cell to cell, destroying cells as it goes. The avirulent form does not penetrate into the tissues but is extruded into the surface area, producing there an accumulation of virus particles. It does not produce extensive disease.

Since in any series of steps of pathogenesis there are various obstacles to be overcome, virulence probably depends on several characters concerned with those obstacles that have a genetic basis. Virulence is then a series of gene-determined characteristics which illustrate the adaptation of that particular parasite to its particular ecological niche. It should be possible to identify different steps in this progression from the avirulent to the virulent. Indeed, there is a series of strains of Newcastle virus varying between the two extremes. The B strain shows capacity to multiply and spread through the adult chicken, but not to kill the chicken readily. It does not have the capacity to enter into and grow within the central nervous system.

Bacterial Infections

In bacterial infections, the work of Burrows (1962) on the plague bacillus, *Pasteurella pestis*, has been one of the most extensive attempts to analyze virulence in detail. He selected two organisms, very alike in their growth on bacterial media, but one was derived from the other and differed in its capacity to produce disease in mice. The MP-6 strain was labeled virulent because it was able to kill mice with an inoculum of from 5 to 100 organisms. The PS strain was labeled avirulent because it took 150,000,000 of the organisms, injected intraperitoneally, to kill. The two strains were analyzed in great detail to see if there were any way in which they could be differentiated on the basis of growth on various media. The need for supplemental factors was examined, their dependency upon hypoxanthine, and the rate at which they mutated from one form to another, and no difference was found. Then the two forms were injected into mice, and they differed in the rate at which they were phagocytized. They both started early to be phagocytized with the same rate, but the virulent form was somehow able to protect itself whereas the nonvirulent one was rapidly phagocytized and subsequently destroyed. Apparently there was a change by the virulent form, and many attempts were made, with some success, to get it to produce a protective substance in culture. When the medium was very carefully controlled, the initial steps of this conversion could be followed so that the organism was able to grow in the phagocyte without being destroyed.

Bacteria populations may change their antigenic characteristics in the host in response to antibodies of the host. A well-known antigenic change in cholera organisms has been shown by Miller and Sack (1968; Chapter 4) to occur in germ-free mice infected with one strain of *Vibrio cholerae*. The antigenic change is delayed if the antibody response of the mice is suppressed, and is accelerated if the host is antigenically stimulated.

THE EVOLUTION OF A DISEASE TENDS TOWARD DEVELOPMENT OF A SYMBIOTIC RELATIONSHIP

This philosophical statement is attributed to Theobald Smith. It is to a large extent true, but there are exceptions. The best evidence supporting the generalization that disease tends to evolve toward a harmless relationship is the history of myxomatosis in rabbits, which is given in detail in Chapter 25. A symbiotic relationship is an association that benefits both partners. What evidence is there of help to the host by the parasite?

Intestinal Bacteria and Sterile Animals

Escherichia is a common intestinal inhabitant. There are strains of *Escherichia coli* which are capable of producing disease, such as enteric epidemics in nurseries for the newborn or aberrant disease in a particular area of the genitourinary tract, as in pyelitis. It is clear that the balance is not such that the organism is no longer able to cause disease.

Do intestinal microorganisms give anything to the host? There is a continual production of various vitamins by bacteria of the intestinal tract, and it is difficult to define how much of this may be occurring in man. Under normal circumstances man may satisfy his requirement for vitamin K by the use of products of intestinal microorganisms. Adults may get some of their vitamin B_{12} from this source, but infants do not.

That interaction between the parasite and the host is of a fundamental nature has been illustrated best by studies of bacterially sterile animals. These animals have been obtained by Caesarean section or by hatching eggs under sterile conditions with all food sterilized. The animals lack the general development of the lymphatic tissue that is present in normally raised animals. Dubos and Schaedler (1962) took one step further in the analysis, in that they did not raise the animals in complete sterility but separated them from their normal flora. They gave them new intestinal bacteria to take the place of the normal ones and showed that these mice changed in certain characteristics. They no longer reacted to the injection of bacterial endotoxin in the same way as the controls. They reacted to a lesser degree to certain pathogenic agents and they grew more rapidly—they had acquired new cultural characteristics.

Sterile animals do not have the antibodies that are present in normal animals. Normal chickens have antibodies to the B-substance of human blood, but germ-free chickens lack these. Something in the intestinal flora is antigenically similar to the human B-substance. Various "natural" antibodies prove to be antibodies of the usual kind developed in response to a specific antigenic stimulus, but the stimulus has been from an identical or similar antigen which happens to be present in a completely different parasite. Thus a chicken contains antibodies to *Leishmania*, an infecting organism it never acquires, and develops these antibodies shortly after it has hatched, and it passes the antibodies through the egg. But a chicken raised under sterile conditions never develops such antibodies.

It is clear from this discussion, and from Chapter 4, that there is an intimate continuing response between the intestinal flora and the host. How much of this is favorable to either is difficult to define.

COMPLEX INFECTIONS—SWINE INFLUENZA

The introduction of a second parasite may so alter the situation that a new host-parasite balance is produced, and new disease. This is illustrated by the complex infection, swine influenza. It is complex not only in needing populations of two parasitic organisms to cause the disease, but also must have had a complex evolutionary history.

Hemophilus and Influenza Virus (Shope, 1944)

Shope and Lewis described swine influenza virus in 1933 and showed that the disease was caused by the virus working with the bacterium *Hemophilus*. *Hemophilus influenza suis* is very closely related to the human *H. influenza*. If the virus by itself is introduced intranasally into the pig, a mild disease results. This seems not to affect the host symptomatically or pathologically to a great extent, but if *Hemophilus* is also introduced a serious disease occurs. The animal is prostrated and at autopsy extensive macroscopic and microscopic changes in the lungs are seen. There is a greater multiplication of the virus in the lung. The *Hemophilus* has in some way stimulated this multiplication. Some evidence that the increased multiplication may be caused by substances given off by the bacterium was obtained in analysis of the combined infection in the chick embryo (Bang, 1943).

Shope next attempted to immunize pigs against the infection in order to identify the important component in the immunization. The isolation units of the Rockefeller Institute where this work was done were very effective. There was a central corridor with windows to the isolation units. It was necessary for a person entering the unit to go through a door and to put on coveralls, a mask, and gloves. The gloves were dipped into phenol solution before the isolation unit was entered. An individual clothed in these garments supplied food in a container. After the feeding the container was removed from another door and immediately disposed of. Under these conditions no cross-infection took place among any of the units.

It was under these conditions that Shope then proceeded to immunize his pigs against *Hemophilus*. At the time when he did the immunization experiments the colony of pigs at the Rockefeller Institute had been exhausted and Shope had to purchase some from outside. A neighboring farmer had pigs with no antibodies to swine influenza so they were purchased and put in the isolation units, kept there awhile, and bled again. They had no infection and so were injected with *Hemophilus*. Some of these pigs then developed a complex disease and Shope isolated the virus of swine influenza. These pigs had had no antibodies, they were in isolation, and so could not acquire the virus from outside. There was no infection going on at the same time anywhere nearby. How did they get infected? This phenomenon occurred repeatedly. Shope was driven to seriously raise the question of whether spontaneous generation was occurring in the unit.

Swine Lungworms and Earthworms

The final story was even more peculiar, but more believable. The pigs which had been bought outside had lungworms. The lungworm gives off its eggs from the lungs of the pig into the sputum. The eggs in the sputum are swallowed and left with the feces on the ground of the farm. Earthworms swallow this material and the eggs. The larvae of the lungworm develop partially in the earthworm, then remain dormant for long periods thereafter. Several years later another pig grubs around in the dirt, finds an earthworm and swallows it. This releases the lungworm, which migrates through the tissues and settles in the lungs. It is a fantastic story but not unusual for parasitic worms. What is even more fantastic is that lungworms can carry the swine influenza virus. Shope took lungworms and eggs from pigs which had recently had influenza, and infected earthworms with the eggs. He kept the earthworms for several years in barrels in the Institute and at his home. Then he fed the earthworms with their lungworms to pigs in isolation. He showed that pigs release the virus from the worm into the pulmonary tissues under certain antigenic stimuli, or change of weather, or introduction of other helminthic antigen. The virus infection thus produced can spread to other pigs. Presumably the disease becomes severe only when *Hemophilus* is also present. There is no evidence that the *Hemophilus* goes through the lungworm.

So here is a peculiar association of virus and bacterium that produces a severe disease, but the virus is resident for a long period of dormancy in a third host. The degree to which various complexities of interaction between host and parasite and varieties of associated parasites may work is only beginning to be realized.

CONCLUSIONS

Infectious disease, which is the resultant of interaction of host and parasite, is subject to the evolutionary forces acting on both components. New disease may arise when a parasite of one host reaches another host (aberrancy) and is successful in establishing itself in that new host species. Selection of more successful helminths was illustrated by a roundworm, *Nippostrongylus*, of rats, which by selection became adapted to the hamster. A comparison of the various myxoviruses of the respiratory tract of man and other animals indicates that these have been exchanged among those hosts in the past. In ecological terms virulence is recognized as a complex series of characteristics of the parasite which facilitate establishment in a host and often disturbance of the environment (host). This was illustrated with a virus, Newcastle disease virus of chickens, and *Pasteurella pestis*, the plague bacillus.

Infection of man and animals with the bacterium *Escherichia coli* in the intestine was presented as an example of an infection which is close to a symbiotic state. Finally the complexity of ecological relationships which may produce disease was illustrated with swine influenza, which is caused by a bacterium and virus acting in concert, with the virus transmitted through the earthworm by parasitic swine lungworms.

REFERENCES

Allee, W. C., Park, O., Emerson, A. E., Park, T., and Schmidt, K. P. 1949. *Principles of Animal Ecology.* Saunders, Philadelphia, Penn.

Andrewes, C. H., and Worthington, G. 1959. Some new or little-known respiratory viruses. *Bull. World Health Organ.* **20**: 435–443.

Bang, F. B. 1943. Synergistic action of *Hemophilus influenza suis* and the swine influenza virus on the chick embryo. *J. Exptl. Med.* **78**: 9–16.

Bang, F. B., and Luttrell, C. N. 1961. Factors in the pathogenesis of virus diseases. *Advan. Virus Res.* **8**: 199–244.

Burrows, T. W. 1962. Genetics of virulence in bacteria. *Brit. Med. Bull.* **18**: 69.

Dubos, R. J., and Schaedler, R. W. 1962. The effect of diet on the fecal bacterial flora of mice, and on their resistance to infection. *J. Exptl. Med.* **115**: 1161–1172.

Haley, A. J. 1958. Host specificity of the rat nematode *Nippostrongylus muris*. *Am. J. Hyg.* **67**: 333–349.

Hanson, R. P., and Brandly, C. A. 1955. Identification of vaccine strains of Newcastle disease virus. *Science* **122**: 156–157.

Hsü, H. F., and Hsü S. Y. L. 1960. The infectivity of four geographic strains of *Schistosoma japonicum* in the rhesus monkey. *J. Parasitol.* **46**: 228.

Miller, E., and Sack, R. B. 1968. Serotypic changes of *Vibrio cholerae* in germ-free mice. *Advan. in Germ-Free Research and Gnotobiology.* Chemical Rubber Co., Cleveland, Ohio.

Shope, R. E. 1944. Old, intermediate and contemporary contributions to our knowledge of pandemic influenza. *Medicine* **23**: 145.

Smith, T. 1934. *Parasitism and Disease.* Princeton Univ. Press, Princeton, New Jersey.

CHAPTER 25

Experimental Epidemiology

FREDERIK B. BANG

Why in this time of abundance of funds and worship of the experimental approach, is there little research with experimental epidemics? What is the cause of an epidemic?

Aldous Huxley wrote in *The Devils of Loudun*: "To think about events realistically in terms of multiple causation is hard and emotionally unrewarding. How much easier, how much more agreeable to trace each effect to a single and, if possible, a personal cause! To the illusion of understanding will be joined, in this case, the pleasure of hero worship, if the circumstances are favorable, and the equal or even greater pleasure if they should be unfavorable, of persecuting a scapegoat."

We are searching for at least a few scapegoats as the causes of epidemics. How are these causes found?

MATHEMATICAL MODELS

There are discussions in Chapters 1, 5, and 27 of attempts to relate numbers of parasites to numbers of hosts through time. Mathematicians such as Lotka (1924) and Bailey (1957) assigned symbols to various components of the interactions and made certain assumptions concerning the time involved and the number of parasites given off per host. They constructed formulae that describe oscillations about a point. By altering the time scale, the oscillations can be changed into a rise and fall of numbers, so that at first the population of the host rises and later the parasite. The host population declines because of the effect of the parasite, and later the parasite declines because of the decline of the host. This same kind of fluctuation but in a different order of magnitude could also represent sick animals and bacteria or viruses. There is thus a recurrent epidemic series, or a continuous variation in the number of the sick animals. Laboratory examples of such curves are found in the studies of Nicholson and Utida on cultures of host and parasite (Chapter 6), but these situations are perhaps too simple.

Certain aspects of more natural situations may be examined. One of the first to try to understand natural epidemics of this sort was a British army officer, Ross. He was not satisfied merely with demonstrating how malaria was transmitted, but wanted to understand the numerical situation involved in the repeated flaring up and dying down of the infection (Ross, 1916).

For a number of years at the Johns Hopkins School of Hygiene and Public Health, Reed and Frost taught certain mathematical aspects of epidemics. Abbey (1952) examined their theory and tested it for the fit of the theoretical epidemic curve to measles and a few other recurrent epidemic diseases. In this work a different sort of assumption was made. The number of infecting organisms was not used, but something called infectiousness was measured. The mathematical model was:

$$C_t = S_t (1 - qC_t)$$

where t = unit of time

        C = number of cases

        q = probability of not having contact with a case

  $1-q$ = probability of having contact with at least one case

        S = number of susceptibles

The role of chance, and the repeated addition of new nonimmunes to the system, caused a variety of fluctuations, and the ups and downs of natural epidemics could be explained in these terms.

HERD STRUCTURE

Topley and Wilson (1955) discussed herd structure in relation to herd immunity. The structure includes not only the host species and its spatial arrangement, but the presence and distribution of alternate hosts and perhaps of arthropod vectors. It includes, as well, all those environmental factors that favor or inhibit the spread of infection from host to host. This herd structure, apart from the susceptibility or resistance of the individual hosts, may play a decisive part in the immunity of the herd.

There are three possible outcomes of an experimental epidemic: the herd is wiped out; the agent dies out and the herd recovers; there is continued existence of both.

MICROEPIDEMIOLOGY IN TISSUE CULTURES

Analysis of experimental epidemics can start with the simplest, the microepidemiology of viruses in tissue culture. The balance between host cell and number of parasitic viruses may be studied for a single virus species and genetically identical host cells. Black and Melnick (1955) described the microepidemiology of herpes simplex and poliomyelitis viruses. Herpes virus spreads predominantly from cell to cell by direct contact, and polio virus spreads by being liberated from the cell and disseminated through the culture.

To justify the term microepidemiology, a more complex situation is needed and this is found in tissue culture infections of equine encephalomyelitis. These are viruses transmitted by mosquitoes from one vertebrate host to another. The host species are a wide variety of vertebrates—snakes, birds, and mammals including man—and the virus is one of the most virulent. It multiplies rapidly, destroying cells, thus killing chicken embryos and mice in 12 to 18 hours. However, in tissue cultures infected with this virus, all three of the possible outcomes of infection of a population can be demonstrated. The infection may be active but brief and relatively unimportant to the host colony. The infection may be brief and completely destructive of the host. Third, infection in a continuous cell line may be chronic, reproducing itself, sometimes in waves of destruction which are incomplete. Regeneration of cells follows, so that the cells are able to grow to a relatively high population again, only to be destroyed without the addition of new virus, but with an increase in numbers of the virus particles present. The chronic infection sometimes ends in death and sometimes there is total recovery. But in a number of cases these chronic infections have been kept going for several years without addition of antibody and without any apparent change in the genetics of the host. There appears to be no change in the susceptibility of the host cells. The explanation for this situation probably lies in a series of successive unbalanced relationships between the numbers of susceptible cells and the number of virus particles (Bang and Gey, 1952).

Interferon is a substance given off by cells, particularly infected cells, which passes to other cells in the culture and protects them against infection. Interferon may be involved in such chronic infections. The analysis of these chronic infections of cell colonies has become popular, and a number of new things have been demonstrated. Deinhardt *et al.* (1958) have shown that with Newcastle disease virus only one of a population of about forty cells may be infected. Other cells are apparently protected against infection with this virus, probably by the presence of interferon.

An experimental epidemic can be considered, then, as having the following components:

Host population with:	reacting with	Parasite population with:
Variation in numbers		Variation in numbers
Variation in genetics		Variation in genetics
Variation in spatial arrangement		Variation in spatial arrangement

Another dimension must be added to this, that of time. Development or change of this complex of host and parasite involves genetic selection and results in evolution of the disease.

CLOSED LABORATORY EPIDEMICS WITH ANIMAL HOSTS

The Spread of Parasitic Amoebae in Populations of Hydras

Stiven (1964) analyzed the demography of a microparasite in experimental epidemics. The value of the intrinsic rate of increase, r, of a population depends on the biology of the species and on the environmental conditions (Chapter 5). The intrinsic rate of increase is used in the Pearl-Verhulst equation for logistic increase of population size:

$$\frac{dN}{dt} = rN \frac{(K-N)}{K}.$$

The strength of the force opposing growth described in this formula by $(K-N)/K$, increases as numbers, N, increase with time, t. K is a constant defined for the particular circumstances.

Stiven applied this idea to the growth of an epidemic, using as a model

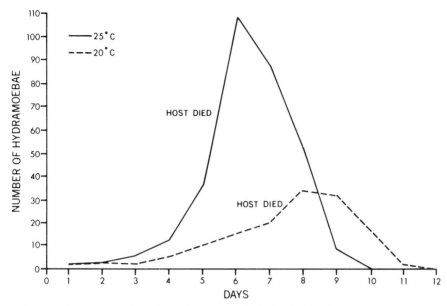

Figure 25.1. Numbers of *Hydramoebae* on one *Hydra* host at 25°C and another at 20°C. Times of deaths of hosts are indicated. (After Stiven, 1964.)

species of *Hydra* and their parasite, *Hydramoeba*, which kills the host in a few days, although some *Hydra* species have some immunity. Figure 25.1 shows growth of two parasite populations at different temperatures, each on a single host.

A newly infected host carrying a known inoculum of amoebae was placed in a host population of known size. Amoebae showed tendencies to move to uninfected hosts after 48 hours. Newly infected hosts were removed daily and replaced by uninfected ones; or they were not removed, so that availability of new hosts limited the growth of the epidemics.

Epidemic growth was measured in terms of infected hosts. Values of *r* for the epidemics were found for various conditions of temperature, inoculum size, and host population density. The epidemic growth was exponential. When hosts were limited, epidemic growth slowed down as in the logistic curve. Logistic curves were computed from a model and parameters derived from population data, and observed data were fitted to them (Figure 25.2). The asymptote was 100 percent infected hosts, so $(K-N)/K$ was in terms of remaining uninfected hosts.

Stiven concluded that the potential for increase of the epidemic was related to the natural rate of increase of the parasite, though these two rates were not the same. For instance, the rate of increase of the epidemic was influenced by the fact that some parasites died during spread to new hosts. The growth of the epidemics also depended on the conditions that Stiven varied—host

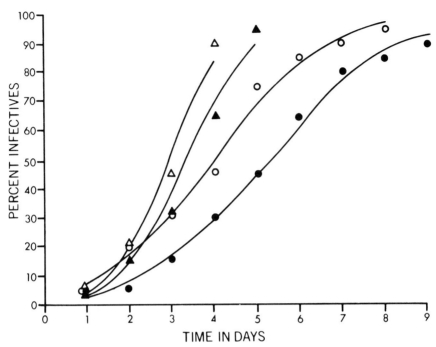

Figure 25.2. Growth of epidemics of *Hydramoebae* on two species of *Hydra* host, shown by triangle or circle, and with two sizes of inoculum, with open symbols representing an inoculum of five, and solid symbols one. Infected hosts are allowed to accumulate. Lines were computed from the logistic model and points were observations. (After Stiven, 1964.)

density, inoculum size, and temperature, and also on host immunity where it occurred.

This was an interesting attempt to describe a particular epidemic disease in fairly simple ecological and mathematical terms. However, when all these experimental epidemics are considered the experiences of Webster and Schneider should be remembered (Chapter 23). They so standardized the conditions under which mouse typhoid was produced that the disease lost much of the natural variability that they had set out to study.

Ectromelia of Mice

Some experimental epidemics in mice were studied for almost half a century. There were two mainstreams of effort, one in the United States at the Rockefeller Institute, by Webster and then by Schneider, the other in England by Topley, Greenwood, and Wilson, then by Fenner in Australia. Both groups of experimenters started with bacterial infections, which here will be discussed after their work with viruses.

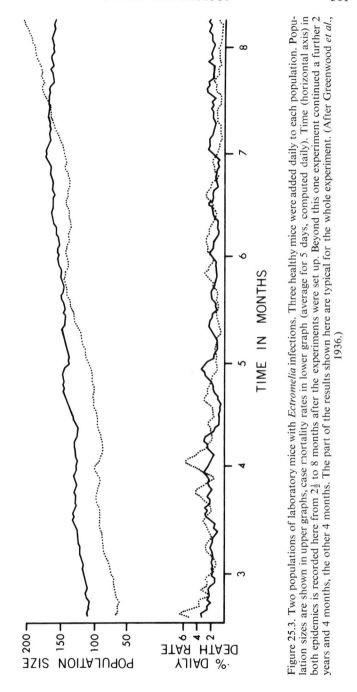

Figure 25.3. Two populations of laboratory mice with *Ectromelia* infections. Population sizes are shown in upper graphs, case mortality rates in lower graph (average for 5 days, computed daily). Time (horizontal axis) in both epidemics is recorded here from 2½ to 8 months after the experiments were set up. Beyond this one experiment continued a further 2 years and 4 months, the other 4 months. The part of the results shown here are typical for the whole experiment. (After Greenwood *et al.*, 1936.)

Ectromelia is a poxlike virus that naturally infects mice, and Fenner's study of its pathogenesis has served as a model for research on other viruses. The infection is acquired through cuts in the skin, and possibly by inhalation of virus into the lungs. The virus multiplies at a great rate in the liver and in certain other target organs. Fenner (1949) first injected the virus intraperitoneally, where it multiplied luxuriantly. Then the virus disseminated to the liver, with concomitant viremia, and subsequent pocklike lesions in the skin. The strains of virus vary in their virulence. Fenner introduced a pair of infected mice into a colony of 50 mice. The mice were kept in small cages that were replaced every day, so that the source of infection was almost entirely the mice themselves. The virulent strain swept through the colony and killed a large proportion of the mice; a less virulent strain killed fewer mice.

One long-term experiment with ectromelia was started by Greenwood *et al.* (1936) on November 18, 1930, and continued to March 19, 1934. The herd consisted initially of 20 infected and 25 uninfected mice. Every day the dead were removed and 3 nonimmunes were added. A second experiment was started with a herd of 10 infected and 25 normal mice. There was daily addition of 3 mice, and the experiment continued for 1 year and 9 months. In nature the added nonimmune animals might be the young mice born into the population. Some results are shown in Figure 25.3 where case death rates for the whole mouse population are presented, and the waves of epidemics are smoothed by calculating 5-day averages for each day. Sixty percent of all deaths showed lesions characteristic of the disease. It was possible to calculate rates of expectation of life and of expectation of acquiring the disease and of dying from it. Epidemics continued as long as mice were available for infection. Results not given in the figure showed that the chance of an individual's acquiring the infection was influenced by vaccination.

Bacterial Infections in Mice

Work on mouse pasteurellosis, caused by *Pasteurella muris*, was more extensive and continued longer. The English work primarily was concerned with infection transmitted by natural contacts among the mice (Topley, 1926). Individual infection lasted a long time, so expectation of life was calculated on the basis of disease lasting 60 days in an individual. The rate of addition of the nonimmunes to the population influenced periodicity of the outbreaks (Figure 25.4). When the rate of addition was decreased, the epidemic peaks were at wider intervals. The effect of immunization by exposure seemed to be less than in the case of ectromelia. The immunity did not last as long. The sick often partially recovered and then relapsed and died. It was suggested that several strains of varying virulence were present, but this was not clearly demonstrated. In these experiments there was no opportunity for selection of resistant strains of mice because the groups were not breeding.

It is difficult to compare the experimental situation with the variable conditions of a natural epidemic.

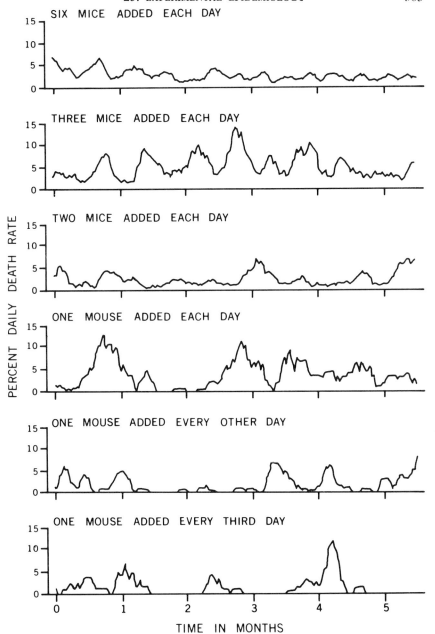

Figure 25.4. Daily death rates in six continued epidemics of *Pasteurella muris* in laboratory mice. The pattern changed with change in the numbers of new mice added at intervals. (After Topley, 1926.)

Respiratory Diseases in Chickens

In recent years there has been some examination of epidemics of respiratory disease in chickens. The mild respiratory infection of chickens, Newcastle disease, can be studied in this way. Andrewes and Allison (1961) studied this model with a virulent strain. One naturally thinks of aerosol spread as a possible mechanism of transmission. If a batch of infected chicks was separated from a susceptible group by a single wire barrier, about half of the second batch acquired the infection. If the wire were doubled so that there were two screens with 2 inches of space between, then by chance about 12 of 13 became infected. The distance separating two wire screens was increased to 9 inches and 12 inches, and none of 14 chicks on the opposite side became infected. If the barrier were solid metal 5 inches high, none of 12 susceptibles became infected. If the barrier were glass and 5 inches high, then more became infected than were spared, but if it were glass 8 inches high, none of 15 became infected. Thus the interposition of a relatively simple barrier between susceptibles and infected animals prevented a respiratory infection from going from one group to the other. It is hard to imagine that aerosol spread of the infection was the mechanism of contagion in this experiment. The effect of social behavior of the chickens on the epidemiology in this situation was both amusing and instructive. If chicks were in cages in a line a, b, c, with simple wire separations, and the group in a were infected and became sick, then the healthy chicks of b and c tended to group where their cages met. Under these conditions there was little spread of infection. If, however, there were no chicks in c, those in b tended to huddle against cage a, and infection was transmitted.

Our own studies with Newcastle disease virus have used a less virulent strain (Thieu and Bang, unpublished). It has been repeatedly shown that transmission rate is affected by the density and the total number of the flock. It is particularly interesting that not all susceptible animals develop the disease, even in the midst of an outbreak, and that an epidemic may stop without any clear cause. The outcome of a particular experiment could not be certainly predicted.

EPIDEMICS INDUCED IN NATURE— MYXOMATOSIS IN RABBITS

This has been a brief introduction to the subject of closed laboratory epidemics. Now an experiment in the field will be described, the epidemiology of myxomatosis in wild rabbits (Fenner and Ratcliffe, 1965). *Oryctolagus cuniculus* is the common laboratory rabbit for which myxomatosis is a fatal disease. It was recorded in Britain in the thirteenth century and later became a pest. Much later it was imported into Australia where it was an even greater pest. *Sylvilagus*, the natural host of myxoma virus, is a genus of cottontail rabbits. It has many species, all of which are confined to the New

World. The myxoma virus is related to the fibroma virus which causes a mild or benign tumor in rabbits, and confers immunity to myxomatosis.

Domestic rabbits had been brought into Australia many times and did not become established in the wild. However, Thomas Austin, a landowner, had a small shipment of wild rabbits brought in on the clipper *Lightening* in 1859. Six years later, 20,000 rabbits had been killed on his estate. About half the continent has climate and soil suitable for the rabbit, and they spread throughout this at the rate of about 70 miles a year.

Early Attempts to Introduce Disease

The rabbits became a plague, damaging the grazing ground of sheep, which are essential to the Australian economy. Therefore since 1860 there have been strenuous and continued attempts to remove the rabbit. Pasteur responded to an offer of a prize for removal of rabbits by experimental infection. He proposed to introduce chicken cholera, but this was never tried, perhaps because it is a dangerous disease for other animals.

In a laboratory in South America there was an outbreak of a severe and fatal disease in *Oryctolagus*. It was found that the infection was derived from contact with *Sylvilagus* rabbits, which had a very mild disease with minor skin lesions. Later, Martin in England showed that the disease would spread through small colonies of *Oryctolagus* with a 99.5 percent mortality rate. Between 1936 and 1943 several attempts were made to get the infection into Australian wild rabbits. These failed and no one really understood why.

The Australian Outbreak

After the war the Australian work started again, and there were failures, but in December 1950 there was a sudden outburst of infection and the disease spread throughout Australia. The results were incredible. The disease appeared to skip from one place to another, sometimes great distances. How was it being spread? It had previously been shown that mosquitoes are able to transmit a bird pox virus. Work under Fenner's leadership showed that the myxoma virus could be transmitted by mosquitoes, but that multiplication of the virus did not occur within the mosquito. The mosquito acted as a flying pin, and the virus could persist on the proboscis for a month or more.

Mortality early in the history of the infection is shown in Figure 25.5. Of some 5,000 rabbits almost none survived. After an outbreak, new rabbits repopulated the area and bred, and then a second outbreak of disease occurred. But the mortality rate dropped from 99.20 percent to 90 percent. Had the virus changed or had the host become more resistant? Fenner had the wisdom to isolate strains of the virus and preserve them. He subsequently tested the strains for virulence. Virulence was graded from I to V on the basis of mortality rate and incubation period, the rate of development of symptoms. The most virulent strains (Grade I) had a mortality rate of 99.9 percent and the least virulent (Grade V), of 50 percent.

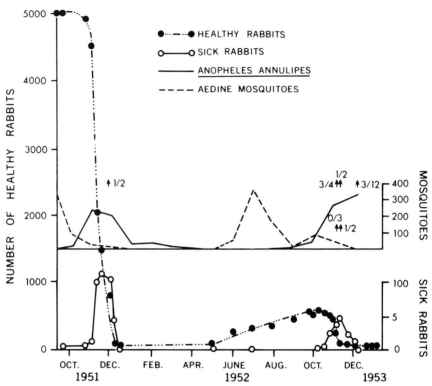

Figure 25.5. Population counts of rabbits and mosquitoes at Lake Urana. The arrows indicate the occasions on which myxoma virus was recovered from batches of mosquitoes; and the numbers of virus isolations out of those attempted are given. (After Myers *et al.*, 1954.)

The strains that were isolated in nature over several years became less and less virulent. How did this occur? The disease was transmitted by the mosquito. The opportunity for the mosquito to acquire infection would be greater if the infection lasted a longer time. So among the variants that would be present in any population, those that produced longer lasting infection were more frequently transmitted by the mosquito; that is, there was natural selection for the less virulent. Localization of the virus in the skin also had selective advantage.

Whether or not the rabbits had also changed was tested. Most infectious agents in nature provoke humoral immunity that is transmitted to the young, and this is true for myxomatosis. Young rabbits, that themselves had not experienced the infection, were held in the laboratory until they had lost their passive immunity, and were tested with a virulent virus from a frozen supply. It was shown that after a few years, the rabbits had become more resistant, although the exact mechanism of the resistance was not determined. Early in

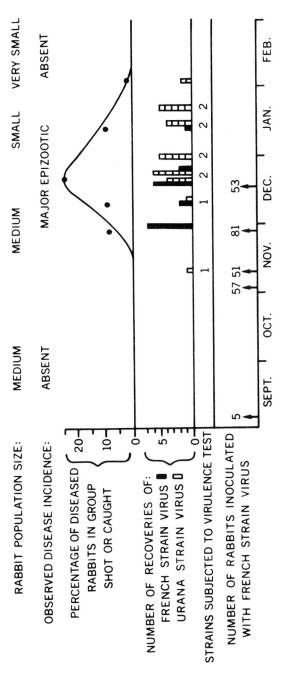

Figure 25.6. A French strain of myxoma virus was introduced into a wild rabbit population. Graph shows the incidence of disease, size of rabbit populations, schedule of inoculation with the French strain, and type of virus recovered. (After Fenner *et al.*, 1957.)

the history of the disease in Australia, the Grade III virulent strain killed 90 percent of the rabbits. After 7 years of epidemics, it killed only 25 percent of the nonimmunes. So there was a continuing evolution of the pair of organisms—the host and the parasite.

Attempts were made to change the balance. Into an area where a less virulent strain of virus had developed, some rabbits were released that had been inoculated with a more virulent virus. The virus could be recognized in the field by its marked pathological effects. For a few days after the introduction, relatively few of the mild strains could be recovered from sick rabbits; then the mild strain again replaced the virulent strain (Figure 25.6).

Although myxomatosis has helped the sheep farmer greatly in Australia, it is no longer the spectacularly destructive disease that was introduced some 10 years ago. It persists as a continuously smoldering disease, at frequent intervals flaring up into minor epidemics. Farmers often reinfect rabbits with virulent strains and then release them.

Myxomatosis in Britain

The epidemics in Europe were somewhat different. Another strain of myxomatosis was introduced by a French farmer because he wanted to rid his own land of rabbits. It spread throughout Europe, and in some way crossed the English Channel, perhaps by means of the mosquitoes, or by men who hate the rabbit. The spread in England took another course. It advanced in a steady wave that was much slower than the progression in Australia. In order to test whether mosquitoes were transmitting the virus, cages of rabbits were suspended above the ground where there was no direct contact with other rabbits, but where they could be bitten by mosquitoes. It was shown that mosquitoes could transmit the virus, but that in nature they did not do so, at least in the epidemics. However, there are plenty of rabbit fleas in Britain and they were transferring the infection. In Australia there are no rabbit fleas. The rabbit had been introduced without them. In transmission of the plague bacillus, dissemination of the vectors' fleas depends upon death of the rat host; so it is also with transmission of myxomatosis in rabbit fleas in Britain. The infected flea leaves a dying host and searches for a new living host, and this no doubt accounts for the slower spread of the infection, though there is also some movement of adult fleas from one living host to another.

There is a second interesting possibility arising from transmission by fleas that has not yet been actually proved. In Britain there has not been such a spectacular loss of virulence of infection as in Australia (Table 25.1). This is thought to be related to the fact that the rabbit host must die before most of the infected fleas are disseminated, and so there is no selective advantage for a less virulent disease.

The biology of the fleas has proved to be interesting. In order to lay eggs, the female flea must feed on the blood of a rabbit that is pregnant (Rothschild, 1965). This means that increase in the numbers of the host is accompanied by increase in numbers of the flea parasite.

"I am fascinated with the present concern for ecology and environmental conservation and the disregard for the most important area of conservation—human beings. I can't avoid the suspicion that it is an avoidance—either conscious or unconscious—to obscure and disguise the fact that our society continues to be not only insensitive about human conservation but actually cruel to human beings."

Dr. Kenneth B. Clark
President, American Psychological Association
Address to New York University School of
Education Alumni Association
October 22, 1970

TABLE 25.1

Comparison of the Virulence of Field Strains of Myxoma Virus in Australia and Britain Several Years After Its Introduction[a]

Country and year	Virulence grade					
	I	II	IIIA	IIIB	IV	V
	Mortality in percent[b]					
	(99)	(95–99)	(90–95)	(70–90)	(50–70)	(50)
Australia						
1950–1951	100[c]					
1958–1959		25	29	27	14	5
1963–1964		0.3	26.0	34.0	31.3	8.3
Britain						
1953	100					
1962	4.1	17.6	38.8	24.8	14.4	0.5

[a] Fenner and Ratcliffe (1965).

[b] Figures given in parentheses in the following line are percentage case mortality rates in standard rabbits.

[c] These figures represent the percentage of virus strains recovered that year, which were found to fall in that particular virulence grade.

Like the mosquitoes, the fleas acquire virus, not from rabbit blood but from infectious lesions. Virus stays on the mouthparts and does not multiply. It is lost by contact with another host, or by eventual virus inactivation. Experiments proved that unfed fleas can retain infectious virus for several months after rabbits have deserted their burrows.

Myxomatosis in Britain has continued to be more effective than in Australia in suppressing rabbit populations. The rabbits are estimated to be at one-tenth their pre-myxomatosis level. This decline has had marked effect on the ecology of the British countryside. Food habits of the predators have changed, and some grassland has become scrubland, which is being succeeded by woodland.

CONCLUSIONS

In the search for understanding of epidemics, a variety of models have been examined—theoretical and mathematical, viruses in tissue culture, amoebic infections of hydra, bacterial and viral infections of mice, and field outbreaks of myxomatosis of rabbits. No matter how virulent the infectious agent, all have in common three possible outcomes. The herd may be eliminated, or may completely recover, or most commonly, persistent recurrent disease is established in the herd. The complexity of the numbers and distribution of the host and of the parasite is so great that few specific predictions can be made, but in all of them, the regulatory effect of reduction of host by agent led to reduction of disease, and increase of host favored increase of transmission.

Even with the most virulent infections known, in small populations of cultured cells or in very large wild populations, continued infection in the population can be established. This depends on factors no more complicated than continued availability of new host individuals. It is not necessary to have variation in susceptibility and in virulence in order to explain the fluctuations of infection rate in a continued infection. However, it is true that hosts may recover from infections and factors involved in this can be studied, and strains of varying virulence do produce different attack rates. In nature there is usually change toward less virulent infectious organisms and more resistant hosts.

In experimental epidemiology the experimenter sets the stage so that he may study the factor he wants to study; in limiting variation he may magnify or even suppress another fact. Experimental epidemiology involves complex ecology.

Greenwood said: "We must put ourselves in school again to learn more of the grammar of the language of epidemiology before we can aspire to write the simple perfect truth. Even within the narrowest concept of the subject, no single group of investigators can complete the experimental study of epidemics. When, generations hence, it may be completed, this will not solve the problems of human epidemics; it will do no more than indicate where solutions of these problems should be sought."

REFERENCES

Abbey, H. 1952. An examination of the Reed-Frost theory of epidemics. *Human Biol.* **24**: 201–233.

Andrewes, C. H., and Allison, A. C. 1961. Newcastle disease as a model for studies of experimental epidemiology. *J. Hyg.* (Cambridge) **59**: 285–293.

Bailey, N. T. J. 1957. *The Mathematical Theory of Epidemics.* Hafner, New York.

Bang, F. B., and Gey, G. O. 1952. Comparative susceptibility of cultured cell strains to the virus of eastern equine encephalomyelitis. *Bull. Johns Hopkins Hosp.* **91**: 427–461.

Black, F. L., and Melnick, J. L. 1955. Microepidemiology of poliomyelitis and herpes B infections. Spread of the viruses within tissue cultures. *J. Immunol.* **74**: 236–242.

Deinhardt, F., Bergs, V. V., Henle, G., and Henle, W. 1958. Studies on persistent infections of tissue cultures. III. Some quantitative aspects of host cell-virus interactions. *J. Exptl. Med.* **108**: 573–589.

Fenner, F. 1949. Mouse-pox (infectious ectromelia of mice): a review. *J. Immunol.* **63**: 341–373.

Fenner, F., Poole, W. E., Marshall, I. D., and Dyce, A. L. 1957. Studies in the epidemiology of infectious myxomatosis in rabbits. VI. The experimental introduction of the European strain of myxoma virus into Australian wild rabbits. *J. Hyg.* (Cambridge) **55**: 192–206.

Fenner, F., and Ratcliffe, F. N. 1965. *Myxomatosis.* Cambridge Univ. Press, London.

Greenwood, M. 1932. *Epidemiology, Historical and Experimental.* Johns Hopkins Press, Baltimore, Md.

Greenwood, M., Bradford Hill, A., Topley, W. W. C., and Wilson, J. 1936. *Experimental Epidemiology.* Medical Research Council Special Report Series 209. His Majesty's Stationery Office, London.

Lotka, A. J. 1924. *Elements of Physical Biology.* Williams and Wilkins, Baltimore, Md. Republished, 1956, as *Elements of Mathematical Biology*, Dover Publications, New York.

Myers, K., Marshall, I. D., and Fenner, F. 1954. Studies in the epidemiology of infectious myxomatosis of rabbits. III. Observations on two succeeding epizootics in Australian wild rabbits on the Riverine Plain of southeastern Australia, 1951–1953. *J. Hyg.* (Cambridge) **52**: 337–360.

Ross, R. 1916. An application of the theory of probabilities to the theory of *a priori* pathometry. *Proc. Roy. Soc.* (*London*), *Ser. A*, **92**: 204–231.

Rothschild, M. 1960. Myxomatosis in Britain. *Nature* **185**: 257.

Rothschild, M. 1965. The rabbit flea and hormones. *Endeavour* **24**: 162–168.

Stiven, A. E. 1964. Experimental studies on the epidemiology of the host-parasite system, *Hydra* and *Hydramoeba hydroxena* (Entz.). II. The components of a simple epidemic. *Ecol. Monographs* **34**: 119–142.

Topley, W. W. C. 1926. Three Milroy Lectures on Experimental Epidemiology. *Lancet* **1**: 477, 531, 645.

Topley, W. W. C., and Wilson, G. S. 1938. *The Principles of Bacteriology and Immunity*. 2nd ed. Wood, Baltimore, Md. Currently published as Wilson, G. S., and Miles, A. A. (eds.) 1964. *Topley and Wilson's Principles of Bacteriology and Immunity*. 5th ed. Williams and Wilkins, Baltimore, Md.

CHAPTER 26

Viral-Induced Tumors in Populations

HARVEY RABIN

Tumors occur throughout much of the plant and animal kingdoms and should be recognized as a general biological phenomenon. In this chapter various vertebrate hosts are included—mammals including man, amphibia, and birds. An attempt is made to present certain tumors in populations from the point of view of infectious disease. These considerations include both the problem of proving viruses as the causal agents of some tumors and studies of the natural history of tumor viruses and tumors in populations.

THE NATURE OF THE VIRUSES AND THE TUMORS
Tumor viruses are "real" viruses with no obvious peculiarities which would relegate them to a special systematic group. Indeed, tumor viruses are known from all the major groups, helical and cubical, DNA and RNA viruses. Likewise, the tumors which they induce are "real" tumors. They conform histologically to accepted structure and they grow and can invade, metastasize, and kill their hosts.

A virus is recognized as a tumor virus by its ability to induce a tumor when inoculated into a susceptible host. That it may also be capable of non-tumorgenic activity should not confuse the status of the virus as a tumor virus. Certain "common" viruses under some circumstances can induce tumor formation.

THE PROBLEMS OF CAUSATION
In microbiology the method of showing a causal relationship between an agent and a disease has been the fulfillment of Koch's postulates. These may be stated as follows:
1. Isolate an organism from the infected animal.
2. Grow it in pure culture.
3. Reinoculate the cultured organism into a suitable experimental animal and reproduce the original condition and, if possible, reisolate the organism.

Much of the difficulty in tumor virology stems from the inability to completely fulfill these postulates, as the following example will show.

Lucké Adenocarcinoma of the Leopard Frog, *Rana pipiens* (reviewed by Rafferty, 1964)
The Lucké adenocarcinoma has been studied mainly in frogs from Vermont and Wisconsin. In the field the relative frequency of frogs with such tumors is about 2.3 percent. Lucké did a series of inoculations with cell-free extracts

TABLE 26.1

The Appearance of Tumors in Frogs Inoculated or Not Inoculated with Cell-Free Extracts of Adenocarcinomas[a]

Treatment	At months post-inoculation						Total	
	0–3		4–6		6			
	Number	Percent	Number	Percent	Number	Percent	Number	Percent
Inoculated with cell-free extract	8/351	2.3	12/104	11.5	44/222	20.7	66/677	9.5
None	16/683	2.3	10/466	6.0	7/104	6.7	33/953	3.5

[a] Lucké in Rafferty (1964).

of tumors, and the results suggested that an agent was being transmitted. Part of his data are summarized in Table 26.1, and at least three noteworthy points can be seen. The difference in the percentage of tumor bearers in the two groups was small, about threefold. The uninoculated group also developed tumors, indicating that at least some of the frogs already had an agent that was the same as or similar to the one presumably present in the cell-free inoculum. The frequency of tumor formation increased with time in both groups, but did so faster in the inoculated group. This threefold difference in tumor formation was for many years the basis for the notion of viral etiology of this tumor.

TABLE 26.2

Attempts to Reproduce the Experiment by Lucké with Adenocarcinomas in *Rana pipiens*[a]

Author	Occurrence of tumors in inoculated frogs		Occurrence of tumors in uninoculated frogs		At months post-inoculation
	Number	Percent	Number	Percent	
Duryee	9/54	17	Low	Unspecified	1.4
Roberts	—	17	—	17	4
Rafferty	29/168	17.3	5/72	6.9	2–4
Rafferty	28/111	25.2	32/115	27.8	5–11

[a] Rafferty (1964).

There have been three other attempts to repeat this experiment. The results of these tests are in Table 26.2. Of the three attempts to confirm Lucké's findings, two showed no difference in tumor formation in inoculated and uninoculated animals on prolonged incubation. Rafferty's study did show the acceleration response of earlier tumor formation in the inoculated group, but the reason for this acceleration is not clear.

Two types of viruses have been isolated from Lucké tumors in tissue culture.

One of these is a herpes-like virus and the other is of uncertain taxonomic position. So far neither of these has been shown to be oncogenic.

All these data were inconclusive, as there was no series of experiments showing a significant difference between the inoculated and uninoculated, but the data did suggest a difference. Tweedell (1967) did an experiment which at last clarified the issue. He inoculated cell-free extracts of tumors into embryos and then raised the embryos through metamorphosis to adulthood. From 55 to 87 percent of such animals developed tumors. There is now good reason to believe that the virus in Tweedell's extracts resembles the herpes-like virus isolated from Lucké tumors in tissue culture.

The study of the Lucké tumor illustrates the inconsistency of the behavior of experimental materials and the difficulty in proving that viruses isolated from tumors may be the etiological agents of the tumors.

Rous Sarcoma Virus

Another aspect of the difficulties involved in fulfilling Koch's postulates can be seen in the "nonproducing" infection of the Rous sarcoma virus, a tumor virus of gallinaceous birds.

The Rous virus is assayed in cultures of chick embryo fibroblasts. Temin and Rubin (1959) noted that tissue cultures from different embryos often yielded significantly different results, with some having as much as a forty-fold reduction in susceptibility. From resistant cultures a virus was isolated which interfered with the multiplication of Rous virus. This virus was called the resistance-inducing factor (RIF). Later, similar interfering viruses were isolated from stock preparations of Rous virus, and these were termed Rous-associated viruses (RAV). These were subsequently shown to be avian leukosis viruses. It was determined that there was usually more RAV than Rous virus in stock preparations made by extracting tumors. The relationship of RAV to Rous virus in terms of virus multiplication, was shown in an experiment by Hanafusa et al. (1963). Stock preparations of Rous virus which contained RAV and Rous viruses, were diluted to the point where they contained only a few virus particles. Such dilutions of virus were then inoculated into cultures of chick embryo fibroblasts. The high dilution resulted in the infection of only a few cells, and these with only one virus particle each, either Rous virus or RAV. Cells infected with RAV are essentially indistinguishable morphologically from normal chick embryo fibroblasts. However, cells infected with Rous virus are clearly recognizable, as they give rise to colonies of transformed cells which are easily distinguished from normal chick embryo fibroblasts.

When such colonies of Rous virus-transformed cells were isolated and assayed for virus, no infectious Rous virus was found. Such cultures were termed nonproducer (N-P) cultures. When RAV was added to such N-P cultures, both infectious Rous virus and RAV were produced, and the culture was a producer culture. These data were interpreted to mean that the Rous virus itself is somehow defective and dependent upon RAV as a helper virus for completion of its natural cycle. This meant that the Rous virus genome became enclosed in an RAV capsid (outer coat) by the process of transcapsidation. More recently it has been shown that the N-P cell does produce a virus after all. But this virus has a different host range from the Rous-RAV transcapsidated particle, and so it was difficult to isolate. This apparently "native" Rous virus has been termed Rous sarcoma virus-o (Vogt, 1967; Weiss, 1967).

Several other animal virus tumor systems are characterized by difficulty in demonstrating an infectious virus. These include simian virus 40, certain adenoviruses, and murine sarcoma virus. The nature of the virus-cell relationships in many of these systems is not clear at present. That is, it is not known whether defectiveness or repression, for example as altered particles, is responsible for the inability to readily isolate virus.

The possibility that many tumors apparently harbor noninfectious virus which is directly involved in the etiology or maintenance of the tumor is very intriguing. That this may be the case in human tumors is especially noteworthy, as no one as yet has proved an etiological role for a virus in a human tumor. The major point is that without knowledge of the N-P state many tumors would mistakenly be regarded as being virus-free.

Knowledge of the virus-cell relationship can also be relevant to the understanding of the natural history of a virus. Kenzy and Neuzil (1953) showed that the frequency of antibodies to Rous virus in chickens was directly related to the amount of clinical lymphomatosis in the flock. Lymphomatosis is a helper virus to Rous sarcoma virus. Based on what is now known about Rous sarcoma virus and transcapsidation, this correlation between lymphomatosis and antibody to Rous virus might well have been predicted. The question of how much Rous virus genome there is in nature remains unanswered.

THE NATURAL HISTORY OF POLYOMA VIRUS

Once a virus is proved to be the etiological agent of a tumor, and enough of its relationship to the cell is known for it to be followed as an antigen and as an infectious agent, its natural history can be investigated. Among the tumor viruses that have been better studied from this viewpoint is the polyoma virus of mice, reviewed by Rowe (1961).

Laboratory Colonies of Mice

The first step was the investigation of the frequency of hemagglutination-inhibition (HI) antibody in laboratory and commercial breeding colonies. Of eight laboratory colonies studied, seven showed animals with HI antibody, with relative frequencies ranging from 3 to 86 percent. In addition, there was a higher frequency among older mice. A laboratory colony from Europe was infected as were several from the United States, and this suggested a wide distribution of the virus. The degree of infection from laboratory to laboratory in general correlated directly with the degree of exposure to experimental work with polyoma virus. However, antibodies were also found in three out of seven commercial breeding colonies, in frequencies of 1 percent to 42 percent, and these had no known exposure to polyoma virus. There was no particular host strain associated with the presence of antibody.

TABLE 26.3

House mice with Hemagglutination-Inhibition Antibody to Polyoma Virus in New York City[a]

Area	Street	Number positive/Number tested
Harlem	1	1/4
	2	0/16
	3	116/423
	4	10/26
	5	0/3
	7	0/17
	9	2/26
	10	3/19
	27	1/11
	28	0/12
	30	0/16
	33	0/25
	40	0/8
	41	0/20
	42	0/11
	43	15/62
Queens		0/109
Bronx		0/65
Lower Manhattan		0/37

[a] Rowe (1961).

Natural Populations of Mice

Two types of natural populations were also studied, urban and rural. Table 26.3 shows some of the results of the study in New York City. There was a somewhat focal distribution of the virus in Harlem. When street 3 was studied house by house a similar pattern emerged, and a suggestion of a focal pattern was found even when a single building on this block was studied floor by floor. In rural areas seven barns were studied with 14/67 (21 percent) of the mice in one barn positive, while no mice (0/99) were found positive in the six other barns. This again suggested a focal pattern of distribution.

No other rodents have been found with HI antibody to polyoma virus, and none of a series of various other vertebrates was found with HI antibody.

The best transmitters of the infection seem to be the newborn. This can be seen in Table 26.4. However, without artificial inoculation of the agent, the chances of infection in the newborn are small. The efficiency of the newborn

TABLE 26.4

Transmission of Polyoma Virus Among Laboratory Mice[a]

Type of contact with infected young mice	Frequency of conversions with time in infected animal room			
	At 3–4 weeks	At 5–6 weeks	At 8–15 weeks	At 5–8 months
Mothers of inoculated newborn	21/21	—	—	—
In cage with infected weanlings	1/276	0/29	5/46	—
In same room	0/314	0/205	5/122	24/69

[a] Rowe (1961).

as a transmitter may result from the intimacy of the contact, and the fact that the newborn is highly susceptible and produces high titers of virus. On the basis of this observation and the facts that the virus is very stable on exposure to the general atmosphere, that it leaves the body in urine, feces, and saliva, and that it can infect by the intranasal route, and to a much lesser degree by ingestion, a scheme is proposed for the spread of the virus through a colony (Figure 26.1).

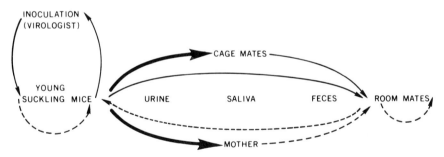

Figure 26.1. Modes of spread of polyoma virus through a laboratory colony of mice, including artificial inoculation. (Adapted from Rowe, 1961.)

Among the most important things to determine is the effect that the virus has on the biology of the host. In this case, the obvious effect is the number of tumors caused. Throughout one year at the Jackson Laboratories in Bar Harbor, Maine, all mice moribund from any cause were bled for determination of HI titer and were autopsied for tumors. The data were examined for any correlation between antibody to polyoma virus and any type of tumor. There was no clear indication that tumor-bearing mice had any excess antibody to polyoma virus, suggesting that this virus was not a major cause of tumors in these mice.

To summarize, polyoma virus is widely disseminated with a somewhat focal distribution in laboratory, commercial, and natural colonies of mice. It grows very well in newborns, which can serve as the major source of infection in a colony. Virus is discharged in urine, saliva, and feces, and can infect by inhalation and possibly by ingestion. Its role in the causation of tumors under natural circumstances is probably very modest.

THE BURKITT LYMPHOMA (Burkitt, 1964)

The association of virus with tumors in humans is perhaps the most intriguing area of tumor-virus research at the moment, and situations in man can be found which are reminiscent of those just described in other species. One of the most interesting human tumors is the Burkitt lymphoma. This condition is found in a belt across the central part of Africa. It occurs mostly in children, with 75 percent of the cases in the age group of less than 13 years and 99 percent in that of less than 19 years. All races are affected. There is a rather definite inverse relationship between frequency of cases and altitude, with cases becoming less frequent at higher altitudes. The altitude factor in turn is seemingly correlated with latitude; the further from the equator the lower is the altitude at which the number of cases diminishes. This suggests that temperature is the limiting factor. These epidemiological observations led to the interpretation that the Burkitt lymphoma is of viral etiology, that the virus is arthropod borne, and that the infection is a zoonosis. Direct research using the electron microscope and tissue culture has shown that virus is present in the tumor cells. Interestingly, the virus closely resembles the herpes-like virus which apparently is the causative agent of the Lucké carcinoma. However, the Burkitt isolates have not, as yet, been shown to be oncogenic.

CONCLUSIONS

Two kinds of discoveries are relevant to the question of whether human tumors are caused by viruses. Girardi *et al.* (1964) showed that certain common viruses, adenoviruses, are oncogenic for newborn rodents. While most other common viruses have so far failed to produce tumors in similar circumstances, the intriguing possibility exists that some common viruses may be oncogenic for man. Second, viruses which may be primarily associated with other animals may be potentially important for man. The implication of a zoonosis was raised concerning the Burkitt lymphoma. Viruses of other host species such as SV40 of monkeys and possibly Rous sarcoma virus of birds are capable of transforming human cells *in vitro*.

This discussion does not deny that other agents such as radiation and chemical carcinogens do have roles in oncogenesis. Indeed, the effects of combining various types of carcinogens are being examined. Likewise the genetics and physiology of the host are doubtless important.

REFERENCES

Burkitt, D. 1964. A lymphoma syndrome dependent on environment. Part II. Epidemiological features. In *The Lymphoreticular Tumours in Africa*. F. C. Roulet (ed.) Karger, New York, pp.119–136.

Girardi, A. J., Hilleman, M. R., and Zurickey, R. E. 1964. Tests in hamsters for oncogenic quality of ordinary viruses including Adenovirus Type 7. *Proc. Soc. Exptl. Biol. Med.* **115**: 1141–1150.

Hanafusa, H., Hanafusa, T., and Rubin, H. 1963. The defectiveness of Rous sarcoma virus. *Proc. Natl. Acad. Sci. U.S.* **49**: 572–580.

Kenzy, S. C., and Neuzil, P. V. 1953. Studies in avian neoplasia II. The incidence of Rous virus-neutralizing antibodies in serum collected from flocks experiencing losses due to lymphomatosis. *Am. J. Vet. Res.* **14**: 123–128.

Rafferty, K. A. 1964. Kidney tumors of leopard frog: a review. *Cancer Res.* **24**: 169–186.

Rowe, W. P. 1961. The epidemiology of mouse polyoma virus infection. *Bacteriol. Rev.* **25**: 18–31.

Temin, H. M., and Rubin, H. 1959. A kinetic study of infection of chick embryo cells *in vitro* by Rous sarcoma virus. *Virology* **8**: 209–222.

Tweedell, K. S. 1967. Induced oncogenesis in developing frog kidney cells. *Cancer Res.* **27**: 2042–2052.

Vogt, P. K. 1967. A virus released by "nonproducing" Rous sarcoma cells. *Proc. Natl. Acad. Sci. U.S.* **58**: 801–808.

Weiss, R. 1967. Spontaneous virus production from "non-virus producing" Rous sarcoma cells. *Virology* **32**: 719–723.

CHAPTER 27

The Ecology of Schistosomiasis

FREDERIK B. BANG

Schistosoma japonicum is a trematode parasite that infects humans in the Orient. The ecological work to be discussed was a study on Leyte in the Philippines that continued for about 10 years. It was sponsored by the World Health Organization, the Philippine Government and the International Cooperative Administration, and was led by the late Dr. T. P. Pesigan of the Philippine Department of Health, and Dr. Farooq and Dr. Hairston, World Health Organization advisors. Their work is a model for ecological studies of parasitic disease. The results apply directly to the particular infection in that location, and quantitative relationships would be altered for other species of schistosomes in other areas of the world. The principles should apply elsewhere.

The principles of attempting to get quantitative data about vectors, infections, and the survival of hosts and agents have also been applied to the problem of measuring transmission of the two filarial worms, *Wuchereria bancrofti* and *Brugia malayi*, the causes of elephantiasis in many areas of the world (Beye and Gurian, 1960).

This account will be concerned only with large differences in orders of magnitude among various factors of interest, and will be limited to schistosomiasis.

THE LIFE CYCLE OF THE PARASITE AND THE RELATION TO ANIMAL INFECTION

The Ecological Niche of the Parasite in the Snail, the Vector

Schistosomiasis is transmitted from one human to another by a vector, a small snail. In the Philippines it is the amphibious *Oncomelania*. The habitat of the snail in the rural Philippines is the rice fields and unplowed grass and swamp lands, including open areas around houses.

The mammalian host, which may be man, releases parasite eggs in the stools. The egg has within it a developing miracidium, and this miracidium becomes a free-swimming multicellular organism and it infects the snail. After a period of from 42 to 63 days in the snail the parasite goes through a developmental stage from which cercariae emerge. One successful miracidium produces many cercariae. The cercaria has a forked tail and it swims in water and hangs on the surface until it comes in contact with an appropriate mammal.

The Ecological Niche of the Parasite in the Mammal, the Definitive Host

The cercaria penetrates the skin of the mammal, which may be man, migrates through the tissues, goes through the lungs, and finally settles in the intestinal veins that lead to the liver. There the adult worms mate. The female pushes into the intestinal venules and deposits tremendous numbers of eggs. The eggs then work their way into the feces and are released. This parasite, like most other parasitic worms, differs then from viruses, rickettsiae, and bacteria in that there is no multiplication of parasitic forms within the definitive host. This fact is important in the approach that Hairston and Pesigan used in their efforts to understand the dynamics of the populations involved in the epidemiology of this infection. The study was approached particularly from the worm's point of view, especially concerning its chances of surviving two periods in its life when it is exposed to great hazards—the intervals between the time when it leaves the mammal and enters the snail and when it leaves the snail and enters another mammal. These are the two periods of great mortality that a parasite population must overcome for long-time survival in an area.

This life cycle has been known for many decades and it is fully accepted wherever one goes, and yet there is no real control of this disease anywhere in the world.

The effect of the disease on man may be illustrated by one particular infected family in Leyte. They lived in a highly endemic area and the youngsters bore the massive brunt of the disease. A young boy of 12 to 14 years had a large abdomen with an enlarged liver. His spleen was enlarged because of the back-pressure of the cirrhosis induced by the infection. Because of the infection he was only half or two-thirds the size he should have been. His father was epileptic, almost surely as a result of the infection. Other members of the family were infected also.

Other mammals that were frequently infected in the study area were field rats, dogs, and pigs. Their role, like man's, depended on their contact with water and also on their defecation in snail and miracidium habitat.

THE EXTENT OF THE INFECTION ON LEYTE

How prevalent is an infection in a particular area and what is its importance to the human population? These are public health problems. There was a change of ideas here, from "numbers of people infected" to "How big a burden does this particular infection impose upon the population?" We are starting to say something new from the public health point of view. What is the population of worms in the area? What is the population that is parasitic on the mammal hosts, and what is the population that is parasitic on the snails, and how are the populations maintained? Is it possible to reduce the worm population—is not this the real aim of public health?

The Numbers of Infected Humans

It is fairly easy to estimate how many humans there are in an area. Since this

TABLE 27.1

Contribution of Hatchable Eggs of *Schistosoma japonicum* by Age Groups of People in an Inland Village on Leyte[a]

Age (years)	Number of people	Number of people passing eggs	Number of hatchable eggs per person per day	Total hatchable eggs per day
0–4	121	13	225	2,925
5–9	136	70	594	41,580
10–14	107	82	2,282	187,124
15–19	76	62	424	26,288
20–29	105	80	385	30,800
30 and over	223	162	238	38,556
	768	469		327,273

[a] Hairston (1962).

is an infection that is acquired at different times in the development of the individuals, the age distribution of the human population should be known. Table 27.1 lists the humans in half a square kilometer that was studied intensively by Hairston, Pesigan, and their group. The infection is greatest in the age group 10 to 14 years: eighty-two out of 107 individuals, almost 80 percent, carried the infection. The total human deposition of eggs in that small area of half a square kilometer was more than 327,000 per day and most of it came from the 10 to 14 age group (Table 27.1).

The first indication of the importance of getting quantitative data came from an older study by Hairston and Pesigan. A large area was sampled and

TABLE 27.2

Contribution of Hatchable Eggs of *Schistosoma japonicum* by Different Species of Mammal Hosts in an Inland Village on Leyte[a]

Species	Population[b]	Number passing eggs	Number of hatchable eggs per animal per day	Total hatchable eggs per day
Dog	74	24	2,333	55,992
Pig	176	80	153	12,240
Cow	0	—	—	—
Water buffalo	58	0	—	—
Goat	1	0	—	—
Field rat	3,075	1,968	222	436,896
				505,128

[a] Hairston (1962).
[b] Crudely estimated for field rats.

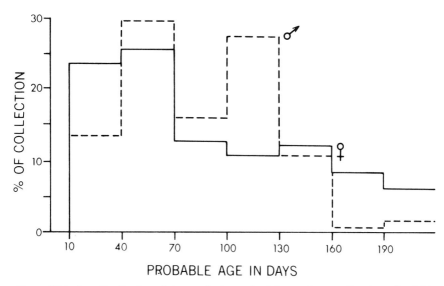

Figure 27.1. Age distribution of *Oncomelania quadrasi* from tube sampling in a rice field. The method failed to collect snails less than 10 days old, and there is an apparent deficiency in the 10- to 39-day age group—a deficiency of about 25 percent of all snails aged 10 days and older. The "hump" at around 130 days reflects a wave of breeding in June. (After Pesigan *et al.*, 1958.)

the numbers of mammalian hosts in the area were estimated. From this study it was concluded that the human host produced the greatest numbers of eggs there. However, Table 27.2 shows that it was estimated that in the Leyte area the field rats produced most eggs.

CENSUS, LIFE TABLES, AND INFECTION OF SNAILS

How does one estimate the numbers of snails in particular areas? How is age determined? How is survival in the field determined? What is the effect of infection on longevity? Surprisingly, these questions can be answered.

Methods of Sampling *Oncomelania quadrasi*

This is an amphibious snail that can be found on the ground in wet areas and can be sampled more easily than the entirely aquatic African vectors of schistosomes. A circle of wire was dropped on the ground and the snails within it were picked out with forceps. Or a piece of pipe was pushed into the mud to obtain a piece of habitat that was carried to the laboratory, where the snails were sorted and counted. With appropriate precautions, sample circles could be randomly distributed and used for estimation of total numbers of snails in the whole area. By this sampling procedure it was possible to estimate numbers of snails at successive times, with fair degrees of accuracy.

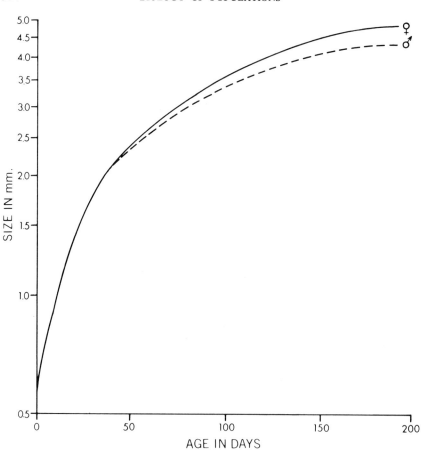

Figure 27.2. Growth curves for male and female *Oncomelania quadrasi* from field and laboratory data. (After Pesigan *et al.*, 1958.)

A collection of snails may be grouped into different size classes (Figure
27.1). Assuming there were no periods of increased mortality, each peak of
numbers of snails of a particular size represents a particular time at which
there were peaks of egg laying. Ages can be estimated from observations of
growth of size of snails in the field and in the laboratory (Figure 27.2). If a
particular population is sampled in, say, July, and again in September, the
same age groups can be identified in the two samples, especially if there are
peaks of numbers that represent peaks of reproduction. If the sampling procedure is adequate and accurate, the decline in numbers of an age group
represents mortality of the group during the time interval. So age-specific
mortality rates can be obtained. With these kinds of figures Hairston and
Pesigan estimated that female *Oncomelania* had an average length of life of

65.8 days and the male 47.6 days. Thus growth curves for the snail were obtained, also mortality rates and beginnings of life tables.

Calculation of the Numbers of Infected Snails

The number of infected snails present in a particular area can be calculated according to the following reasoning. It has to be the result of the number of eggs deposited multiplied by the probability of hatching, multiplied by the probability of being near enough to a snail host, multiplied by the chance of establishing an infection in that host once it has been penetrated.

S (infected snails) per day

$$
\begin{aligned}
= \; & E \text{ (eggs) per day} \\
& \times \; h \text{ (probability of hatching)} \\
& \times \; d \text{ (probability of being deposited near snails)} \\
& \times \; p \text{ (probability of penetrating snail)} \\
& \times \; e \text{ (probability of establishing infection).}
\end{aligned}
$$

Can figures be given for all parts of this equation? This also has been done.

According to Hairston there were about 5,600,000 snails in the study area. This estimate was based on series of counts made over many years. A direct count over many years established that about 438,000 of the snails were infected, a rate of about 1 in 10. Of these infected snails, 280,000 were females and 158,000 were males. This difference was mainly due to the fact that the female has a longer life. The figure needed was the number of snails that became infected per day, so the total number found to be infected had to be divided by the length of life in days. Length of life of the female in nature was approximately 60 days and of the male 40 days. But this figure had to be corrected for mortality caused by the infection. An infected female lived 51 days and an infected male 37 days. Thus the number infected per day was 5,500 (280,000/51) for female snails and 4,200 for males, and 9,700 for the combined sexes. This is crude arithmetic using data from the field, from half a square kilometer.

What were the chances of survival from the time of release of the egg from the mammal, to successful infection of the snail? Table 27.1 gives numbers of humans living in the area and numbers of schistosome eggs given off per person per day. The figures have to make allowance for the fact that the 10- to 14-year age group put out more hatchable eggs per person than did any other age group. It may be concluded from these data that a total of 327,000 eggs was being deposited from humans in that area per day. Other mammals were depositing schistosome eggs in that area too—dogs, pigs, goats, and field rats. These mammals proved to be contributing about 500,000 eggs per day. So the formula became:

9,700 S per day=827,000 $hdpe$.

The probability of hatching, h, was estimated from detailed studies to be 0.5; p, the probability of penetrating, was 1.0; e, the probability of establishing an infection, was about 0.5.

$$9,700 = 827,000 \times 0.5 \times d \times 1.0 \times 0.5$$

and d = approximately 0.05.

This was the probability of the egg being deposited near a snail. This means that the chance of a miracidium getting into a snail ($d \times p$) was 0.05×1 or $1/20$.

This was an amazingly high ratio of success considering the prolific nature of the infection, the tremendous numbers of eggs deposited, and the tremendous numbers of snails that were present. Hairston was rather amused and startled by this because in an initial review of the general situation he had estimated that about $1/30$ to $1/40$ of the area that he studied did have snails in it.

CALCULATION OF THE NUMBER OF ADULT WORMS IN ALL MAMMALS

Another part of the story is the success of a parasite in getting back into a mammal. What were the chances of a particular worm surviving during this time in nature?

Flukes produced per day in the mammalian hosts were equal to the number of cercariae shed per day from snails, multiplied by the probability of their reaching a mammal, multiplied by the probability of maturing in that mammal.

F (flukes) per day = C (cercariae shed) per day

$\times \ c$ (probability of reaching a mammal)

$\times \ m$ (probability of maturing in a mammal)

Estimations had been made of the numbers of snails in an area, as discussed above, and also of the proportions infected. Infected snails were found to give off cercariae at approximately the rate of 2.5 per snail per day and to shed about 54 in a lifetime.

How many humans were present and what was the rate at which they acquired infection over a period of time? It was assumed that the rate at which they acquired infection was moderately constant. They were always going into the infected areas. The children, who were most heavily infected, were frequently exposed and in a fairly random manner. It was found that about 54 percent of the people in an area showed infection each year by egg deposition. However, the figure needed is those that became infected each day. In order for a mammal to demonstrate infection by passage of eggs he must contain both a male and a female fluke. Thus for an individual of the 54 percent to show infection, he must have at least 2 worms. What average number of worms per infected person gives 54 percent of the whole population with at least 2 worms of different sexes? Such a calculation involves use of a Poissonian distribution curve, which cannot be treated in detail here. A 0.54 chance of infection with 2 or more worms requires an average infection in the whole population of 2.3 worms per person and a 0.73 chance for each member of the population of acquiring at least 1 worm. The number of persons in the population, 768, is multiplied by 2.3 to give 1,700, the total

number of new worms acquired each year. This is divided by days in the year
to give 4.6, the new infection rate per day. Thus of the total population of
cercariae in the area only approximately 4 per day were successful in becoming
established in a human host. A tremendous number of cercariae were deposited, almost a million per day. The chances of infecting man were 0.000004, or
4 in a million.

What happened to the rest of the cercariae? What were the chances of
reaching a rat and penetrating? How many rats were infected? Of 55 rats
trapped within the particular area, there were 34 that were infected. Hairston
is the first to recognize the need for better data on the rats. However, these are
the available data. The 34 infected rats contained an average of 20.4 flukes
each. To make some use of this information, the total population of rats and
the dynamics of the rat population should be known. What was the mean
length of life of a rat in the area being investigated? Again the data were not
adequate and Hairston had to use a death-rate of 0.5 in 42 days, that was
calculated for some rats in Hawaii. That is, the half-life was about 42 days and
the mean life was 60 days. There were 3,075 rats in the half a square kilometer.
If a rat acquired 20 flukes over a period of 60 days then each rat acquired an

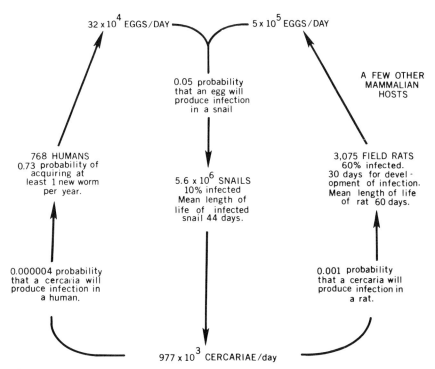

Figure 27.3. Summary of estimations of the population dynamics of schistosome infection
in half a square kilometer in Leyte.

average of 0.3 flukes per day. It took about 30 days for the rat to develop its infection and during that time some rats died. With modifications for this and for other details which tended to cancel out, it was calculated that there were 1,036 rats being infected with 1 worm each per day. The 1,036 adult flukes per day were recruited from the 997,000 cercariae that were released daily into the aquatic environment. Therefore the probability of one of the released cercariae becoming successfully established in a rat was 0.001, or 1 in a thousand.

Some of the results of the investigation are summarized in Figure 27.3. Several items should be noted. In this particular area the chances of the infection derived from man getting into the snail were very high (1/20). The numbers of snails that were available to get infected were high and the rate of mortality of the infective agent when going from the mammalian host to the snail was low. The survival of the cercariae reaching man was very low and the chance of survival of cercariae reaching the rat was 200-fold greater. The conclusion is clear. This was a disease of the rat in that particular area. It was maintained by the rat in this highly infected area where tremendous numbers of humans were infected. If one were to wipe out all the human infections and to introduce a new group of uninfected humans to the area, these people would within a few years acquire the same load of infection. The chances of the cercariae surviving in the rat were so good that the infection would be maintained in the absence of infected humans. Obviously the humans played some role in maintaining the population of rats. So the importance of combining all sorts of methods of controlling the total situation was clearly demonstrated. The ecologist here could work with profit not only with the medical people but with the agriculturalists.

AN ESTABLISHED INFECTED COMMUNITY

If the infection survived at the same level as it had for 10 or 20 years previously as in Leyte, one successful worm produced one successful worm for the next generation. A decrease from generation to generation would be a decline in the infection, an increase would occur during a spread of the infection. This allowed Hairston to make a check on the data by another route of reasoning. He found that the figures did correspond fairly closely, and this helped to confirm the reality of the numerical events discussed above.

However, other factors must be taken into account. There apparently is an optimum density for the parasites. Densities of more than two or three female schistosomes per host reduce the numbers of eggs deposited by each female. Multiple infections of snails may well lead to higher mortality rates of the snails. Thus there are density-dependent factors regulating the parasite population. On the other hand, if the parasite population is reduced to a point where the chance of infection of the vertebrate host by both a male and a female worm is reduced significantly, then the success of the parasite is abruptly decreased.

CONCLUSIONS

The quantitative data emphasized: the importance of rats on Leyte in maintaining the endemic infection; the need for great reduction in the numbers of rats and snails, if the disease in man is to be controlled.

Such quantitative study of a vector, human disease, reservoir situation is a new method. It is open to criticism at various points. It also raises the questions: does it apply to the situation in other areas; or to what extent does it apply? But it is a model for future considerations. With this kind of thinking, not necessarily with the amount of detailed data, it will be possible to determine the load of infection in one area and the importance of the various hosts in the continuation of a particular disease.

REFERENCES

Beye, H. K., and Gurian, J. 1960. The epidemiology and dynamics of transmission of *Wuchereria bancrofti* and *Brugia malayi*. *Indian J. Malariology* **14:** 415–440.

Hairston, N. G., Hubendick, B., Watson, J. M., and Olivier, L. J. 1958. An evaluation of techniques used in estimating snail populations. *Bull. World Health Organ.* **19:** 661–672.

Hairston, N. G. 1962. Population Ecology and Epidemiological Problems. In *Ciba Foundation Symposium on Bilharziasis*. G. E. W. Wolstenholme and M. O'Connor (eds.) Churchill, London.

Hairston, N. G. 1965. On the mathematical analysis of schistosome populations. *Bull. World Health Organ.* **33:** 45–62.

Pesigan, T. P. *et al.* 1958. Studies on *Schistosoma japonicum* infection in the Philippines. In three parts:
1. Pesigan, T. P., Farooq, M., Hairston, N. G., Jauregui, J. J., Garcia, E. G., Santos, A. T., Santos, B. C., and Besa, A. A. General considerations and epidemiology. *Bull. World Health Organ.* **18:** 345–455.
2. Pesigan, T. P., Hairston, N. G., Jauregui, J. J., Garcia, E. G., Santos, A. T., Santos, B. C., and Besa, A. A. The molluscan host. *Bull. World Health Organ.* **18:** 481–578.
3. Pesigan, T. P., Farooq, M., Hairston, N. G., Jauregui, J. J., Garcia, E. G., Santos, A. T., Santos, B. C., and Besa, A. A. Preliminary control experiments. *Bull. World Health Organ.* **19:** 223–261.

CHAPTER 28

Arthropod Populations as Vectors and Reservoirs of Disease

LLOYD E. ROZEBOOM

Many arthropod populations are vectors of disease in that they directly transmit infectious agents from host to host. In the process they support or do not support agent multiplication. Beyond this there are several special situations in which the arthropod population insures long-term or inter-epidemic survival of the parasite and so is a temporary or permanent reservoir of disease. Many intrinsic and environmental factors that affect vector and vector-reservoir populations profoundly influence disease transmission and thereby help determine the nature of epidemics of disease. Chapter 15 discussed rapid genetic changes in vector populations caused by selection by insecticides. Change in mosquito behavior will be further discussed here. All these topics are of general biological interest because they involve evolutionary mechanisms and also are of considerable practical importance.

PATHOGEN-VECTOR RELATIONSHIPS

Free-living arthropods acquire vitamins from their food and from organisms in the environment. Species which imbibe plant juices or animal blood also require another source and such arthropods often possess highly specific symbiotic organisms. The symbionts live in special pouches associated with the digestive tract or they may be intracellular, and elaborate mechanisms operate to assure the transfer of the symbiont from generation to generation. One theory of the origin of arthropod-borne organisms that are pathogenic to man is that the symbiotic forms gave rise to the intracellular rickettsiae and even to some of the viruses. Mechanisms for hereditary survival of the arthropod host had already been established; hence the arthropod in such cases would serve as a true reservoir as well as the vector. However, most of the other disease organisms may have evolved in the vertebrate host, with the arthropod becoming involved secondarily as a vector and temporary or obligate host. There are, therefore, all degrees of dependence of the many pathogenic and parasitic organisms on the arthropod vector. These relationships may be presented concisely as follows:

Mechanisms of Transmission

MECHANICAL TRANSMISSION

The pathogen is transferred by simple contamination of the external surface of the body or the mouthparts, or through the digestive tract.

BIOLOGICAL TRANSMISSION

The agent undergoes development in the vector.

Propagative transmission: the agent multiplies without a special cycle.

Cyclo-developmental transmission: obligatory development through part of the life cycle takes place in the arthropod to produce the infective stage, but there is no multiplication.

Cyclo-propagative transmission: there is an obligatory cycle of parasite development culminating in increase in numbers of the infective stage.

Basic Epidemiological Patterns of Arthropod-Borne Diseases of Man

DIRECT CYCLE FROM MAN TO VECTOR TO MAN

Mechanical transmission: e.g. contamination of flies by enteric bacteria and then deposition of the bacteria on human food (also mechanical transmission of myxomatosis virus among rabbits by mosquito mouthparts).

Propagative transmission: e.g. dengue virus by *Aedes aegypti* in urban epidemics.

Cyclo-developmental transmission: e.g. filarial worms.

Cyclo-propagative transmission: e.g. malaria.

INDIRECT CYCLE WITH MAN BECOMING SECONDARILY INVOLVED DURING AN EPIZOOTIC

Mechanical transmission: e.g. tularemia by bites of *Chrysops discalis*.

Propagative transmission: e.g. plague by fleas from rodents; jungle yellow fever and arboviruses by mosquitoes.

Cyclo-developmental transmission: e.g. filariasis—subperiodic *Brugia malayi* by mosquitoes from monkeys.

Cyclo-propagative: e.g. sleeping sickness trypanosomes by *Glossina*; Chagas disease by *Triatoma* bugs.

INDIRECT CYCLE, WITH MAN AS AN INCIDENTAL VICTIM (i.e. as an indicator species)

The disease is maintained in reservoir hosts and reservoir vectors: e.g. Rocky Mountain spotted fever; scrub typhus.

THE VECTOR AS A DISEASE RESERVOIR

In most of the above situations the arthropod vector serves as a temporary host for the pathogenic organisms. Most of the blood-sucking insects have rather short life spans so that even when the insect is an obligate host it cannot be used by the parasite for interepidemic survival. Situations in which the agent survives in vector populations for long periods will now be discussed.

Papatasi fever virus is transmitted by one of the frailest of insects, *Phlebotomus papatasi*. The fever occurs during the season of the year when the flies are active. The adult flies do not hibernate and the virus does not survive in man. Barnett and Suyemoto (1961) isolated strains of the virus in nature from adult flies, including males, which do not suck blood. This supports observations by Moshkovsky *et al.* (1937) and Petrischeva and Alymov (1938) that the larvae may become infected and pass the virus on to the adults. The larvae become infected either by transovarial passage of the virus or by feeding on virus-infected bodies or feces of adult flies.

The mosquito-transmitted arboviruses do not pass from the infected female through the egg to the next generation. The mechanism of winter survival of western and eastern encephalitis viruses is not well known. It may involve new introductions by migrant birds, recrudescence in birds, survival in snakes or other reptiles; or the virus may survive in certain hibernating adult mosquitoes. Laboratory experiments show that such survival is possible but on the other hand the species of mosquitoes which are vectors do not need a meal of blood in the fall to go into hibernation. Where winters are very mild there may be midwinter feeding and a two-step survival cycle. This problem remains unsolved. In the tropics species of sabethine mosquitoes, which inhabit the forest canopy, are capable of surviving the hot, dry months of the year. They may harbor the virus of yellow fever until the onset of the rainy season permits resumption of transmission by *Haemagogus* mosquitoes among monkeys.

In outbreaks of urban plague, the rat flea, *Xenopsylla cheopis*, transmits the bacilli from rats to man. The mechanism of transmission involves blockage of the upper portion of the digestive tract, and in some cases it possibly is mechanical by contamination of the mouth parts. The blocked flea does not live for more than a few days although some may become unblocked and live a normal life span. The flea evidently is not a reservoir in this situation. However, in sylvatic plague the burrows of the wild rodent hosts are inhabited by species of fleas which are remarkably long-lived. These insects can harbor plague bacilli for from several months to a year or longer.

It is among the ticks and the mites that the arthropod serves as a true reservoir as well as a vector. Several strains or species of spirochetes are found in rodent hosts and are transmitted among these hosts by soft ticks of the genus *Ornithodorus*. There is both stage-to-stage and transovarial passage of the spirochetes through the ticks. New lines of infection are begun when the ticks feed on infected rodents but even without reinfection the spirochetes can maintain themselves in the ticks for several generations, and possibly indefinitely.

The rickettsiae of Rocky Mountain spotted fever can survive indefinitely by stage-to-stage and transovarial passage through hard ticks, especially *Dermacentor andersoni* and *D. variabilis*. New lines of infection occur when the ticks feed on ground squirrels or other rodent hosts which harbor the rickettsiae.

The virus of Colorado tick fever is transferred in *Dermacentor andersoni* from larva to nymph to adult, or from nymph to adult, after the immature stages have acquired the virus by feeding on infected rodents. However, there is no evidence that the virus is passed from one generation to the next so that in this case the tick cannot serve as the only reservoir. It is claimed that certain other tick-borne viruses can be transferred from mother to offspring through the eggs.

The rickettsiae of scrub typhus appear to be capable of indefinite survival in the mites *Leptotrombidium akamushi* and *L. deliensis*. Only the larval stage is parasitic. It acquires the infection from the rodent host, and the rickettsiae are passed through the free-living, nonparasitic nymphal and adult stages, and through the eggs to the succeeding generation.

It should be noted that although the vertebrate host is considered to be the true reservoir in all these arthropod-borne infections, there is no direct passage from mother to young. The organisms must depend upon new infections by way of a vector in order to survive in these so-called vertebrate reservoirs.

SPECIATION IN MOSQUITOES

Because of their importance as disease vectors the mosquitoes have been taxonomically studied more thoroughly than most other families of insects. Perhaps it is for this reason that a number of species groups, or species complexes, is known to exist. Member populations of such groups may resemble one another closely in morphology, but differ in habits and show varying degrees of reproductive isolation. They may have sympatric (overlapping) or allopatric (discontinuous) distributions. These studies have contributed much information which has been used by students of evolution in the formulation of theories of speciation.

Mechanisms of Speciation

A reference bibliography for this topic is in Grant's (1963) book, "The Origin of Adaptations". Mayr (1942) defined species as groups of actually or potentially interbreeding natural populations, which are reproductively isolated from other such groups. Geographically isolated populations of one species are prevented from interbreeding, but would be able to breed if brought together. The barriers to interbreeding among species can be genetic incompatibility, and behavioral, physiological, and morphological characteristics.

The evolutionary steps in species formation are based on genetic variability. A species is able to inhabit a given place because it has undergone adaptation, but mutation and other genetic variability continue to occur. If the range becomes discontinuous as the result of geological catastrophe, or because of migration or by other environmental change, one or both of two separated groups may undergo further adaptations. Ford (1964) discussed reproductive isolation and genetic divergence in a continuously distributed butterfly species, *Maniola jurtina*, in a place where there was no habitat discontinuity. As two separated groups continue their ways, they may diverge in appearance and habits, and eventually evolve into separate species with firmly established reproductive isolation.

In summary, the parts of the process of speciation are:

Individual genetic variability particularly from mutation and recombination.

Separation and isolation of populations.

Selection of the most fit genomes, leading to—

Adaptation of populations to new environments.

Development of reproductive barriers:

Mating barriers.

Sterility on cross-mating.

The *Anopheles maculipennis* Complex

This group of mosquitoes has been studied in detail because when malaria was prevalent in Europe, it was noted that in many places there was no direct correlation between the presence of malaria and *Anopheles* population density. People spoke of anophelism without malaria. Many theories were proposed to explain why there was no malaria where *Anopheles* densities were high. Roubaud suggested that some anophilines had more teeth than others, and so were able to bite through the hides of cattle and other animals, and thus enjoy a competitive advantage over the weaker-toothed man-biting mosquitoes! Counting mosquito teeth became a popular activity! Falleroni noticed that two different kinds of eggs were produced by female *Anopheles maculipennis*, and this led to extensive research on egg types, biology, and malaria transmission. It was shown that instead of one kind of *A. maculipennis* there were seven. From crossbreeding experiments Bates (1940) concluded that there were five species and four subspecies:

> *Anopheles maculipennis* Meigen
> *Anopheles messeae* Falleroni
> *Anopheles melanoon melanoon* Hackett
> *Anopheles melanoon subalpinus* Hackett and Lewis
> *Anopheles labranchiae labranchiae* Falleroni
> *Anopheles labranchiae atroparvus* Van Thiel
> *Anopheles sacharovi* Favre

The last three were shown to be the important malaria vectors, because even when domestic animals were present, a proportion of the females entered houses and bit people. These populations were associated with the more highly malarious coastal districts as the larvae are found especially in brackish water.

The *Anopheles gambiae* Complex

Anopheles gambiae provides the major reason why malaria control is very difficult in tropical Africa. It breeds in the puddles, pools, and other waters which accumulate during the rains. The adults may be intensely domestic, entering houses to feed on man, but some populations have been observed to live in a wild state, that is, in relatively uninhabited areas. A form of *A.*

gambiae was found to be a brackish water breeder on the west coast of Africa, and later another brackish water form was discovered on the east coast. The two brackish water forms are considered to be distinct species, and have been named *A. melas* and *A. merus*. The fresh-water *A. gambiae* appears to be composed of several populations which actually may be cryptic species. These were first discovered in the course of genetic experiments on insecticide resistance, in which it was necessary to crossbreed different strains. Davidson and Jackson (1962) found that the 15 strains they attempted to cross could be placed in two groups, A and B. Member populations of Group A are interfertile, and member populations of Group B are interfertile. However, when a member of Group A is crossed with a member of Group B, the F_1 males are sterile. There is no evidence of clear-cut geographical separation of the two groups. Both are found in western and eastern parts of Africa. Populations of both groups may occur close together, and if the sterility barrier prevents an exchange of genetic material they should be considered to be specifically distinct. Indeed, Paterson (1963) calls them species A and B. Both of them are malaria transmitters; they enter houses to feed on man, and dieldrin resistance is found in both. A third fresh-water form was found by Paterson, Paterson, and van Eeden (1963) in Swaziland and Southern Rhodesia. This species C is exophilic, and bites cattle in preference to man. Antimalarial spraying of houses had no effect on its populations. House spraying, by eliminating the endophilic species A and B and leaving species C unscathed, appeared to bring about a change in habits of the *A. gambiae* population. Actually there seems to have been a replacement of the endophilic by the exophilic species. It will be interesting to watch this situation. Perhaps species C, in sole possession of the breeding habitats, will increase in population density and bite man as well as animals in significant numbers out of doors.

Other Mosquito Species Groups

Other examples of species groups of mosquitoes could be given. One of special interest is the *Aedes scutellaris* group of the South Pacific, in which the 18 or so species and subspecies probably originated through isolation on islands. Another is the *Culex pipiens* complex, with two main divisions, *C. pipiens* and *C. fatigans*, which themselves may represent species complexes. Crossbreeding has demonstrated varying degrees of incompatibility among many local populations, and Laven (1959) insisted that the basis for this is cytoplasmic inheritance. Introduction of incompatible genes into a natural population may be an effective means of control. This could be a great help in reducing the transmission of Bancroftian filariasis in the cities of South Asia.

BEHAVIORAL CHANGES BROUGHT ABOUT BY INSECTICIDES

In the early malaria control projects DDT was applied as a residual deposit in houses. Not only did the adult anophelines disappear, but larval densities in the breeding places were greatly reduced. Later, larval densities increased even though adults could not be found in houses. Then it was realized that a change in adult behavior was probably taking place. Insects resistant to insecticides might have greater exposure to insecticide, and this might enhance irritability and cause other behavioral change. This "behavioral resistance" has been described for several malaria vectors. Furthermore, even species noted for their strongly anthropophilic and endophilic habits may also have wild-type populations away from houses (exophilic). Species C of *Anopheles gambiae* has already been mentioned. In central Brazil the dangerous *A. darlingi* also shows a high degree of exophily.

MOSQUITO SPECIATION AND PUBLIC HEALTH CONTROL MEASURES

Species formation in mosquitoes, as in other animals, occurs by selection of characters that adapt populations to the environment. When man began constructing shelters, these were exploited by certain mosquitoes which became endophilic malarial vectors. By use of residual insecticides, this shelter has now been removed from the mosquitoes' environment. If wild-type populations still exist in highly endophilic species, it appears that it is not too difficult for these species to thrive in an environment without human habitations. Malaria transmission by outdoor-biting *Anopheles* is at present a serious problem in some places, and it is likely that further change to exophily will be encouraged by the selective action of insecticides.

QUANTITATIVE EVALUATIONS OF VECTOR POPULATIONS IN NATURE

To assess the importance of an arthropod in a disease transmission in a given area, there must be estimates of population densities, host preferences, survival, dispersion, and rates of infection.

Population Density

As indicated in Chapters 1 and 5, densities of most animal populations are difficult to measure. Estimates for arthropods can be made from systematic catches, grid counts, or capture–recapture experiments. Biting-densities of the vector may be far more meaningful, especially when combined with infection rates of the vector. For instance, in the swamp-forest environment in Palawan in the Philippines, Rozeboom and Cabrera (1965) observed an average biting rate for female *Mansonia bonneae* mosquitoes of 8.2 per man-hour. If this rate were maintained for 6 hours each day, in 1 month a person would be bitten about 1,500 times by this species. Also, one of every 200 *M. bonneae* caught in the forests and villages carried infective larvae of the filarial parasite, *Brugia malayi*. During the 6 months of the year when the mosquito is abundant, anyone living in the area could expect to be bitten 40 or 50 times by infective females.

The numbers of mosquitoes taken per man-hour from bait or shelters, or per trap-night, are useful in evaluating relative densities, or in following the progress of control measures.

Dispersal

Dispersal of mosquitoes and other diptera has been studied by direct observation of flight activity, by capture of adults at known distances from breeding places, and by release-and-recapture experiments. In former years when malaria control depended upon attack on the larvae in their breeding places, it was considered axiomatic that the effective flight range of *Anopheles* mosquitoes was 0.5 to 1 mile. It is now known that some vector species engage in long migratory flights, and that even those which remain within a restricted area have considerable daily movement.

Survival

The daily survival rate of *Anopheles* malaria vectors is basic in the mathematical models of malaria epidemiology which have been proposed by Macdonald (1957). Longevity of mosquitoes and other arthropods can be recorded in the laboratory under known conditions of temperature, humidity, and food. Under favorable conditions, adult mosquitoes live for a month or two. But it does not follow that they live this long in nature, and to arrive at some estimate of survival of wild mosquitoes indirect means are used. Laboratory experiments usually show that there is little mortality of adult females during the first 2 or 3 weeks following emergence. Thus, during this period of their life span when these females would act as disease vectors, mortality in the field is assumed to be independent of age. That is, the causes of death are exerted equally against all age groups. This is a basic assumption in Macdonald's postulates, but is not necessarily true.

An indirect method of estimating daily survival rates of a natural popula-

TABLE 28.1

A Theoretical Cohort of a Vector with a Daily Death Rate of 10 Percent

Day (x)	Deaths per day d_x	Survivors at day x l_x	Daily survival rate per 1 p	Rate of survival from day 0, per 1
0	—	100	—	—
1	10	90	0.90	0.90
2	9	81	0.90	0.90^2
3	8.1	72.9	0.90	0.90^3
4	7.2	65.6	0.90	0.90^4
5	6.6	59.6	0.90	0.90^5
n	—	—	—	0.90^n

tion of mosquitoes involves determining the proportion surviving between two events in the life cycle.

Table 28.1 is a life table of a theoretical cohort of 100 with a daily mortality rate of 10 percent. Assuming that event 1 occurred at day 0 and event 2 at day 4, of the original number, 65.6 survived to event 2. If the daily survival rate, p, were constant,

$$65.6 = p^4 \times 100$$
$$p = \sqrt[4]{0.656}$$
$$= 0.90 \text{ or } 90\%$$

and the daily mortality rate $= 10\%$.

If $p = $ daily survival rate,

    $M = $ proportion of population surviving to day n,

    $n = $ number of days from one event to the succeeding event,

then $M = p^n$
$$p = \sqrt[n]{M}$$

These expressions are basic to the models established by Macdonald for the epidemiology of malaria and they apply also for other arthropod-borne diseases.

Two events in the life cycle of female mosquitoes that can be recognized are emergence of the adult and oviposition. Polovodova (1941) and Detinova (1962) showed that it is possible to determine whether a mosquito has deposited eggs by examining the ovariole ducts for the presence of relic swellings. It is necessary to know what proportion of an emerging female population has survived to deposit eggs, and how many days elapse from emergence to oviposition.

For several reasons it is considered more accurate to determine survival from blood meal to blood meal, rather than from emergence to oviposition. Freshly blooded (fed) females are collected and dissected. The ovarioles will show how many egg batches have been laid; these in turn reveal how many blood meals were taken. Then the proportion of parous to nulliparous females can be considered to be the result of survival during the intervals between those feeds.

Others have measured the proportions surviving through stages of development of parasites that they carry.

Three examples follow to demonstrate the use of these indirect methods of estimating survival.

Survival of *Anopheles funestus*

Gillies and Wilkes (1963) obtained freshly blooded females in villages in East Africa (Table 28.2). Three days was the average interval between blood meals.

$$\therefore p = \sqrt[3]{0.732}$$
$$= 0.9012$$

TABLE 28.2

Survival of *Anopheles funestus* Obtained in East African Villages[a, b]

Season	Pregravid (ovaries still in resting state)	Nulliparous	Parous	Nulliparous + Parous
Cool	1,694	1,645	4,447	6,092
Hot	1,403	1,329	3,694	5,023
Total	3,097	2,974	8,141	11,115
Percent	—	26.8	73.2	—

[a] Gillies and Wilkes (1963).

[b] 14,212 freshly fed female mosquitoes, *Anopheles funestus*, were collected, and of these 11,115 were gravid. Females take approximately 3 days from feeding to laying eggs and feeding again. It is accepted for a rough estimate that the percentage of parous females in the total gravid population, represents percentage survival of the parous females over the 3-day interval.

percent daily survival rate, p, = 90.1
and percent daily mortality rate = 9.9.

Survival of *Anopheles gambiae*

Davidson and Draper (1953), working also in East Africa, collected mosquitoes from houses and divided them into two groups. One group was dissected immediately and the number carrying sporozoites was counted and this percentage of the total collected was called the immediate sporozoite rate. The other group was held for completion of the extrinsic incubation period.* The extrinsic incubation was represented by n in the calculation, and was shown to be 13 days in that particular place. The females at collection had actually already passed through one day of n, so the time interval became $n-1$. The number that developed sporozoites in the protected environment of the laboratory at the end of n represented the proportion at capture that was in the earliest stage of infection. This proportion gave the delayed sporozoite rate.

Total survival rate,

$$p^{n-1} = \frac{\text{immediate sporozoite rate}}{\text{delayed sporozoite rate}}$$

and the daily survival rate,

$$p = \sqrt[12]{\frac{\text{immediate sporozoite rate}}{\text{delayed sporozoite rate}}}$$

Values of p in 3 different months are given in Table 28.3. It should be noted that there is rather close agreement between daily mortality rates for East African *Anopheles funestus* and *A. gambiae*, as determined by the two methods.

* The extrinsic incubation period is the time required for the development of a parasite in the vector host.

TABLE 28.3

Proportions of Female Mosquitoes, *Anopheles gambiae*, with Malaria Sporozoites in earliest (Delayed) and Completed (Immediate) Stages of development[a, b]

| Month | Sporozoite rates | | | | Daily % mortality rate |
	Immediate i	Delayed d	i/d	p	
September	0	0	—	—	—
October	4.6	11.2	0.411	0.93	7.0
November	5.0	12.6	0.397	0.93	7.0
December	5.1	20.0	0.255	0.89	11.0

[a] Davidson and Draper (1953).

[b] The decline in sporozoite infection rate per total sample is taken to represent death of females occurring in the field during the 12-day interval.

Survival of *Culex fatigans*

Another method of calculating survival was applied by Laurence (1963) to *Culex fatigans* in Vellore, India. It consisted of determining the proportions of *Wuchereria bancrofti*-infected females that had survived long enough in the field to permit advanced stages of the parasite to develop (Table 28.4).

He also measured p from the time of emergence to oviposition. Of 200 females examined for parity, 108, or 54 percent, showed follicular relics. If n were assigned a value of 4 days, p became 0.857; if n were 3 days, p became 0.814, and daily mortality rates were 14 percent and 19 percent.

There are too many inaccuracies and assumptions in these procedures for complacency. One of the most vexing problems is establishing a true value for n, the time intervals between events. Environmental factors and intrinsic biological variables affect the feeding-oviposition cycles of mosquitoes and

TABLE 28.4

Calculation of Daily Survival Rates of Female Mosquitoes, *Culex fatigans*, from Proportions Carrying Successive Stages of Development of the Parasite *Wuchereria bancrofti*[a]

| | | Stage of parasite development | | | |
	Total infected	Microfilaria	Sausage stage $< 250\,\mu$	Sausage stage $> 250\,\mu$	Infective stage III
Number of females	101	37	22	32	10
Percent of total infected that had developed to and beyond that stage	—	—	63.4	41.6	9.9
n	—	—	2	4	10
p	—	—	0.796	0.803	0.794
Daily % mortality rate	—	—	20.4	19.7	20.6

[a] Laurence (1963).

other insects. Also these formulae are based either on the assumption that the age distributions of the populations depend on constant daily rates of emergence of adult females or on the assumption that the daily infection rates of females have not changed during the time the population was under study. Nevertheless, it is helpful to have even approximations of the daily survival rates of vector populations. Garret-Jones and Graf (1964) compiled a table combining estimated survival rates of vector mosquitoes and extrinsic incubation periods of the malarial parasite. With a 13-day extrinsic incubation period, the average number of days of infective life per infected mosquito would be 2.41 days.

Macdonald (1955) pointed out that models of malaria epidemiology are not intended for the demonstration of fully understood happenings, but are scientific tools of great value in experimentation and research. A model shows how epidemiological characteristics would interlock if all assumptions on which it is made were correct. Comparisons of such models with nature shows which of the assumptions are finally acceptable and which can be rejected.

CONCLUSIONS

The topics that have been presented emphasize the complexity of the epidemiology of the arthropod-borne diseases. Even in the most direct type of man-mosquito-man transmission, the dynamics of the vector population present many variables which require intensive study for proper evaluation. Moreover evolutionary processes of selection and adaptation may be intensifying some problems by bringing about behavioral changes of vector populations.

The main topics treated in this chapter are:

The basic mechanisms of development and transmission of pathogenic agents by arthropods.

The possible impact of insecticidal programs on the behavior, as well as on specific insecticide resistance, of vector populations.

The kinds of studies that are made on vector populations. Emphasis has been placed on natural survival of these populations because of its importance to epidemiology. More accurate measurements of survival are desirable, and there is opportunity here for mathematically inclined investigators to improve present methods.

REFERENCES

Barnett, H. C., and Suyemoto, W. 1961. Field studies on sandfly fever and kala-azar in Pakistan, in Iran, and Baltistan (Little Tibet) Kashmir. *Trans. N.Y. Acad. Sci.* **23:** 609–617.

Bates, M. 1940. The nomenclature and taxonomic status of the mosquitoes of the *Anopheles maculipennis* complex. *Ann. Entomol. Soc. Am.* **33:** 343–356.

Davidson, G. 1955. Measurement of the ampulla of the oviduct as a means of determining the natural daily mortality of *Anopheles gambiae. Ann. Trop. Med. Parasitol.* **49:** 24–36.

Davidson, G., and Draper, C. C. 1953. Field studies of some basic factors concerned in the transmission of malaria. *Trans. R. Soc. Trop. Med. Hyg.* **47:** 522–535.

Davidson, G., and Jackson, E. 1962. Incipient speciation in *Anopheles gambiae* Giles. *Bull. World Health Organ.* **27**: 303–305.

Detinova, T. S. 1962. Age-grouping methods in diptera of medical importance, with special reference to some vectors of malaria. *World Health Organ. Monograph. Ser.* **47**: 216 pp.

Dobzhansky, T. 1951. *Genetics and the Origin of Species.* 3rd ed. Columbia Univ. Press, New York.

Ford, E. B. 1964. *Ecological Genetics.* Wiley, New York.

Garrett-Jones, C., and Graf, B. 1964. The assessment of insecticidal impact on the malaria mosquito's vectorial capacity, from data on the proportion of parous females. *Bull. World Health Organ.* **31**: 71–78.

Gillies, M. T., and Wilkes, T. J. 1963. Observations on nulliparous and parous rates in a population of *Anopheles funestus* in East Africa. *Ann. Trop. Med. Parasitol.* **57**: 204–213.

Grant, V. 1963. *The Origin of Adaptation.* Columbia Univ. Press, New York.

Laurence, B. R. 1963. Natural mortality in two filarial vectors. *Bull. World Health Organ.* **28**: 229–234.

Laven, H. 1959. Speciation by cytoplasmic isolation in the *Culex pipiens* complex. In *Cold Spring Harbor Symp. Quant. Biol.* **24**: 166–173.

Macdonald, G. 1955. *Trans. Roy. Soc. Trop. Med. Hyg.* **49**: 155–156.

Macdonald, G. 1957. *The Epidemiology and Control of Malaria.* Oxford Univ. Press, Oxford, England.

Mayr, E. 1942. *Systematics and the Origin of Species.* Columbia Univ. Press, New York.

Moshkovsky, S. D., Diomina, N. A., Nossina, V. D., Pavlova, E. A., Levchitz, J. L., Pels, H. J., and Roubtzova, V. P. 1937. Pappataci fever. VIII. On the presence of the virus of pappataci fever in *Phlebotomus* born from eggs laid by infected females. *Med. Parazitol. i Parazitarn Bolezni* **8**: 922–937. Abstract in *Trop. Dis. Bull.* 1944. **41**: 565.

Paterson, H. E. 1963. The species, species control and antimalarial spraying campaigns, implications of recent work on the *Anopheles gambiae* complex. *S. African J. Med. Sci.* **28**: 33–44.

Paterson, H. E., Paterson, J. S., and van Eeden, G. J. 1963. A new member of the *Anopheles gambiae* complex. A preliminary report. *Mediese Bydraes* **9**: 414–418.

Petrischeva, P. A., and Alymov, A. J. 1938. On transovarial transmission of virus of pappataci fever by sandflies. *Arch. Biol. Sci.* **53**: 138–144.

Polovodova, V. P. 1941. Age changes in ovaries of *Anopheles* and methods of determining age composition in mosquito populations. *Med. Parazitol. i Parazitarn Bolezni.* **10**: 387.

Rozeboom, L. E. 1952. The significance of *Anopheles* species complexes in problems of disease transmission and control. *J. Econ. Entomol.* **45**: 222–226.

Rozeboom, L. E., and Cabrera, B. D. 1965. Filariasis caused by *Brugia malayi* in the Republic of the Philippines. *Am. J. Epidemiol.* **81**: 200–215.

CHAPTER 29

Vertebrate Populations as Reservoirs of Disease

KEERTI V. SHAH

There are more than 150 infections of man that have reservoirs in other vertebrates. The simplest definition of a reservoir is that it is the maintaining species for the parasite (Audy, 1958). It is the host species, or a complex of species, of critical importance for the perpetuation of the parasite upon which the parasite is dependent for its existence. It is generally the host with which the parasite has a long-established and stable association. Such association is mostly unspectacular—there are few overt signs like disease to indicate that the parasite is present. The term *zoonoses* refers to infections of man such as plague, rabies, and arbovirus infections that are acquired naturally from other vertebrates. Specifically human infections are sometimes transmitted to other vertebrates, for example measles, parainfluenza 3 virus, and *Salmonella* infections, and tuberculosis can be transmitted to rhesus monkeys; these infections are called *anthroponoses.*

THE EVOLUTION OF HOST-PARASITE RELATIONS

Man shares some infectious agents with other vertebrates because of his evolution. Mammals have been on the earth for about 125,000,000 years and man about 1,000,000. Infectious agents of man—viruses, rickettsiae, many of his bacteria, protozoa, and larger parasites—must have come from parasites of his animal ancestors and associated animals, rather than from free-living forms. Extensive man-to-man transmission was probably infrequent when man was a rare or scattered species. Infections may have become more specialized and more of them specifically human infections as man started living a pastoral life in settled communities.

When a parasite enters a new host, the chance of successful establishment is slight. The failures of adaptation leave no traces and so are difficult to document. If the parasite is able to infect a new host, it may produce a severe disease in this aberrant host. In a successful adaptation the host and the parasite change after the initial introduction (Chapter 25 and Audy, 1958). With its usually shorter generation time and its rapid increase in numbers, the parasite has the greater potential for change. Two types of forces act on the parasite. One leads to specialization. The parasite becomes more fully adapted, biochemically, morphologically, physiologically, and behaviorally, to the life of the host, leading to host specificity. When a single species is required for the completion of the life cycle of the parasite, the host-parasite relationship is *intraspecific.* As the parasite becomes more adapted to one host species, it becomes progressively less suited to other hosts.

The second type of force brings about changes in the parasite that make it adaptable to additional hosts. Eventually more than one species may be required for the completion of the life cycle of the parasite, and the relationship is *interspecific, obligatory*. In such instances the two species required for the maintenance of the parasite frequently form adjacent links in a food chain, or coexist in the same environment. Another type of relationship is *interspecific, optional*, in which infection of additional hosts by the parasite, while it occurs, is not necessary for the maintenance of the parasite. Examples of these relationships are given in Figure 29.1.

THE CHARACTERISTICS OF A RESERVOIR

The reservoir of human infection may consist of one species or of a group of species, and a number of characteristics can be demonstrated. The reservoir population is susceptible to infection, so that the parasite is able to multiply in it. Natural infection with the parasite can be shown, and the infection occurs at the right time of the year, and is frequent enough to maintain transmission. The host population is numerous enough to provide an adequate supply of nonimmunes. The parasite usually causes little or no morbidity or mortality in the reservoir species, this situation being the outcome of a long-established and stable association of obvious survival value for the host and the parasite. A well-known exception to this rule is rabies virus which kills most of its wild-life hosts, except some bat species; vampire and fructivorous bats apparently maintain rabies asymptomatically and transmit infection for long periods. The parasite has one or more modes of transmission within the reservoir species, and for dissemination to man. Also, the period of infectiousness is of critical importance; for infections transmitted by blood-feeding arthropods, this is the time during which the parasite is present in the bloodstream in high enough concentration to infect arthropods. The period of infectiousness may vary from only a few days, as for some arboviruses, to several months or even years as for *Brucella abortus*, which is disseminated intermittently in the milk of infected cattle. The virus of Bolivian hemorrhagic fever is excreted in the urine of the rodent reservoir, *Calomys callosus*, for over 100 days. Rio Bravo virus, which is related antigenically to group B arboviruses, has been recovered from the saliva of an asymptomatic Mexican freetail bat for periods as long as 2 years (Constantine and Woodall, 1964).

INTEREPIDEMIC SURVIVAL OF THE PARASITE

The question of interepidemic survival of arboviruses in the temperate zone has received much attention, and aspects of it have been discussed in Chapters 1 and 28. Since transmission by arthropods does not take place in the cold winter months, and vertebrate hosts are viremic for only a short period, how does the virus survive from one epidemic season to another? A number of mechanisms have been proposed. The virus may survive in the hibernating mosquito, or in the case of ticks, in the long-lived adult, or by transovarial

The evolution of intraspecific relationships

Parent virus → man, measles
→ ruminants, rinderpest
→ dog, distemper

Other examples of intraspecific relationships:
cholera, amebiasis, poliomyelitis.

Obligatory interspecific relationships

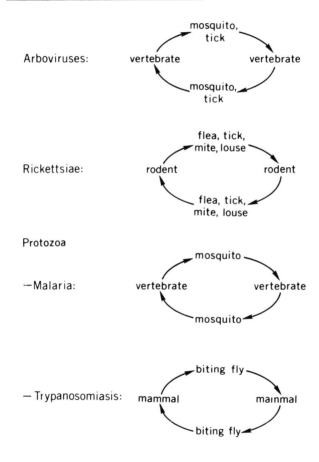

Arboviruses: vertebrate → mosquito, tick → vertebrate → mosquito, tick → vertebrate

Rickettsiae: rodent → flea, tick, mite, louse → rodent → flea, tick, mite, louse → rodent

Protozoa

—Malaria: vertebrate → mosquito → vertebrate → mosquito → vertebrate

— Trypanosomiasis: mammal → biting fly → mammal → biting fly → mammal

Optional interspecific relationships

Rabies: mammals

Ornithosis: birds, mammals

Figure 29.1. Types of relationships of parasites to species of hosts.

transmission (Chapter 28). Other possibilities suggest a major role for vertebrates. It has been shown experimentally that a small proportion of wild birds
infected with western equine encephalitis may have a recrudescent viremia as
long as 10 months after infection (Reeves, 1961). If such a phenomenon
occurred in nature, it could initiate the transmission cycle of the next year by
infecting fresh mosquitoes. The possibility of garter snakes providing the
overwintering mechanism was discussed in Chapter 1. Western equine
encephalitis virus has been isolated from naturally infected frogs in Saskatchewan (Burton *et al.*, 1966). A similar role has been suggested for bats
(Sulkin *et al.*, 1966). A number of insectivorous bat species, peripherally
inoculated with Japanese or St. Louis encephalitis virus, developed prolonged
viremia without disease. During artificially induced hibernation at low temperatures viremia was not demonstrable, but it reappeared at the termination
of hibernation. Goldfield and his colleagues, quoted by Shope (1965), reported
isolations of eastern and western equine encephalitis viruses from rodents in
winter months, indicating the possibility of a transmission cycle in rodents for
overwintering. Johnson (1967) suggested that these transmission cycles in
rodents may be independent of arthropods. The hypothesis that viruses may
be reintroduced by migratory birds from time to time has been proposed
several times. It is thought that Murray Valley encephalitis virus is not maintained in southeastern Australia, where it frequently produces human disease,
but is brought in by a migratory bird-mosquito cycle from tropical Australia
and New Guinea, where it is enzootic (Miles, 1960).

A DISEASE OCCURS IN SPECIFIC HABITATS

During field studies of tick-borne encephalitis, relapsing fever, and tularemia,
Pavlovsky developed the concept of the "natural nidality" or focal localization of disease (Pavlovsky, 1966). He emphasized that a disease itself tends to
have a natural habitat, in the same way as do species—that it occurs in a land-
scape which is characteristic for it. For example, the broad-leaved taiga
forests of the far east are the natural "nidi" for tick-borne encephalitis, and
the wooded plains and steppes of Siberia for tick-borne typhus. Pavlovsky's
enunciation of this concept in 1939 led to extensive ecologically oriented
research in the Soviet Union, to determine the structure of natural nidi for
various infectious agents, and helped in devising protective measures against
them. Though similar concepts were developed earlier for specific diseases like
plague and trypanosomiasis, Pavlovsky's generalization of it was of great
value, particularly for research in the Soviet Union.

TRANSMISSION FROM RESERVOIR TO MAN

The great variation in the manner in which the infecting organism is transmitted from the reservoir to man is illustrated by the following examples:
    a. Man may be an incidental and dead-end host of wildlife infection—an
indicator species. In this category are many of the most important zoonoses,

for instance, encephalitis viruses in the United States, and rickettsial diseases, except louse-borne typhus.

b. Infection may be transmitted from wildlife to a population that is commensal with man, which may then maintain its own transmission cycle, and also infect man. Examples of these are given in Table 29.1. Many human

TABLE 29.1

Examples of Diseases Transmitted to Man by Commensal Animals

Disease	Wildlife reservoir populations	Animal populations commensal with man	Mode of transmission to man
Rabies	Skunk, fox, coyote, raccoon, mongoose, squirrel, badger, rat, bat	Dog and cat	Bite
Plague	Ground squirrel, rabbit, prairie dog, other rodents	House rat	Fleas
Ornithosis	Birds	Turkey, pet birds	Contact
Tick-borne encephalitis	Rodents	Goat	Milk

infections result from man's close association with commensal animals. The commensal population acts as an amplifier of the parasite, for example of rabies virus in dogs and of plague bacillus in rats.

c. Infection may be originally derived from wildlife, but it initiates a man–vector–man transmission cycle, as in urban yellow fever, where the human parasite is identical with the wildlife parasite, but the vector species are different.

d. Transmission to man may be rare, but potentially significant. Some investigators believe that new strains of human influenza are derived from animal reservoirs. Such hypotheses have been proposed for the pandemics of 1918 and 1957.

FACTORS INFLUENCING INFECTION OF MAN

For human infection to occur, the reservoir, the pathogen, the vector populations (if any), and man should come together in space and time in an environment favorable for transmission. These requirements for transmission contribute to the seasonality of many of the zoonoses and also determine the high risk groups in the human population. Since the populations of reservoir species, amplifiers, pathogens, and vectors fluctuate widely, and not always concurrently, the occurrence, size, and severity of epidemics are hard to predict. The ecologist and epidemiologist hope to identify key factors of predictive value (Chapter 5).

Man can be exposed to an established zoonosis in several ways. He may inadvertently enter a focus of wildlife infection, such as plague, or Rocky

Mountain spotted fever, in the course of his work or pleasure. This will result in sporadic and sometimes numerous infections, as in scrub typhus in the military populations. He may also extensively disturb the habitat of a zoonosis in order to clear areas for settlement and farming, and thus be exposed to infections in epidemic proportions.

The types of host-parasite relationships that may evolve from the latter type of situation are exemplified in African trypanosomiasis, transmitted by tsetse flies, *Glossina*. *Trypanosoma brucei*, probably the ancestral form, is prevalent as an extensive wildlife infection in ruminants with no overt disease. It produces a virulent disease in equines and cattle, but does not infect man naturally or even experimentally. On the other hand, *T. gambiense*, probably derived many centuries ago from *T. brucei*, is fully adapted to man. It has no wildlife reservoir and is maintained by transmission in man alone, with occasional infection of his domestic animals. *Trypanosoma rhodesiense*, apparently a more recent derivative from *T. brucei*, is more closely related to it and has a wildlife reservoir. It occasionally infects man and produces a fulminant disease. The human disease caused by *T. rhodesiense* is more severe than that caused by *T. gambiense*.

It is likely that naturally occurring fluctuations in the reservoir vertebrate populations influence the probability of human infection. An increase in the population density of a wildlife reservoir is likely to increase the number of infectious organisms available for dissemination, and to increase the total number of ectoparasites such as mites, ticks, and fleas which may be vectors. It is also likely to increase the possibility of transfer of the organism to another species, perhaps a population commensal with man. Havlik (1954) reported that in Czechoslovakia, increase in vole populations one year was followed by epidemics of tick-borne encephalitis the next.

The questions concerning the nature of the reservoir will now be considered for three specific infections: Kyasanur Forest disease, simian malaria in man, and Bolivian hemorrhagic fever.

THE RESERVOIR OF KYASANUR FOREST DISEASE (KFD)

Kyasanur Forest disease is a tick-transmitted group B arbovirus infection, occurring in a 600 mile square area in Mysore State in south western India. It was first discovered in 1957, in humans as an epidemic febrile illness with occasional hemorrhages, and as a fatal disease among the two species of forest monkeys present in the area, *Presbytis entellus* and *Macaca radiata* (Work, 1958). Retrospective investigations found no evidence of that disease in the human or monkey populations before 1956.

The affected area is situated in a forest on the eastern slopes of the western Ghats mountain range, at about 2,000 feet above sea level. The annual rainfall is about 80 inches, and most of it falls during the monsoon season of June to September. The forest is largely evergreen, with deciduous patches and a heavy underscrub. Villages and hamlets are located in cleared areas. Most of the villagers are farmers and grow rice in terraced fields close to the villages.

The disease is seasonal, with most cases occurring between February and June.

KFD virus is closely related to the tick-borne encephalitis viruses in the Soviet Union and eastern Europe, to Langat virus in Malaya, and to Powassan virus in Canada and the United States. It is transmitted to man and monkeys by the bite of ticks of the genus *Haemaphysalis*. Virus isolations have been made infrequently from ticks of the genus *Ixodes*, which are rarely found on man but are common on small mammals. Species of *Ixodes* and *Haemaphysalis* feed on different individual hosts as larva, nymph, and adult. Transovarial transmission of this virus in naturally infected ticks does not occur, or is very rare. Transstadial transmission has been demonstrated for *Haemaphysalis* and *Ixodes* species.

The complex epidemiology of the disease, including the identification of the reservoir species, has been under intensive investigation by the Virus Research Center, Poona, and the State Health Department of Mysore State (Virus Research Center, 1964). Simplified versions of their findings are given in Tables 29.2 and 29.3. While the reservoir species of KFD have not been fully identified, a number of important conclusions have been reached.

Infection of man occurs almost exclusively by the bite of infected nymphs of the genus *Haemaphysalis*. This is the only stage found on man, and the epidemic season is also the time of abundance of *Haemaphysalis* nymphs. An

TABLE 29.2

Characteristics of Kyasanur Forest Disease Virus Infection of the Common Vertebrates of the Epidemic Area

	Man	Monkey	Cattle	Small mammals	Ground-feeding birds
Susceptibility to infection	+	+	+	+	+
Virus isolation	+	+	−	+	−
Number in epidemic area	20,000	Thousands	Thousands	+1 million	+1 million
Time from birth to maturity in years	15	4	2	0.2	1
Number of offspring per pregnancy	1	1	1	4–10	4–8
Antibody prevalence percent	5	8–10	40	5	10–15
Viremia-duration					
–Days	7–8	7–8	1–2	6–8	6–8
–Titer	High	High	Very low	Moderate to high	Moderate to high
Effect of infection	Ill, but survives	Death common	No effect	No or minor effect with some exceptions	No effect

TABLE 29.3

Prevalence of Attached *Haemaphysalis* and *Ixodes* spp. on the Common Vertebrates in the Kyasanur Forest Disease Epidemic Area

| | *Haemaphysalis* spp. | | | |
	Larva	Nymph	Adult	*Ixodes* spp.
Man		+		
Monkey	+	+	+	+
Cattle			+	
Small Mammals	+	+		+
Ground-feeding Birds	+	+		

uninfected nymph feeding on a viremic human can become an infected adult, but since the adult does not attach to another human host, man–vector–man transmission is unlikely. Man is exposed to infection in the forests around the villages.

The infected *Haemaphysalis* nymphs which transmit disease to man must have become infected by feeding as larvae on viremic hosts. Therefore, from the standpoint of human infection, vertebrates which infect *Haemaphysalis* larvae are the most important source of infection. *Haemaphysalis* larvae are found on small mammals, including rodents and shrews, on birds, and on monkeys.

Despite the high tick-load of adult *Haemaphysalis* carried by cattle, and the high antibody prevalence in them, cattle are not important as virus disseminators. This is because the transient low-titered viremia that the cattle develop during KFD infection is insufficient to infect ticks. However, cattle are probably very important for the maintenance of a large *Haemaphysalis* population. Also, examination of cattle sera for antibodies is one of the most sensitive indicators of past virus activity in an area.

The *Ixodes* are not of direct importance in human infection since they do not bite man. Their role in the maintenance of virus is not clear. It is important to note that a cycle of rodent-*Ixodes*-rodent transmission, if it occurs in nature, may remain undetected. Such a cycle could occur in the monsoon months when the early stages of *Ixodes* are very active, and those of *Haemaphysalis* are not, but evidence to substantiate this hypothesis has not yet been obtained.

The small mammals in the epidemic area appear to meet the criteria for a reservoir, but detailed studies of individual species have revealed an irregular degree of susceptibility to the virus and a low infestation rate of ticks. Ground-feeding birds, some of which have a high tick load and moderately high virus titers in blood after experimental infection, may also play a role. The present indications are that the reservoir may consist of a number of species. The spread of KFD infection to new areas has been very slow, and only to contiguous villages; this indicates that the reservoir is probably "ground-bound" with a limited range of dispersal. It is considered likely that the virus was present in the area as a silent infection before 1956, but the reasons for its sudden appearance as an epidemic illness of man and monkeys in 1956 have not been clarified.

SIMIAN MALARIA IN MAN

Plasmodia infect a wide variety of vertebrates including man, higher apes, monkeys, rodents, birds, and reptiles. While all plasmodia must surely have a common ancestor, they are now so well specialized for growth in particular host and vector species that they are not able to infect species distantly related to their host. The four plasmodia that commonly infect man have no other reservoir except man; this is a basic premise for the World Health Organization policy of malaria eradication. A possible exception is the infection of chimpanzees with *Plasmodium malariae* in some parts of Africa. The chimpanzee strain, originally described as *P. rodhaini* is indistinguishable from *P. malariae* and they are now considered to be synonymous. However, it is likely that the mosquito vector of *P. malariae* in chimpanzees is sylvatic, and that human infection from a chimpanzee source is probably uncommon and confined to people working and living in the tropical forest (Garnham, 1966). Apes in Africa harbor other parasites morphologically indistinguishable from *P. falciparum* and *P. vivax*, but these do not infect man and are distinct species.

The possibility that simian plasmodia may naturally infect man has been extensively investigated in recent years (Coatney, 1968). Monkeys in Malaysian forest maintain at least eight species of plasmodia, and parasitemia rates as high as 80 percent occur in some of these species. Of the many natural vectors of simian malaria, some bite man on the ground and monkeys in the trees with equal facility. The first human case of naturally acquired simian malaria was reported by Chin *et al.* (1965). The infecting parasite was *P. knowlesi*, whose natural host is the crab-eating macaque, *Macaca irus*. The patient was an American surveyor who acquired the infection during a 4-week stay in Malaysia. The infection was transmissable to other humans, and to several species of monkeys, by inoculation of blood parasites. In the course of experimental human infection, counts as high as 20,850 per milliliter of blood were obtained.

The above data indicate that simian malaria is occasionally transmitted to man from its natural reservoir. However, there is no evidence at present of natural transmission of simian plasmodia in a man-mosquito-man cycle. Experimentally, vectors of human malaria, like *Anopheles freeborni* and *A. quadrimaculatus*, transmit the simian plasmodium *P. cynamolgi* from man to man. Initially the efficiency of such transmission is low, but it appears to increase with passage in man.

THE RESERVOIR OF BOLIVIAN HEMORRHAGIC FEVER

Epidemics of Bolivian hemorrhagic fever have been recognized in the Beni

Province of Bolivia since 1959, and in 1963 the etiological agent was identified and named Machupo virus. The habitat is "grassland broken occasionally with 'islands' of forest and laced with tree-lined rivers and streams" (Kuns, 1965), and the area is tropical with 1,500 millimeters of annual rainfall.

Some of the epidemiological features of the disease suggested that the virus was transmitted from a reservoir host (Mackenzie, 1965). These features were: lack of transmission from man to man by contact, clustering of cases in same household over periods of many months, and appearance of new "infected" houses close to already infected households. Virus isolation was attempted from a number of bat and rodent species. The only species from which virus was isolated was the mouse-like rodent, *Calomys callosus*, but the ectoparasites of this rodent did not harbor virus. Extermination of the rodent brought about a dramatic decrease in cases (Kuns, 1965).

Calomys callosus is a pastoral species distributed through the grassland. It readily invades households in villages that are located in the clearings of the climax forest. Johnson (1965) has suggested that the virus was recently introduced into this species from an unidentified source. Infected *Calomys callosus* develop a chronic infection, mostly asymptomatic, with excretion of virus in the urine for a period of months. Human infections are thought to result from direct contamination of food, water, and air.

Viruses antigenically related to, but distinct from, Machupo virus have been isolated in Argentina, Trinidad, and Brazil. In Argentina, Junin virus is the etiologic agent of Argentinian hemorrhagic fever, a disease similar to Bolivian hemorrhagic fever. Tacaribe virus, repeatedly isolated from *Artibeus* bats in Trinidad, and Amapari virus isolated from rodents and mites in Brazil, are as yet not known to be associated with human disease.

CONCLUSIONS

If the parasite is established in a stable and long-term association in a vast reservoir, its eradication is virtually impossible. However, for some infections, risks of human infection can be very significantly reduced. When the organism is transmitted to man directly from the reservoir species, as with arthropod-borne encephalitis viruses in the United States, the hopes of reducing disease lie in development of vaccines for exposed people, and in reducing the contact of man with the arthropod vectors. In other instances like rabies and plague, where most cases occur from contact of man with his commensal animals, and not with the vast wildlife reservoirs, marked reduction in disease may be accomplished by protection of the commensal animals by vaccination, as in rabies, or by reduction or elimination of the commensal animal, as in plague. If most human cases are derived from man–vector–man transmission of an infection originally derived from a wildlife reservoir, as in urban yellow fever, this urban transmission cycle can be effectively broken by controlling the vector.

In all instances, knowledge of the ecology of the parasite populations and their reservoir, amplifier, and vector populations, is required for the institution of rational measures for control of human infection.

REFERENCES

Audy, J. R. 1958. The localization of disease with special reference to the zoonoses. *Trans. Roy. Soc. Trop. Med. Hyg.* **52**: 308–334.

Burton, A. N., McLintock, J., and Rempel, J. G. 1966. Western equine encephalitis virus in Saskatchewan garter snakes and leopard frogs. *Science* **154**: 1029–1031.

Chin, W., Cantacos, P. J., Coatney, G. R., and Kimball, H. R. 1965. A naturally acquired quotidian-type malaria in man transferable to monkeys. *Science* **149**: 865.

Coatney, G. R. 1968. Simian malarias in man: Facts, implications and predictions. *Am. J. Trop. Med. Hyg.* **17**: 147–155.

Constantine, D. G., and Woodall, D. F. 1964. Latent infection of Rio Bravo virus in salivary glands of bats. *Public Health Rept. (U.S.)* **79**: 1033–1039.

Garnham, P. C. C. 1966. *Malaria Parasites and Other Haemosporidia.* Blackwell, Oxford, England.

Havlik, O. 1954. Significance of increases in rodent population in the epidemiology of Czechoslovak tick encephalitis. *Cesk. Epidemiol., Mikrobiol., Immunol.* **3**: 300 (see 1955 *Bull. Hyg.* **30**: 574).

Johnson, H. N. 1967. Natural ecology. In *Methods in Virology*, Vol. I., K. Maramorosch and H. Koprowski (eds.) Academic Press, New York.

Johnson, K. M. 1965. Epidemiology of Machupo virus infection III. Significance of virological observations in man and animals. *Am. J. Trop. Med. Hyg.* **14**: 816–818.

Kuns, M. L. 1965. Epidemiology of Machupo virus infection. II. Ecological and control studies of the hemorrhagic fever. *Am. J. Trop. Med. Hyg.* **14**: 813–816.

Mackenzie, R. B. 1965. Epidemiology of Machupo virus infection and pattern of human infection, San Joaquin, Bolivia, 1962–1964. *Am. J. Trop. Med. Hyg.* **1**: 808–813.

Miles, J. A. R. 1960. Epidemiology of the arthropod-borne encephalitides. *Bull. World Health Organ.* **22**: 339–371.

Pavlovsky, E. N. 1966. *Natural Nidality of Transmissible Diseases.* (Translated from Russian.) Univ. of Illinois Press, Urbana, Illinois.

Reeves, W. C. 1961. Overwintering of arthropod-borne viruses. *Progr. Med. Virol.* **3**: 59–78.

Shope, R. E. 1965. Transmission of viruses and epidemiology of viral infections. In *Viral and Rickettsial Infections of Man.* F. Horsfall and J. Tamm (eds.) Lippincott, Philadelphia, Penn.

Sulkin, S. E., Allen, R., and Sims, R. 1966. Studies of arthropod-borne virus infections in Chiroptera. III. Influence of environmental temperature on experimental infection with Japanese B and St. Louis encephalitis viruses. *Am. J. Trop. Med. Hyg.* **15**: 406–417.

Virus Research Center. 1964. *Kyasanur Forest Disease*, 1957–1964. Indian Council of Medical Research Publication.

Work, T. H. 1958. Russian spring-summer virus in India. Kyasanur Forest disease. *Progr. Med. Virol.* **1**: 248–277.

CHAPTER 30

Senescence as a Biological Problem

FREDERIK B. BANG

That man and other animals age and show their age is so remarkable and so commonplace a phenomenon that it is part of the poetry of life. It is likely that the scientists' attempts to understand and to delineate senescence have not equaled the attempts of artists such as Rembrandt (Plate 30.1). Nevertheless, senescence must be understood in more prosaic terms, and so one turns to definitions.

AGING DEFINED AS THE INCREASING FORCE OF MORTALITY

Most of the biologists who have attempted to define senescence have used the statistical concept that senescence is that increasing force of mortality which comes with increasing age. There is a decrease in the ability to resist the traumas of life; with increasing age there is loss of resilience of some sort.

Population Curves

If a group of individuals were dying at a constant rate the survivorship curve from the l_x column of life tables would look like A in figure 30.1. If with increasing age there were a greater chance of survival, such as occurs in some fish, there would be a curve as in B. If a constant number were dying per unit time the curve would look like C; such a pattern of survival would be seen for chickens on an ocean liner if the same number were sacrificed for dinner every day. If the death rate increased with age the curve would be as in D, and this is frequently found for animal populations. Figure 30.2 gives survivorship curves for man, Irish wolfhounds, and Mouflon sheep.

Gompertz (1825) showed that for man an increasing rate of death with age seemed to occur at a particular rate itself, indicating that there might be some uniform factor involved in the process of senescence. There are other ways of looking at human life-table data. Figure 30.3 shows numbers dying plotted against time, and from this Benjamin (1959) pointed out that perhaps there is a time, that he put at roughly 76 years for man, about which there is an even Poissonian distribution of deaths. These deaths may be due simply to the natural termination of the life span. With English data he showed that the peak of deaths from old age, as he determined it, was changing its value. In the past hundred years in England the peak changed from age 72 to 76 years in males, and from 73 to 80 in females.

Plate 30.1. Four of Rembrandt's self-portraits, showing the progression of aging: in 1632 at the age of twenty-five; at thirty-four; at fifty-three; and probably in his final year, at sixty-three. (The Burrell Collection in the Glasgow Art Gallery and Museum; the National Gallery, London; and Andrew Mellon Collection in the National Gallery of Art, Washington, D.C.; and the Wallraf-Richartz Museum, Cologne.)

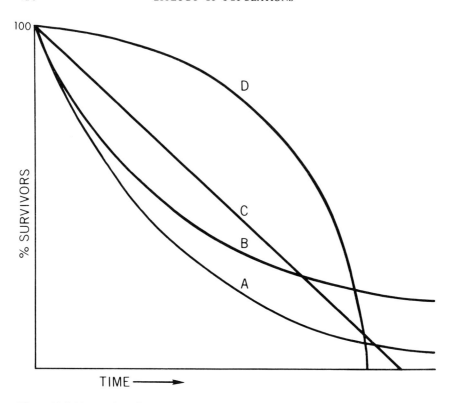

Figure 30.1. Types of survivorship curves.   A: Constant death rate.   B: Decreasing death
rate.   C: Constant number of deaths per unit time.   D: Increasing death rate.

THEORIES OF THE CAUSE OF SENESCENCE

People have continually struggled with theories as to what is involved in the
process of senescence. Perhaps the rate of living might in some way be related
to senescence. Pearl (1928) studied a transparent water flea, *Daphnia pulex*,
and measured the rate at which its heart beat, which was easy to observe. In
males there were approximately 4.3 beats per minute, and in females 3.7. The
male had a mean life of 37.8 days and the female 43.8. Heart rate multiplied by
length of life came to the same figure for each sex, 162.5 and 162.1. So the
rate of living may have had some influence on the longevity, perhaps by means
of a store of energy which might be exhausted at different rates.

Strehler (1962) in his book on the problems of aging dealt with a similar
idea. Figure 30.4 shows the inverse of absolute temperature plotted against
the logarithm of rate of aging of various invertebrates. There is a relation
between the rate of aging, as he has calculated it, and temperature. This
apparently involves the same idea that somehow energy is used up over a
period of time.

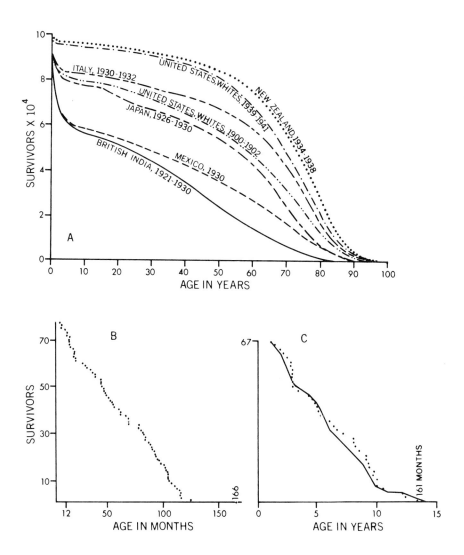

Figure 30.2. Survivorship (l_x) curves. A. Man in various countries. B. Mouflon sheep after the age of 12 months, in the London Zoo. C. Irish wolfhounds after 12 months. The solid line represents total population of 67, the dotted line, another 38 whose exact age of death was not known. (After Comfort, 1964.)

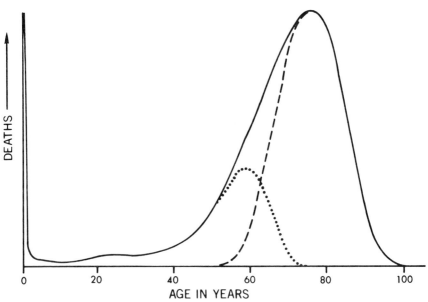

Figure 30.3. Deaths of males (d_x) plotted against age (x), from a life table constructed from English census data of 1950–1952. Solid line represents total deaths. All deaths after age 76 are called "senescent deaths." The mirror image of the post-age-76 part of the curve is drawn as a dashed line from age 50 to 76, and is said to also represent senescent deaths. When senescent deaths represented by the dotted line are subtracted from the total deaths, the remainder, represented by the dotted line, are called "anticipated deaths." (After Benjamin, 1959.)

There are, of course, many other ideas. One is that there is an inescapable accumulation of damage with time, so that there is a logarithmic increase in the rate of mortality. Some believe that the accumulated damage is mutation in somatic cells. There are about 10^{14} somatic cells in the human body. Of these 10^{12} are known to be expended every hundred days, or 10^{10} every day. If this is so, and the mutation rate is 10^{-6} per cell generation per locus, and if there are 10^4 loci in each cell, then the replacement cells arising daily carry 10^8 new mutations. Do some of these accumulate and cause increasing damage? Moreover, repeated infections might cause accumulation of damaging viral material in cells.

DEGENERATIVE DISEASE IN WILD ANIMALS AND NEANDERTHAL MAN

Schultz (1956) caught wild primates and examined their skeletons for lesions, such as caries (Plate 30.2), that accumulate in individuals as time passes. He repeatedly found signs of direct trauma and also signs of degenerative lesions in various joints, changes which can only be called osteoarthritis. This sort of

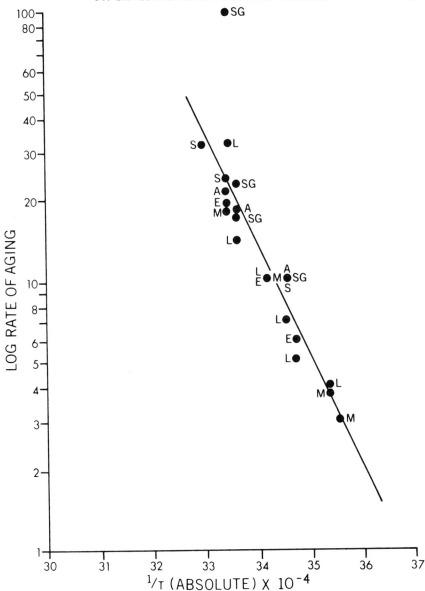

Figure 30.4. Logarithms of relative rates of aging of several arthropod species at various temperatures. Rate of aging was calculated as the reciprocal of mean longevity and normalized to 20°C. SG= *Drosophila* sp.; A and L= *D. melanogaster*; E= *Ptinus tectus* (beetle): M= *Daphnia magna* (crustacean "water flea"). (After Strehler, 1962.)

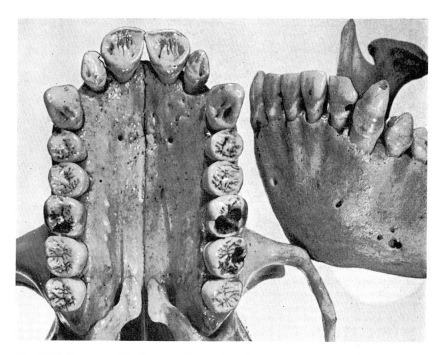

Plate 30.2. Carious cavities in upper right and left first molars, left second molar, and both lower lateral incisors of an adult female, wild-shot orangutan. (Schultz, 1956.)

thing has been found also in various fossils, even from the times of the pre-dominance of reptiles, and later in sabertoothed tigers. Straus and Cave (1957) examined skeletons of Neanderthal man, who is pictured as walking stooped. Straus said that it may well be that he walked bent over, but that in a skeleton he examined Neanderthal was stooped because he had too much osteoarthritis to be able to straighten his back.

MEASUREMENTS IN MAN
Accommodation of the Eye
There are many measurements of change of function with age, particularly for man. Friedenwald (1952) took a human survivorship curve with survivors plotted on a logarithmic scale. With it he plotted another curve, that of sur-vivorship of vision in a cohort adjusted so that mortality of individuals was not included (Figure 30.5). Figure 30.6 shows that the range for which the human eye can accommodate also declines with age. This capacity to focus on

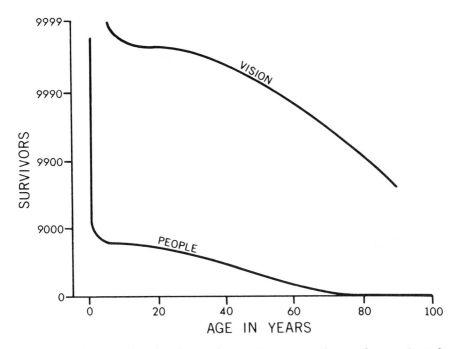

Figure 30.5. Survivors plotted against age in years. Lower curve: humans from a cohort of 10,000. Upper curve: survival of vision in a cohort of 10,000 people with sight, assuming that none of the people died. Based on United States census date of 1910. (After Friedenwald, 1952.)

objects over a range of distances can be measured for humans of all ages except those below 8–10 years who cannot cooperate in the tests. Friedenwald (1952) said in conclusion, "The insoluble problem of senescence is as to whether these changes represent inherent changes of mortal flesh, or the cumulative effects of potentially avoidable disease, or potentially avoidable subclinical damage produced by a non-ideal environment." Even in specific events such as change of the eye, it is not possible to tell whether avoidable damage has come from without or whether there is inherent inevitable change in structure.

The Diffusion Capacity of Lungs

Cohn *et al.* (1954) were interested in the capacity of oxygen to get from air into capillaries in the lungs. They modified a simple formula: diffusion of oxygen is equal to $V/p - p_l$, where V is the quantity of oxygen absorbed per minute, p is the tension of the oxygen in the alveoli, and p_l is the tension of the oxygen in the vessels that come from the lung (Riley 1965). These are measurements that can be made with some accuracy. Figure 30.7 shows the relationship between the diffusing capacity of oxygen into the lungs, and age. There

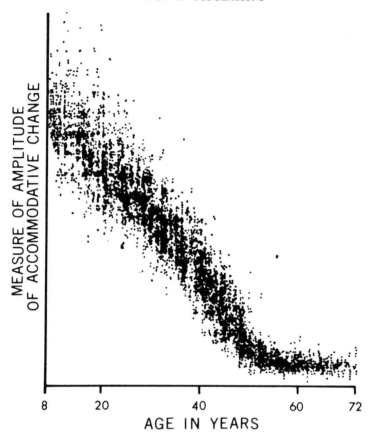

Figure 30.6. The amplitude of accommodative change in the lens of the eye in man, plotted against age, from measurements in 4,000 cases. (After Friedenwald, 1952, after Duane.)

is with increasing age a decreasing capacity of the oxygen to gain entrance into the lung.

The Type and Amount of Collagen

The last example from man comes from other interesting studies of human lungs. The amount of collagen that is present at autopsy in the lungs of men and women was calculated from the amount of hydroxyproline, the amino acid specific for collagen (Briscoe *et al.*, 1959). Both men and women showed a gradual increase in the amount of collagen with age regardless of the cause of death. Scar tissue, or collagen, or whatever it may be called, accumulates with time. It is possible that this is not only an accumulation, but a change in the material itself.

Verzár (1963) measured weights that could be lifted through specific

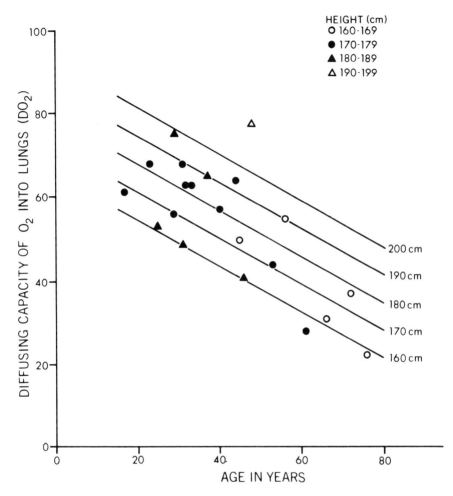

Figure 30.7. The relationship between diffusing capacity of oxygen into the lungs, and age and height in men. (After Cohn *et al.*, 1954.)

distances by tendon (collagen), which contracts when it is heated in a water bath. Tendon from a rat or a frog showed capacity to raise heavier weights as age of the donor increased. With increase of age the tendon became stiffer and more contractile. The way in which the molecules change with age is not known, but while contracting under the influence of heat, young collagen gave up more of several compounds, including hydroxyproline. With extensive contraction very little material was released, and contractability under the influence of heat was changed. Whether the fact that older collagen raised greater weights is part of the inherent change of aging is not known.

There are many more measurements of physiological change with age, particularly in man, and many show that there is a gradual increase in the rate of occurrence of various lesions or various types of disease. The inseparability of disease and the general phenomena of aging will be emphasized again.

THE GENETICS AND EVOLUTION OF LONGEVITY

Populations support other populations that are above them in the pyramid of numbers. The much higher rates of productivity of the lower populations usually involve shorter length of life. Microorganisms, particularly, have very short lengths of life and thus they are able to rapidly change the genetic constitution of their populations. Length of life is part of the adaptation of a species to its niche, and death of individuals is necessary for much of the process of adpative change.

Selection operates before the end of reproduction. Diseases that may not influence individuals until after reproduction, such as osteoarthritis, may therefore have no selective influence. However, if genetic factors responsible for vigor in early life and high fertility are associated by pleiotropism and close linkage with loss of vigor in postreproductive life, then loss of vigor in old age can be maintained in the population.

A specific example of this kind is the susceptibility of Scottish sheep to scrapie (Parry, 1962). This is a fatal disease of the central nervous system. According to Parry, clinically affected individuals contain self-replicating, pathogenic particles that apparently are not naturally disseminated to other sheep. The disease is recessive, appearing in all individuals homozygous for the autosomal allele, s, but in most cases not until they are $2\frac{1}{2}$ to $3\frac{1}{2}$ or more years old. An ss ram, breeding first at 8 months may serve 30 to 40 ewes for three or four seasons before the onset of symptoms of disease. Among the stock herds there is a considerable amount of inbreeding, and by introduction of an ss ram, some flocks have acquired relative frequencies of the s allele estimated to be 0.8 and more, and attack rates as high as 74 percent. Unfortunately the s allele has pleiotropic effects that have been highly preferred by "show" and breeders' standards. Ss and ss individuals have rapid growth in infancy, an early adolescence, exceptional development of the musculature of the trunk and limbs, high fertility, and fine wool. Thus this disease of older individuals has been kept at a high frequency—as a polymorphism—by the great selective advantage given to homozygous recessive and heterozygous individuals, because of their vigor and their build in early life.

Humans have genetic factors involving longevity. Pearl's famous statement contains some truth—that the best way to have long life is to choose long-lived ancestors (Pearl and Pearl, 1943). This is supported by studies of the time of death of monozygotic and dizygotic twins (Kallmann, 1953). Monozygotic twins were shown in this study to be 35 to 36 months apart in their deaths; the dizygotic twins died on the average almost twice as far apart. Thus apparently the capacity to die at the same time is more closely associated where genetic similarity is greater. This is not an astonishing conclusion, considering the multiplicity of factors that are undoubtedly involved in the sequences that lead to death.

Hybrids in Drosophila and Laboratory Mice

It has long been known that mean length of life in *Drosophila* may be 50 to 60 days in one laboratory strain, and 90 to 100 days in another. If the strains are crossed, the F_1 generation has a mean length of life greater than both parents, but no greater than that of wild-caught *Drosophila*. The original strains had been inbred, and the resulting homozygosity was associated with loss of vigor. Similar studies with mice have shown that there are various disease phenomena that appear earlier in the life of the homozygous mouse.

The most likely explanation of hybrid vigor is that many of the recessive pleiotropic effects of major genes depress fecundity, fertility, and longevity. These effects tend to become recessive because heterozygotes with genetic arrangements that result in deleterious effects being hidden have better survival and reproductive rates. Thus, selection tends to make unfavorable effects of a gene recessive, and also favorable effects dominant. The homozygote shows deleterious recessive effects as loss of vigor (Ford, 1964).

EXPERIMENTAL STUDIES OF SENESCENCE

Rats and Nutrition

The classic studies of McCay and his associates (1943) showed that the rate at which laboratory rats die is influenced by their nutrition. If during early life and adolescence rats are given a perfectly nutritious diet with a very low level of calories, the mean length of life increases from 800 to 1,400 days. This immediately raises questions, for in many cases there is apparently nothing inherent in the life of the cell which means that it has to die within a certain number of days or in a certain number of years. G. O. Gey (personal communication) has kept certain cells from individual rats alive in tissue culture for four and five times the life span of the individual animal—in fact one talks of the immortality of cell lines. Malignant cells derived from a woman who had carcinoma of the cervix in Johns Hopkins Hospital, the HeLa cells, have been kept through many cell generations for some 15 years since she died. Yet it is well known that some normal diploid cells in tissue culture do seem to run an appointed course of time and then die out. However, there is something involved in the individual animals which causes increased mortality rates beyond the constitution of individual cells.

McCay concluded that inadequate calorie intake during adolescence delayed maturation and thereby senescence. An important question again was whether any chronic disease was involved, disease being defined as specific change in tissues that can be described and measured. One factor called a degenerative disease of the rat, is bronchiectasis. Some pathologists have called it the arteriosclerosis of the rat. It is a disease of the bronchi causing destructive changes in the cartilage of the bronchi. At the termination of the bronchi it leaves big holes in which pus accumulates. It has been shown that this is an infectious viral disease, acquired early in life, at that time causing a mild pneumonia, but leading in later life to bronchiectasis. It is a very common disease, present in laboratory colonies all the time.

What did McCay's rats actually die from? Bronchiectasis was present but also sarcomas and lymphosarcomas. The long-lived rats were studied to see if they had delayed onsets of these particular diseases. Unfortunately data accumulated so far are not adequate for conclusions. For instance at 620 days there were 9 out of 10 rats with bronchiectasis in rats delayed in growth, and also in those not delayed. Simms and Berg (1957) have made more extensive studies by sampling and autopsy of rats of all ages. They have found myocardial degeneration and other muscle degeneration, periarteritis, and chronic nephrosis, all of which have anatomical symptoms that can be measured histologically. They measured the age of onset of these lesions, and developed curves that showed the onset of these specific degenerative disease processes, and their progress from early life toward death. However, actual causes of death were not established.

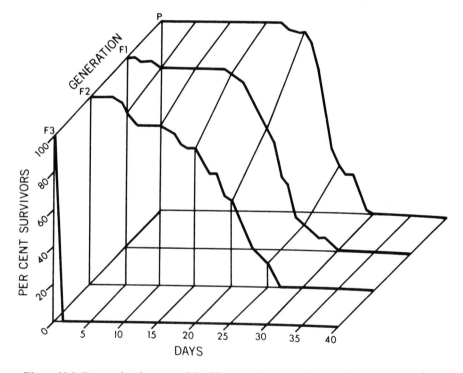

Figure 30.8. Progressive decrease of the lifespan of a strain of the rotifer, *Philodina citrina*, raised in each generation from eggs laid by old parthenogenetic individuals. The third generation is essentially non-viable. (After Lansing, 1952.)

Rotifers and the Inheritance of Longevity

An animal which is not usually connected with the topic of aging is the rotifer, a little invertebrate with cilia arranged at the top in a circular fashion. It is transparent and various processes can be seen in action. Parthenogenetic rotifers, whose use simplified the genetic aspect of a study, do show senescence. A senile rotifer has a low egg production, or produces no eggs, and its ovaries shrink. It has a decreased rate of movement of cilia and slow passage of food through the intestine. It changes from a clear highly refractile individual to an opaque one. Survivorship curves for *Philodina citrina* are shown in Figure 30.8. (Lansing, 1952). If in every generation only eggs from old animals are allowed to develop, then the fourth generation has such a short life span that the animal "shuts itself off," and does not reproduce, and the experiment ends. Laboratory selection favoring the opposite character, by allowing only eggs from young parthenogenetic individuals to develop, produces rotifers with long life spans. Apparently there is accumulation with parental age of material in the eggs, perhaps in the cytoplasm, that is manifested in increased mortality rate, or rates of degeneration, of the animals developed from those eggs.

Paramecium and Longevity in Relation to Conjugation, Autogamy, and Nuclear Breakdown

Paramecium is a free-living freshwater ciliate known as the "slipper animacule." It has been a favorite subject of study for a long time and it has been argued that this animal shows senescence. It frequently divides into two individuals, and so it is difficult to measure the rate of senescence in individuals. If the rate of division in a clone decreases, perhaps that is senescence. If the clone no longer divides, the individuals die. Thus the investigators may refer to the vitality and senescence of a clone. Rejuvenation can occur through conjugation, in which two paramecia join and exchange micronuclei. The macronucleus is responsible for continual influence on the cytoplasm. Then it was shown that there are certain strains of paramecia that survive without conjugation. A process of autogamy, a breakdown of nuclear material, rejuvenates the individual. If the animal is well fed, the process of breakdown of the micronucleus and formation of a new micronucleus, does not occur. If the animal is starved before it has become too senile, then rejuvenation by autogamy takes place. The mechanism of rejuvenation by autogamy is unknown, though there have been various suggestions as to what its nature might be (Sonneborn, 1954, 1955). Apparently, there are processes within the macronuclear region that decline with time and reach a final stage. It is necessary to reorient the nuclear mechanisms in some way, so that various undetermined processes, and cell division, can start again at their original rates.

CONCLUSIONS

The physical study of senescence describes it, but does not tell whether the process is repeated trauma, an accumulation of damage of some other sort, or is inherent in the animal itself. Senescence can be considered to be that increasing force of mortality that comes with age. But this is a statistical concept that measures what happens but does not explain why. The study of senescence in several species has not been free of the continually concomitant question of disease, and perhaps it is impossible to separate them.

There is no doubt that when disease occurs before the end of reproduction it is a selective factor. When it occurs after reproduction it is difficult to judge whether it affects the evolution of the race, and whether evolution influences the prevalence of the disease. Perhaps it is the accumulation of many "diseases" that is now studied as senescence.

Some General Conclusions

This text has described DNA, the magic thread of life. Its complex code allows genetic continuity in the populations that are in turn responsible for its propagation. The innumerable populations fill their niches to the levels where their increase is totally restrained by environmental resistance. This in part is caused by other populations whose existence and variety make possible the complex mechanisms of ecosystems and the biosphere.

Darwinian fitness, the capacity for DNA continuity, is relative survival from generation to generation and is recognized as the driving force of evolution. As individuals of most populations become older and are no longer important in reproduction, an increasingly rapid force of aging sets in, one that seemingly cannot be reversed in the intact animal.

Man, because of a superior development of behavioral organization and cultural inheritance, has cast aside to varying degrees, for a few centuries, many of the restraints of predators, starvation, infectious disease, and much else of his natural environment. But there has not been time enough for his culture to evolve to the level where there is assurance he càn attune his population density to resources by other means than disasters, and can suppress behavioral ills such as war.

With these depressing thoughts, I have turned repeatedly to books such as the Ciba Foundation Symposium on the future of man (Wolstenholme, 1962). In this J. Huxley initiates discussion on the potential value of expanding man's mind with psychedelic drugs; and J. B. S. Haldane from a lofty pinnacle talks of how man as an astronaut might perhaps evolve into a legless animal when relieved of gravity. However, the book does recognize that the secret of DNA is but child's play as compared to the understanding of child's play.

Where does one turn? Perhaps it will help to understand populations in general and how they behave. There must be improved communication from one human population to another, and search for the barriers to communication and understanding. One may return almost to a medieval faith exemplified by Die Vier Ritter (Plate 30.3). The four horsemen represent four of the scourges that man calls forth to ride all the quarters of the earth. The horsemen have been interpreted as Tyranny, Murder—including international and civil warfare—Famine, and Death—including pestilence. Until man calls

Plate 30.3. Etching by A. Dürer of the Four Horsemen of the Apocalypse: Tyranny, Murder, Famine, and Death.

instead for Righteousness, Truth, Justice, and Peace, the four horsemen will continue to ride the earth.

Perhaps the vision for which people in medicine and public health are working will be a reality when the curve of human survival is almost horizontal until the age of about 100 years and then drops off precipitously.

REFERENCES

Benjamin, B. 1959. Actuarial aspects of human lifespans. In *The Lifespan of Animals*. G. E. W. Wolstenholme, and M. O'Connor (eds.) Ciba Foundation Colloquia on Ageing, Vol. 5. Little, Brown. Boston, Mass.

Briscoe, A. M., Loring, W. E., and McClement, J. H. 1959. Changes in human lung collagen and lipids with age. *Proc. Soc. Exptl. Biol. Med.* **102**: 71–74.

Cohn, J. E., Carroll, D. G., Armstrong, B. W., Shepard, R. H., and Riley, R. L. 1954. Maximum diffusing capacity of the lungs in normal male subjects of different ages. *J. Appl. Physiol.* **6**: 588–597.

Comfort, A. 1964. *Ageing: The Biology of Senescence*. Holt, Rinehart, and Winston, New York.

Ford, E. B. 1964. *Ecological Genetics*. Wiley, New York.

Friedenwald, J. S. 1952. The eye. In *Cowdry's Problems of Ageing: Biological and Medical Aspects*. A. I. Lansing (ed.) Williams and Wilkens, Baltimore, Md.

Gompertz, B. 1825. On the nature of the function expressive of the law of human mortality, and on a new mode of determining the value of life contingencies. *Phil. Trans.* **115**: 513–585.

Kallmann, F. J. 1953. *Heredity in Health and Mental Disorder*. Norton, New York.

Lansing, F. J. (ed.) 1952. *Cowdry's Problems of Ageing: Biological and Medical Aspects*. Williams and Wilkens, Baltimore, Md.

McCay, C. M., Sperling, G., and Barnes, L. L. 1943. Growth, ageing, chronic disease, and life span in rats. *Arch. Biochem.* **2**: 469–479.

Parry, H. B. 1962. Scrapie: a transmissible and hereditary disease of sheep. *Heredity* **17**: 75–105.

Pearl, R. 1928. *The Rate of Living*. Knopf, New York.

Pearl, R., and Pearl, R. de W. 1943. Studies on human longevity. VI. Distribution and correlation or variation in the total immediate ancestral longevity of nonagenarians and centenarians in relation to inheritance factor in the duration of life. *Human Biol.* **6**: 98–222.

Riley, R. 1965. Gas exchange and transportation. In *Physiology and Biophysics*. T. C. Ruch and H. D. Patton (eds.). Saunders, Philadelphia, Penn.

Schulz, A. H. 1956. The occurrence and frequency of pathological and teratological conditions and of twinning among non-human primates. In *Primatologia: Handbook of Primatology*. Vol. I. H. Hofer, A. H. Schulz, and D. Starck (eds.) Karger, New York.

Simms, H., and Berg, B. N. 1957. Longevity and the onset of lesions in male rats. *J. Gerontol.* **12**: 244–252.

Sonneborn, T. M. 1954. The relation of autogamy to senescence and rejuvenescence in *Paramecium aurelia. J. Protozool.* **1**: 38–53.

Sonneborn, T. M. 1955. Heredity, development and evolution in *Paramecium. Nature* **175**: 1100.

Straus, W. L., and Cave, A. J. E. 1957. Pathology and the posture of Neanderthal man. *Quart. Rev. Biol.* **32**: 348–363.

Strehler, B. 1962. *Time, Cells, and Ageing*. Academic Press, New York.

Verzár, F. 1963. The aging of collagen. *Sci. Am.* **208**: 104–114.

Wolstenholme, G. E. W. (ed.) 1963. *Man and His Future*. a Ciba Foundation Volume. Little, Brown. Boston, Mass.

Author Index

Numbers in parentheses indicate the numbers of the references when these are cited in the text without the name of the author.

Numbers set in *italics* designate those page numbers on which the complete literature citations are given.

431

Subject Index

439